Proceedings of the International Symposium on

COMPUTER SOFTWARE IN CHEMICAL AND EXTRACTIVE METALLURGY

Pergamon Titles of Related Interest

Coudurier FUNDAMENTALS OF METALLURGICAL PROCESS
Davenport FLASH SMELTING
Norrie AUTOMATION FOR MINERAL RESOURCE DEVELOPMENT
  (An IFAC Proceedings Volume)
Witts MINERALS PROCESSING TECHNOLOGY 4th Edition
Yoshitani AUTOMATION IN MINING, MINERAL AND METAL
  PROCESSING 1986 (An IFAC Proceedings Volume)

Other CIM Proceedings
Published by Pergamon

Bickert REDUCTION AND CASTING OF ALUMINUM
Jonas DIRECT ROLLING AND HOT CHARGING OF
  STRAND CAST BILLETS
Kachaniwsky IMPACT OF OXYGEN ON THE PRODUCTIVITY OF
  NON-FERROUS METALLURGICAL PROCESSES
Macmillan QUALITY AND PROCESS CONTROL IN REDUCTION AND
  CASTING OF ALUMINUM AND OTHER LIGHT METALS
Plumpton PRODUCTION AND PROCESSING OF FINE PARTICLES
Rigaud ADVANCES IN REFRACTORIES FOR THE
  METALLURGICAL INDUSTRIES
Ruddle ACCELERATED COOLING OF ROLLED STEEL
Salter GOLD METALLURGY
Tyson FRACTURE MECHANICS
Wilkinson ADVANCED STRUCTURAL MATERIALS

Related Journals
(Free sample copies available upon request)

COMPUTERS AND CHEMICAL ENGINEERING
MINERALS ENGINEERING
CANADIAN METALLURGICAL QUARTERLY
ACTA METALLURGICA
SCRIPTA METALLURGICA

Proceedings of the International Symposium on

COMPUTER SOFTWARE IN CHEMICAL AND EXTRACTIVE METALLURGY

Montréal, Canada
August 28-31, 1988

Vol. 11 Proceedings of the Metallurgical Society
of the Canadian Institute of Mining and Metallurgy

Edited by

W.T. THOMPSON

Royal Military College, Kingston, Ontario, CANADA

F. AJERSCH

Ecole Polytechnique, Montréal, Quebec, CANADA

G. ERIKSSON

Ecole Polytechnique, Montréal, Quebec, CANADA

PERGAMON PRESS

New York   Oxford   Beijing   Frankfurt   São Paulo   Sydney   Tokyo   Toronto

Pergamon Press Offices:

U.S.A.	Pergamon Press, Inc., Maxwell House, Fairview Park, Elmsford, New York 10523, U.S.A.
U.K.	Pergamon Press plc, Headington Hill Hall, Oxford OX3 0BW, England
PEOPLE'S REPUBLIC OF CHINA	Pergamon Press, Qianmen Hotel, Beijing, People's Republic of China
FEDERAL REPUBLIC OF GERMANY	Pergamon Press GmbH, Hammerweg 6, D-6242 Kronberg, Federal Republic of Germany
BRAZIL	Pergamon Editora Ltda., Rua Eça de Queiros, 346, CEP 04011, São Paulo, Brazil
AUSTRALIA	Pergamon Press (Aust.) Pty Ltd., P.O. Box 544, Potts Point, NSW 2011, Australia
JAPAN	Pergamon Press, 8th Floor, Matsuoka Central Building, 1-7-1 Nishishinjuku, Shinjuku-ku, Tokyo 160, Japan
CANADA	Pergamon Press Canada Ltd., Suite 271, 253 College Street, Toronto, Ontario M5T 1R5, Canada

---

First printing 1989

Library of Congress Cataloging in Publication Data

International Symposium on Computer Software in Chemical and
 Extractive Metallurgy (1988 : Montréal, Québec)
  Proceedings of the International Symposium on Computer Software in
 Chemical and Extractive Metallurgy, Montréal, Canada, August 28-31,
 1988 / edited by W.T. Thompson, F. Ajersch, G. Eriksson.
  p. cm. -- (Proceedings of the Metallurgical Society of the
 Canadian Institute of Mining and Metallurgy ; vol. 11)
  Includes index.
  ISBN 0-08-036087-4
  1. Chemistry, Metallurgical--Computer programs--Congresses.
 2. Metallurgy--Computer programs--Congresses. I. Thompson, W.T.
 (William T.) II. Ajersch, F. III. Eriksson, G. IV. Title.
 V. Series.
 QD132.I58   1988
 669'.0028'553--dc19   88-38661
                                                                  CIP

In order to make this volume available as economically and as rapidly as possible, the authors' typescripts have been reproduced in their original forms. This method unfortunately has its typographical limitations but it is hoped that they in no way distract the reader.

Printed in the United States of America

PREFACE

*This proceedings volume consists of a series of papers
presented at the International Symposium on Computer Software in
Chemical and Extractive Metallurgy held in Montreal, August
30-31, as part of the 27th Annual Conference of Metallurgists of
the Canadian Institute of Mining and Metallurgy.*

*The program, consisting of presentations from North
America as well as overseas, was intended to provide an overview
of software development activities including data bases in chemical thermodynamics, applications of thermochemical and transport
properties in metallurgical processes, modelling of mineral processing systems, expert systems and training packages for plant
operations. Some of the papers deal with stand alone programs
for use with personal computers while others refer to on-line
systems available through data packet switching networks. The
range of activities covered, in the editors' opinion, is not
exhaustive, but is a reasonably good reflection of present progress in this rapidly developing specialty.*

*The symposium was organized jointly by the Basic
Sciences and Non-Ferrous Pyrometallurgy sections of the Metallurgical Society of CIM. The editors wish to express their
thanks to these groups and in particular Ralph Harris and George
Kachaniwsky who helped in soliciting representative contributions. The editors are also grateful to the authors, session
chairmen, institutes, universities and companies who through
their co-operation and enthusiasm made this symposium possible.*

F. AJERSCH
Ecole Polytechnique

W.T. THOMPSON
Royal Military College

G. ERIKSSON
Ecole Polytechnique

October 1988

PROCEEDINGS OF THE INTERNATIONAL SYMPOSIUM ON

COMPUTER SOFTWARE IN CHEMICAL AND EXTRACTIVE METALLURGY

Conference Technical Program Chairman:

A.D. Pelton
Ecole Polytechnique
Montreal, Quebec

Symposium Organizing Committee:

F. Ajersch	W.T. Thompson	G. Eriksson
Ecole Polytechnique	Royal Military College	Ecole Polytechnique
Montreal, Quebec	Kingston, Ontario	Montreal, Quebec

Session Chairmen:

I Thermochemical Computation and Data Banks

G. ERIKSSON, *Ecole Polytechnique, Montreal, Quebec*
M. BLANDER, *Argonne National Laboratory, Argonne, Illinois, U.S.A.*

II Pyrometallurgical and Process Applications

F. AJERSCH, *Ecole Polytechnique, Montreal, Quebec*
G. KAIURA, *Falconbridge Ltd., Sudbury, Ontario*

III Heat and Mass Transfer

F. MUCCIARDI, *McGill University, Montreal, Quebec*
R. BERGMAN, *University of Toronto, Toronto, Ontario*

IV Expert Systems and Artificial Intelligence

W.T. THOMPSON, *Royal Military College, Kingston, Ontario*
A. VAHED, *Inco Ltd, Sudbury, Ontario*

TABLE OF CONTENTS

page

SESSION I

<u>Thermochemical Computation and Data Banks</u>

Co-Chairmen: GUNNAR ERIKSSON
              Ecole Polytechnique
              Montreal, Quebec

              MILTON BLANDER
              Argonne National Laboratory
              Argonne, Illinois, U.S.A.

CALCULATIONS OF THE THERMODYNAMIC PROPERTIES OF METALLURGICAL SOLUTIONS

Milton Blander
Materials Science Program/Chemical Technology Division
Argonne National Laboratory
9700 South Cass Avenue
Argonne, Illinois 60439-4837

ABSTRACT

Predictive theories for metallurgical solutions are important precursors for computer software in chemical and extractive metallurgy. A limited selection of concepts useful for slags and other ionic systems will be discussed and include the quasichemical theory, the conformal ionic solution theory, and polymer theory. We emphasize theories which usefully predict solution properties of multicomponent ionic systems, such as silicates and molten salts, to illustrate the range of possible uses.

KEYWORDS

Solution theories; molten salt solutions; ionic solutions; conformal ionic solutions; silicate solutions.

INTRODUCTION

The development of computer software for thermodynamic calculations for reactions involving multicomponent ionic metallurgical melts can benefit from the use of theoretical developments that greatly minimize the necessary input data. The most useful theories permit one to perform calculations for multicomponent solutions based on data for the subsidiary binary solutions. Since, for a given number of components, there are a much smaller number of possible binary systems than there are multicomponent systems and since, in addition, there is enough understood about the physics of binary systems to make educated guesses of properties when data are not available, these theories permit one to calculate the thermodynamic properties of a very large number of systems *a priori*.

There are two major types of solutions we will consider: (A) simple solutions for which the excess thermodynamic properties can be represented by polynomials and (B) solutions that are highly ordered and cannot be represented by the usual polynomial representation. In addition, there are two classes of systems in each of the above two categories–additive systems and reciprocal systems. Additive systems contain either one kind of cation and different kinds of anions (e.g., $A^+/X^-,Y^-,Z^-, \ldots$) or different kinds of cations and one kind of anion (e.g.,

A^+, B^+, C^+, ... $/X^-$). Reciprocal systems contain at least two cations and two anions and generally exhibit very large deviations from ideal solution behavior.

Before one can discuss deviations from ideality, one needs to define an ideal solution. For a given salt constituent of an ionic solution, $A_m X_n$, one can write the chemical potential in an ideal solution, μ_i^{id},

$$\mu_i^{id} = \mu_i^\circ + RT \ln X_A^m X_X^n \tag{1}$$

where i is the constituent $A_m X_n$, μ_i° is the standard chemical potential of constituent i and X_A and X_X are cation and anion fractions, respectively; these are defined as $X_A = n_A/\Sigma n_c$ and $X_X = n_X/\Sigma n_a$ where the n's designate the number of moles of the subscript species, the subscript c denotes cations, and a denotes anions. Eq. (1) serves to completely define an ideal solution.

The deviations of real solutions from ideal solution behavior is expressed in terms of excess functions and activity coefficients

$$\mu_i = \mu_i^\circ + \mu_i^{id} + \mu_i^E = \mu_i^\circ + RT \ln X_A^m X_X^n \gamma_i = \mu_i^\circ + RT \ln a_i \tag{2}$$

where $\mu_i^E (= RT \ln \gamma_i)$ is the excess chemical potential, γ_i is an activity coefficient and a_i is an activity of the component i. The molar excess free energy of mixing is

$$G^E = \sum_j X_j \mu_j^E \tag{3}$$

where X_j is the mole fraction of component j.

In what follows, we will present a brief description of different theories that have proven to be particularly useful. Since the subject is complex, applications require reading the references for detail. Because of limited space, this discussion is not exhaustive and some useful theories are not discussed.

CONFORMAL IONIC SOLUTION THEORY

The conformal ionic solution theory is a statistical mechanical perturbation theory and is the most fundamental and successful method for predicting the properties of multicomponent molten ionic systems. Equations up to fourth order have been deduced for the excess free energies of mixing and partial molar excess free energies of mixing of ternary additive ionic systems in which the three salts are of the same charge type (Saboungi and Blander, 1975a).

$$G_m^E = \sum\sum_{i<j} a_{ij} X_i X_j + \sum\sum_{i \neq j} b_{ij} X_i^2 X_j + \sum\sum_{i<j} c_{ij} X_i^2 X_j^2$$
$$+ A X_1 X_2 X_3 + \sum_{i \neq j < k} B_i X_i^2 X_j X_k \tag{4}$$

where X_i, X_j, and X_k are ion fractions of either cations or anions, a_{ij}, b_{ij}, and c_{ij} are binary interaction coefficients and A and the three B_i's are ternary interaction coefficients that are given in terms of the binary coefficients by the expressions

$$A = \left(b_{12}^{1/2} + b_{13}^{1/3}\right)\left(b_{21}^{1/3} + b_{23}^{1/3}\right)\left(b_{13}^{1/3} + b_{23}^{1/3}\right) \tag{5}$$

and

$$B_i = 2(c_{ij}c_{ik})^{1/2} \tag{6}$$

Equations (4), (5), and (6) permit one to calculate the properties of ternary systems from the properties of the three subsidiary binaries. There has been empirical support for a theoretical conclusion by Førland (Førland, 1957) that the substitution of equivalent fractions for ion fractions into the expression for the excess free energy (not into the ideal term) leads to better representations of data in most binary systems (and probably multicomponent systems). Equivalent fractions are defined by

$$y_i = \frac{z_i X_i}{\sum_{j=1}^{3} z_j X_j} \tag{7}$$

where z_i and z_j are the charge on the ith and j^{th} ions, respectively; the ions considered are all cations or all anions. With this modification, one can usually make accurate predictions for mixtures of salts of different charge type. Since the ternary interaction terms (i.e., the terms with A and B_i as coefficients) are relatively small, these equations do not have a clear advantage over empirical methods; however, since these equations are fundamental, their use should be preferred over empirical rules.

The equations deduced from the CIS theory for reciprocal systems (Blander and Yosim, 1963) have proven to be particularly useful (Blander and Topol, 1966). Constituents of reciprocal systems often exhibit very large deviations from ideal solution behavior. In ternary reciprocal systems, two constituents exhibit negative deviations and the other two exhibit positive deviations from ideal behavior. This leads to chemical behavior that is a complex function of composition and to phase diagrams with complex topologies.

Conformal ionic solution theory for ternary reciprocal molten salt systems up to second order terms leads to an expression for the total excess free energy of mixing of the three component salts, AX, BX, and BY,

$$\Delta G_M^E = X_A X_Y \Delta G^\circ + X_Y \Delta G_Y^E + X_X \Delta G_X^E +$$
$$X_B^{E} {}_B^E + X_A \Delta G_A^E - X_A X_B X_X X_Y \frac{\left(\Delta G^\circ\right)^2}{2ZRT} \tag{8}$$

where ΔG° is the standard free energy change for the metathetical reaction

$$AX(\ell) + BY(\ell) \rightleftharpoons AY(\ell) + BX(\ell) \tag{9}$$

The X_i are anion or cation fractions, ΔG_i^E is the excess free energy of mixing of the binary mixture in which the two salts have the common ion i, and Z is a parameter that can be taken as 6. The activity coefficients of a component, e.g., BY, are given by the expression

$$RT \ln \gamma_{BY} = X_A X_X \Delta G^\circ + X_A X_X (X_X - X_Y) \lambda_A$$
$$+ X_X (X_A X_Y + X_B X_X) \lambda_B + X_A (X_A X_Y + X_B X_X) \lambda_Y$$
$$+ X_A X_X (X_A - X_B) \lambda_X - X_A X_X (X_A X_Y + X_B X_X - X_B X_Y) \frac{\left(\Delta G^\circ\right)^2}{2zRT} \tag{10}$$

where the binary excess free energies, such as ΔG_X^E, are given by a quadratic expression

$$\Delta G_X^E = X_A X_B \lambda_X \tag{11}$$

where λ_X is an energy parameter for the AX-BX system. The empirical substitution of higher order expressions for ΔG_i^E in Eq. (8), when available, appears to be a useful extension of Eq. (10). In Fig. 1, we compare the results of the first use of Eqs. 8-11 for the calculation of the complex topology of the phase diagram for the Li^+, $K^+/Cl^-,F^-$ reciprocal ternary system with measurements (Blander and Topol, 1966). The substitution of equivalent fractions in these equations has been shown to lead to good predictions of the phase diagrams of systems with anions and or cations having different charges (Saboungi and Blander, 1975b). One of the best examples of the use of these equations was for deducing the phase diagrams and gas-liquid reactions of fuel cell carbonate electrolytes (Pelton, Bale, and Lin, 1981). In Fig. 2, we exhibit results of calculations for the Li^+, $K^+/CO_3^{2-},OH^-$ system and a measured phase diagram for comparison. These equations have been extended to higher order multicomponent systems (Saboungi, 1980).

ORDERED SOLUTIONS

The most difficult class of solutions to describe are the ordered solutions, which includes, e.g., silicates and chloroaluminates. This difficulty arises from the fact that the enthalpies of mixing tend to be "V" shaped and the entropies of mixing tend to be "m" shaped. These shapes are poorly represented by the usual polynomial representation of deviations from ideal behavior. We have deduced a set of equations with such properties based on empirical modifications of the quasichemical theory (Blander and Pelton, 1983, 1984, 1987; Pelton and Blander, 1984, 1986; Pelton, Eriksson, and Blander, 1988)). For binary systems, the excess free energy of mixing is given by the expression

$$G^E = RT\left(aX_1 + bX_2\right)\left[y_1 ln\frac{K-1+2y_1}{y_1(K+1)} + y_2 ln\frac{K-1+2y_2}{y_2(K+1)}\right] \tag{12}$$

where X_i is the mole fraction of component i, and the values of a and b are equal to 0.34435 times the cationic charge of components 1 and 2, respectively. The quantities y_i are equivalent fractions and K is given by the equation

$$K = [1 + 4y_1 y_2 \left[exp\left(W_{12}/RT\right) - 1\right]]^{1/2} \tag{13}$$

where W_{12} is an energy that is taken to be dependent on temperature and composition and is written as a polynomial in concentration

$$W_{12} = \sum_{k=0} \left(h_k - TS_k\right)y_2^k \tag{14}$$

where no more than four values of k chosen between 1 and 7 were found to be needed and where the component 2 is taken to be more acid than component 1.

All these quantities define the number of the 1-1, 2-2, and 1-2 bonds in the quasichemical model, n_{11}, n_{22}, and n_{12}. The total number of bonds to each component 1 cation is 2a and to component 2, 2b so that if n_i is the number of moles of component i, then

$$an_1 = n_{11} + n_{12}/2$$
$$bn_2 = n_{22} + n_{12}/2 \tag{15}$$

and if the fraction of ij bonds, $X_{ij} = n_{ij}/\Sigma n_{ij}$, then

$$X_{11} = y_1 - X_{12}/2$$
$$X_{22} = y_2 - X_{12}/2 \tag{16}$$

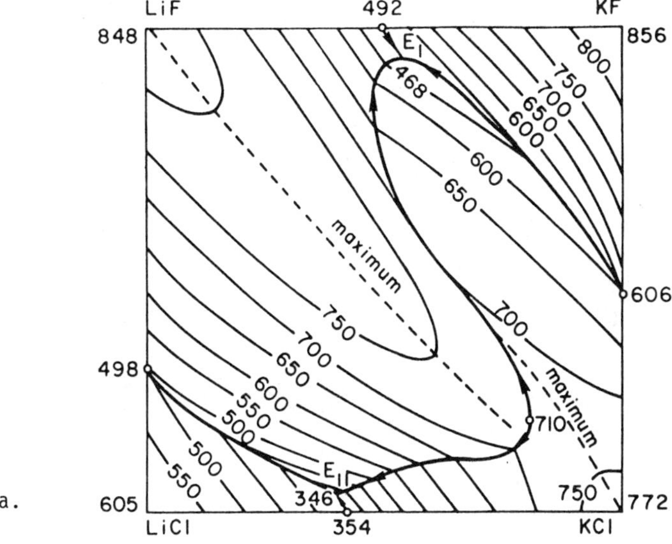

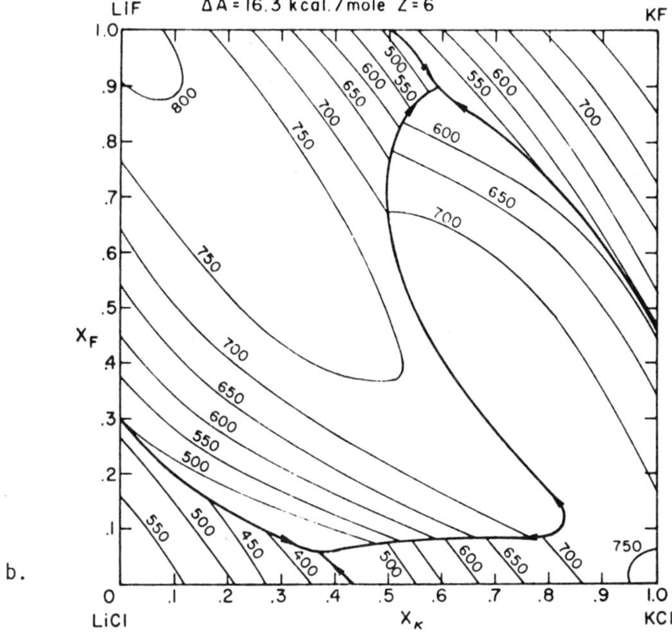

Fig. 1.   Phase diagrams of the Li$^+$, K$^+$/F$^-$, Cl$^-$ system. a. Measured. b. Calculated with the CIS theory.

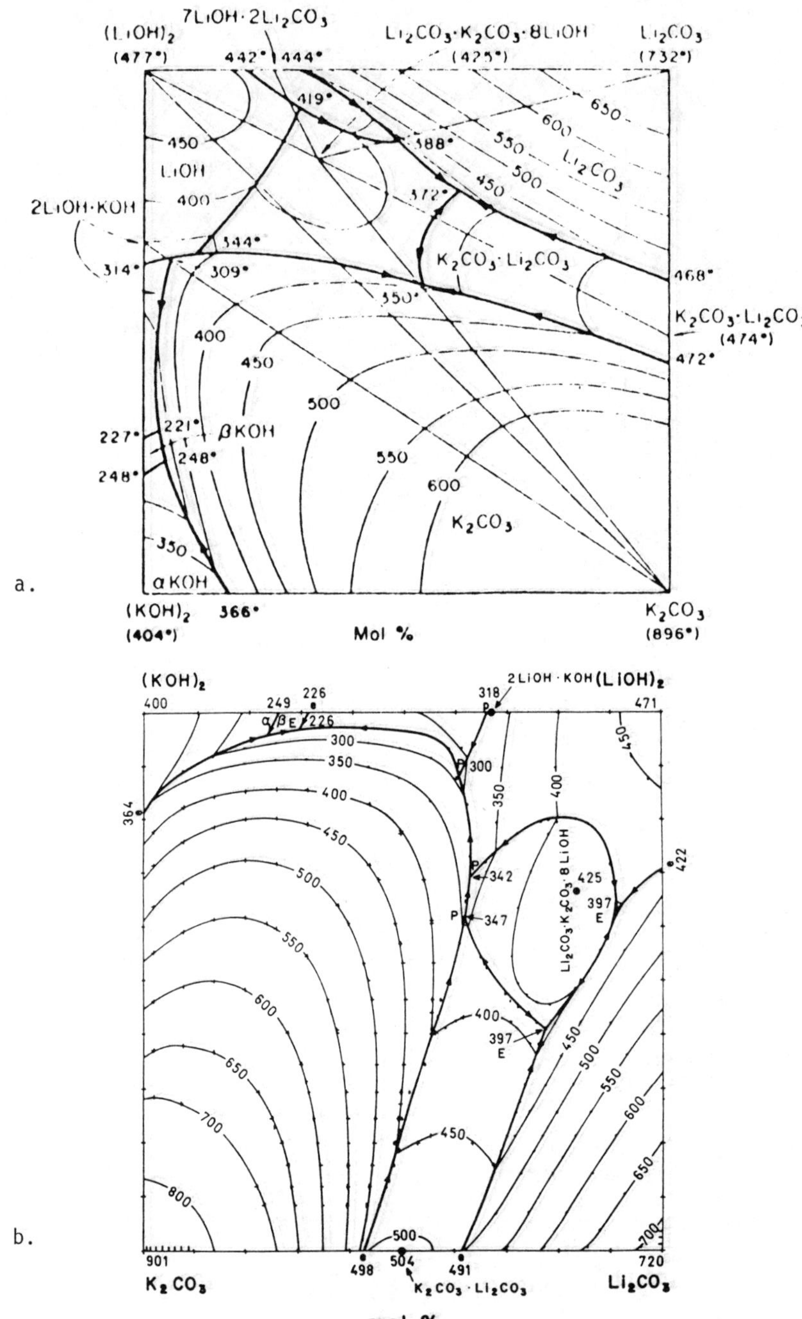

Fig. 2.   Phase diagrams of the Li^+, K^+/CO_3^{2-}, OH^- ternary system. a. Measured. b. Calculated with the CIS theory.

and the distribution of bonds is governed by the equation

$$\left(X_{12}^2/X_{11}X_{22}\right) = 4exp\left(-W_{12}/RT\right) = \alpha_{12}^2 \tag{17}$$

The solutions to Eqs. (16) and (17) can be used to calculate the molar excess free energy from the equation

$$\frac{G^E}{RT} = aX_1 ln\frac{X_{11}}{y_1^2} + bX_2 ln\frac{X_{22}}{y_2^2} \tag{18}$$

The equations are somewhat more complex for ternary systems with the two expressions of Eq. (16) replaced by the three expressions of Eq. (19)

$$X_{11} = y_1 - X_{12}/2 - X_{13}/2$$
$$X_{22} = y_2 - X_{12}/2 - X_{23}/2$$
$$X_{33} = y_3 - X_{13}/2 - X_{23}/2 \tag{19}$$

where now, for example, $y_3 = cX_3/(aX_1 + bX_2 + cX_3)$. In addition there are three expressions replacing Eq. (17) of the form

$$\left(X_{ij}^2/X_{ii}X_{jj}\right) = \alpha_{ij}^2 = 4exp\left(-W_{ij}/RT\right) \tag{20}$$

By substituting Eq. (19) into Eq. (20), one obtains three simultaneous equations with three unknowns that can be solved numerically. The power series representation of the three values of W_{ij} can be approximated in several ways. For systems with only one acid component (e.g., silica) and with the remaining components being basic, a logical representation is the "asymmetric approximation." If we choose SiO_2 to be component 1, then we set

$$W_{12} = c_0 + c_1 y_1 + c_2 y_1^2 + \cdots$$
$$W_{13} = c_0' + c_1' y_1 + c_1' y_1^1 + \cdots \tag{21}$$

and

$$W_{23} = c_0'' + c_1'' t_{23} + c_2'' t_{23}^2 + \cdots \tag{22}$$

where $t_{ij} = y_j/(y_i + y_j)$. The coefficients c_i contain a constant and a temperature dependent term.

The molar excess free energy of the ternary system is given by the expression

$$\frac{G^E}{RT} = aX_1 ln\frac{X_{11}}{y_1^2} + bX_2 ln\frac{X_{22}}{y_2^1} + cX_3 ln\frac{X_{33}}{y_3^2} \tag{23}$$

The complex equations for the partial molar quantities are readily calculated from Eq. (23).

With the equations thus deduced, one can calculate the properties of a ternary system based on the properties of the three subsidiary binaries. This approximation leads to very good representations of ternary data (Blander and Pelton, 1984; Pelton and Blander, 1987).

For binary systems, the parameters h_j and s_j are deduced from a complex optimization procedure that performs a global and simultaneous analysis of all thermodynamic data on a system. This includes liquidus phase diagrams, activity, data, data on miscibility gaps, enthalpies of fusion, free energies of formation of compounds, etc. A small set of resultant parameters (seven at most in the systems considered, including temperature coefficients) are then used to recalculate the input data so as to double check the accuracy of the curve fitting procedure, as well as the efficacy of the use of the equations for representing the data. The results were generally very good. For multicomponent systems, we used the asymmetric approximation given in Eq. (21). When silica is the only acid component, the predictions based on this approximation were in very good agreement with measured data (Blander and Pelton, 1984; Pelton and Blander, 1987). A partial theoretical justification for such a method can be based on theories for ternary systems (Blander and Pelton, 1984).

RESULTS OF THERMODYNAMIC ANALYSES

We have performed analyses of the binary systems containing the components MgO, FeO, CaO, Na_2O, K_2O, MnO, and SiO_2, and of several ternary systems.

We illustrate our calculations for one ternary system below. The analysis of the three binary subsystems and the ternary system CaO-FeO-SiO_2 was performed using as input the liquidus phase diagram, activities of CaO and SiO_2, free energies of formation of $CaSiO_3$ and Ca_2SiO_4, and the miscibility gap in the CaO-SiO_2 system, measured activities of FeO in the CaO-FeO system, and the activities of FeO, the phase diagram, and the free energy of formation of Fe_2SiO_4 in the FeO-SiO_2 system. To illustrate some of the results, we exhibit (1) the calculated phase diagram of the FeO-SiO_2 system in Fig. 3, along with measured values of the invariant points and (2) a comparison of activities of "FeO" measured in the iron saturated molten FeO-SiO_2 system with calculated values in Fig. 4.

Using our asymmetric combining rules, the data for the binary systems were combined and led to the results for a ternary system given in Fig. 5; this figure illustrates the correspondence between calculated and measured values (Blander and Pelton, 1984) of the activities of FeO in the CaO-FeO-SiO_2 system. The differences are well within the uncertainties in the measurements and are partly due to the differences between the binary data used in our analysis and that used to represent the measured data for the ternary. We find that this method essentially permits us to make predictions in ternary systems, based solely on data for the three subsidiary binary systems for cases in which silica is the only acid component. When alumina and silica were both present, a more complex representation was necessary.

The good correspondence of calculations with the complex concentration dependence of activities in the CaO-FeO-SiO_2 system illustrates the fact that our equations properly take into account the kinds of ternary interaction terms known to exist in such systems. This feature lends confidence in the use of our equations for predictions in multicomponent systems (containing only silica as an acid component), based solely upon the subsidiary binaries. If, as it appears, this is generally true, our method provides an important predictive capability.

The ability to predict the thermodynamic properties of slags is important in the calculation of equilibria involving slags. However, the methods that have been developed are limited to oxides, and nonoxide components have to be treated differently. For example, in practical uses, slags often will contain components having anions other than oxide, such as sulfides, sulfates, phosphates, and chlorides. Equations deduced for sulfide capacities could be adapted for such solutes. For example, for the dissolution of CaS to form a dilute solution in the CaO-SiO_2 binary system, the equation for the activity of CaS (Reddy and Blander, 1987)

$$ln\ a_S \cong\ ln\ \Phi_S + \left(1 - \frac{1}{m}\right)\Phi_p \tag{24}$$

deduced from Flory's polymer theory (Hildebrand and Scott, 1950) where Φ_p is the volume fraction of polymeric silicate ($\sim$1), Φ_S is the volume fraction of sulfide (approximated by

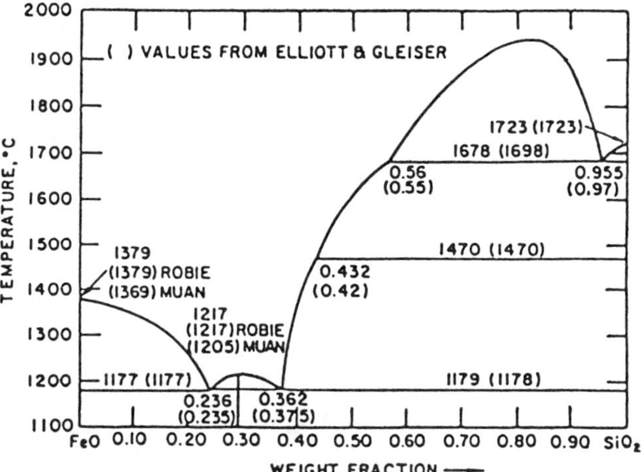

Fig. 3. Calculated phase diagram of the FeO-SiO$_2$ system. Numbers in parentheses are measured values.

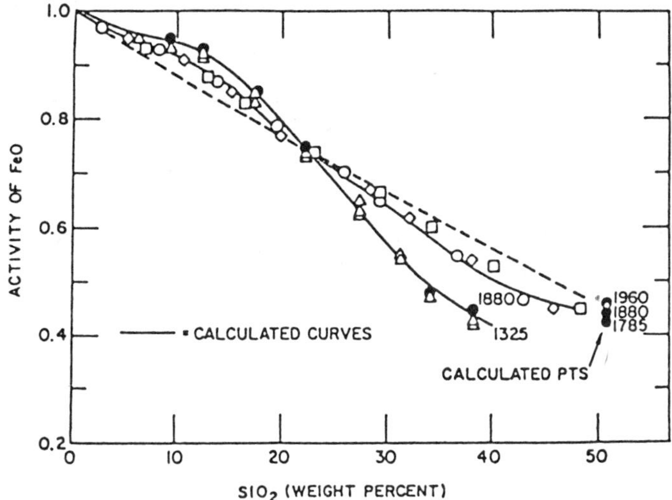

Fig. 4. Activities of "FeO" measured in iron saturaed molten FeO-SiO$_2$ at 1325°C ($\triangle$), 1785°C (o), 1880°C ($\diamond$), and 1960°C ($\square$). The two solid lines represent calculated points at 1325°C and 1880°C. The filled circles along one solid line represent individual calculated points and the three filled circles labeled 1960, 1880, and 1785 represent calculated points at three temperatures and fixed composition that illustrate the calculated temperature dependence.

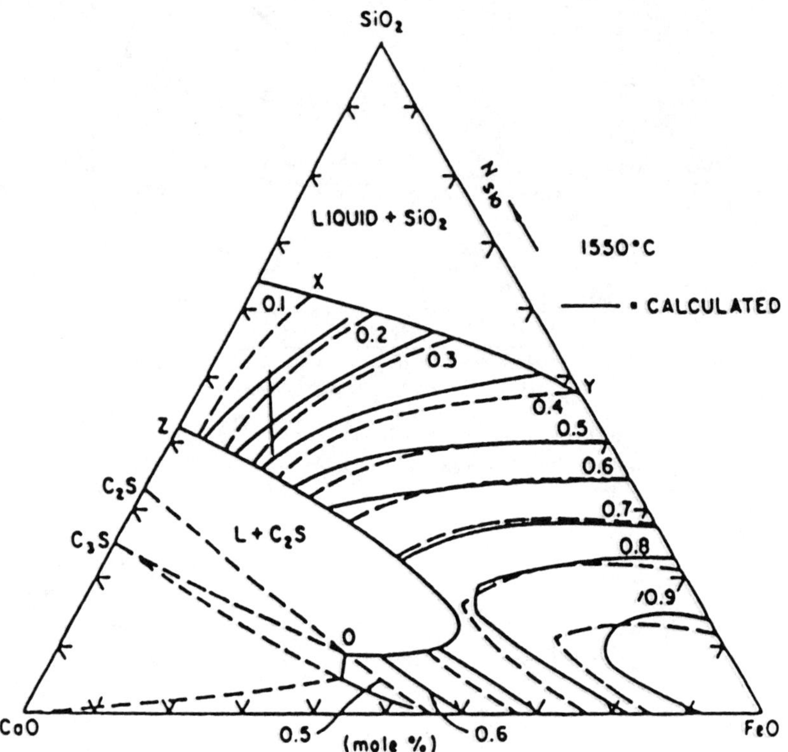

Fig. 5.   Activities of FeO in iron saturated CaO-FeO-SiO₂ at 1550°C. Dashed lines are measured and the solid lines represent our calculations.

$(n_S/(n_S + n_{Si}))$ where n_S is the number of moles of sulfide, n_{Si} is the number of moles of silica and m is the average length of the silica polymer chain. Since the activity of the sulfide is a weak function of the polymer chain length, a linear interpolation between $X_{SiO_2} = 0.333$ (m = 1) and $X_{SiO_2} = 0.500$ $((1/m)\sim0)$ appears to be consistent with data. A similar equation can be applied for other calcium salts. Equation (24) has been used to calculate sulfide capacities in binary melts in excellent agreement with measurements (Reddy and Blander, 1987; Chen, Reddy, and Blander, 1987; Reddy, Blander, and Chen, 1987) and has been extended to multicomponent melts through the use of the Flood-Grjotheim approximation (Reddy and Blander, 1988). For sulfide capacities in a silicate melt, this approximation is given by

$$ln \, C_S = \sum_i X_i \, ln \, C_S(i) \tag{25}$$

where i represents the cation of a basic oxide in the multicomponent melt, X_i is the basic cation fraction and $C_S(i)$ is the sulfide capacity of a binary melt having the oxide of the cation i as the basic component and has the same mole fraction of SiO_2 as the multicomponent melt; C_S is the sulfide capacity of the multicomponent melt. The sulfide capacity for a binary system MO-SiO_2 is given by

$$C_S = (wt\%S)\left(\frac{P_{O_2}}{P_{S_2}}\right)^{1/2} = 100W_S \cdot K_M \cdot a_{MO}\frac{X_{SiO_2}}{\overline{W}}\frac{\Phi_S}{a_{MS}} \tag{26}$$

where wt % S is the sulfur content of the slag in equilibrium with a gas having a given ratio of P_{O_2}/P_{S_2}, W_S is the atomic weight of sulfur, K_M is the equilibrium constant for the reaction

$$MO(\ell) + \frac{1}{2}S_2(g) \rightleftarrows MS(\ell) + \frac{1}{2}O_2(g) \tag{27}$$

a_{MO} is the activity of the oxide, $\overline{W}$ is the average molecular weight of the solution, and (ϕ_S/a_{MS}) can be calculated from Eq. (24).

There are many variants of these relations that can be adapted for other types of dilute solutes in silicates. A theory for more concentrated solutions of nonoxide solutes in silicates is under development.

CONCLUSIONS

In developing computer algorithms for the calculation of chemical equilibria in complex multi-component systems, it is important to use thermodynamically self-consistent equations, based on fundament concepts. For ionic systems, we have provided an all too brief annotated introduction to a number of equations that have proven useful for making predictions in multicomponent systems a priori. We have not discussed the coordination cluster theory (CCT) that applies for a dilute solute in an ionic or metallic solvent (Blander, Saboungi, and Cerisier, 1979; Blander and Saboungi, 1980, 1981). The CCT leads to predictions in a multicomponent solvent based on data for the binary solvent subsystems. There are other concepts and equations at different stages of development that should extend our capabilities further.

ACKNOWLEDGMENTS

This work was performed under the auspices of the Division of Materials Science, Office of Basic Energy Sciences, United States Department of Energy, under Contract W-31-109-ENG-38.

REFERENCES

Blander, M. and Pelton, A. D. (1983). Computer-assisted analyses of the thermodynamic properties of slags in coal combustion systems. ANL/FE-83-19, Argonne National Laboratory, Argonne, IL 60439.

Blander, M. and Pelton, A. D. (1984). Analyses and predictions of the thermodynamic properties of multicomponent silicates. Proc. of Second Int'l. Symposium on Metallurgical Slags and Fluxes, H. A. Fine and D. R. Gaskell (Eds.), TMS-AIME, Warrendale, PA 295-304.

Blander, M. and Pelton, A. D. (1987). Geochim. Cosmochim. Acta, 51, 85-95.

Blander, M. and Saboungi, M.-L. (1980). Acta Chem. Scand. Ser. A, 671-676.

Blander, M. and Saboungi, M.-L. (1981). The coordination cluster theory of solutions. In N. A. Gockcen (Ed.), Chemical Metallurgy–A Tribute to Carl Wagner, TMS-AIME, Warrendale, PA 223-231.

Blander, M., Saboungi, M.-L., and Cerisier, P. (1979). Metall. Trans., 10B, 613-633.

Blander, M. and Topol, L. E. (1966). Inorganic Chem., 5, 1641.

Blander, M. and Yosim, S. J. (1963). J. Chem. Phys., 39, 2610.

Chen, B., Reddy, R. G., and Blander, M. (1988). Sulfide capacities of CaO-FeO-SiO_2 slags. Proc. Third Int'l. Conference on Molten Slags and Fluxes, in press.

Førland, T. (1957). Properties of some mixtures of fused salts. Norg. Tek. Vitenskapsakad., Ser. 2, No. 4, 55.

Hildebrand, J. H. and R. L. Scott (1950). In The Solubility of Non-Electrolytes, 3rd ed., Reinhold Publishing Co., New York, NY 347-351.

Pelton, A. D., Bale, C. W., and Lin, P. L. (1981). Calculation of thermodynamic equilibria in the carbonate fuel cell. Report to the U.S. Department of Energy, Vol. I.

Pelton, A. and Blander, M. (1984). Computer assisted analyses of the thermodynamic properties and phase diagrams of slags. Proc. of Second Int'l. Symposium on Metallurgical Slags and Fluxes, H. A. Fine and D. R. Gaskell (Eds.), TMS-AIME, Warrendale, PA, 281-294.

Pelton, A. D. and Blander, M. (1986). Metall. Trans., 17B, 805-815.

Pelton, A. D., Eriksson, G., and Blander, M. (1988). A quasichemical model for the thermodynamic properties of multicomponent slags. Proceedings of the Third International Conference on Molten Slags and Fluxes, University of Strathclyde, Glasgow, Scotland, June 27-29, 1988, in press.

Reddy, R. G. and Blander, M. (1987). Metall. Trans. B, 18B, 591-596.

Reddy, R. G. and Blander, M. (1988). Sulfide capacities of MnO-SiO_2 Slags. Proc. 116th TMS-AIME Annual Meeting, Denver, CO, in press.

Reddy, R. G., Blander, M. and Chen, B. (1987). Thermodynamic prediction of sulfide capacities in Na_2O-SiO_2 melts. Proc. of the Joint Int'l. Symposium on Molten Salts, G. Mamantov, M. Blander, C. Hussey, C. Mamantov, M.-L. Saboungi, and J. Wilkes (Eds.), The Electrochemical Society, Inc. Pennington, NJ, 156-164.

Saboungi, M.-L. (1980). J. Chem. Phys., 73, 5800-5806.

Saboungi, M.-L., and Blander, M. (1975a). J. Chem. Phys., 63, 212-220.

Saboungi, M.-L. and Blander, M. (1975b). J. Am. Ceram. Soc., 58, 1-7.

HSC - SOFTWARE VER. 3.0 FOR THERMODYNAMIC CALCULATIONS

Antti Roine

Outokumpu Research Centre
P.O. Box 60
SF-28101 Pori, Finland

ABSTRACT

Most of the common thermodynamic problems can be solved easily and quickly using the HSC-program which is developed by **Outokumpu Oy**. The program is integrated with the thermodynamic database which contains basic thermodynamic data, i.e. **enthalpy (H)**, **entropy (S)**, **heat capacity (C)** and molecular weights, for over 2400 substances. The database can easily be expanded or edited. The program runs in all common IBM PC, XT, AT and PS/2 compatible computers.

The present version 3.0 of the HSC-program can be used to calculate:

1) Enthalpy, entropy and free energy values for given compounds and temperatures.

2) Enthalpy, entropy, Gibbs energy and equilibrium constant values for given chemical reactions and temperatures.

3) Vapor pressures for pure elements and compounds.

4) Heat and material balances as well as adiabatic temperatures for the chemical processes or for the experimental data.

5) Theoretical equilibrium compositions of phases using the free energy minimization method (**GIBBS-program**).

The HSC-program can extrapolate the thermodynamic data of the liquid state to the lower temperatures if required and also to the higher temperatures, if such data is not available in the HSC-database for these higher temperatures.

KEYWORDS

Thermodynamics; computer software; IBM PC compatibles; heat balance calculations; enthalpy; entropy; heat capacity; multiphase equilibrium calculations; free energy minimation.

INTRODUCTION

Traditionally, thermodynamic calculations have been based on experimental data which is listed in different thermodynamic data books and articles in various scientific journals. The difficult searching stage and complicated calculations have made the whole process time-consuming.

During recent years, several international data banks with thermodynamic databases and versatile programs for different applications, have been founded. For many simple thermodynamic problems it is, however, inconvenient and also expensive to make contact with these data banks using modems.

The main idea of the HSC-program is to enable most of the standard but time-consuming thermodynamic calculations to be solved by most common PC computers, so that the user can concentrate on new and interesting problems without wasting his time on the solution of old ones.

The present version of the program can calculate enthalpy, entropy and free energy values for given compounds and temperatures, equilibrium constants for chemical reactions and heat and material balances for experimental data, as well as the theoretical equilibrium compositions of several phases.

The HSC-software is integrated with the thermodynamic database which contains basic thermodynamic data, i.e. enthalpy (H), entropy (S) and heat capacity (C) values for over 2400 compounds. This database can easily be expanded or edited.

HSC-software is menu-driven (interactive) and **is intended to be used without an operation manual**. The program asks clear questions and gives all of the valid alternative answers to these questions. One important feature of the program is that the results of the calculations can be transferred to common spreadsheet programs such as MS-Excel and Lotus 123. This gives the possibility of carrying out further calculations and graphics in a spreadsheet environment.

The HSC-program will run on any IBM compatible PC, XT, AT or PS/2 computer. A hard drive is not necessarily needed, but is highly recommended. The program automatically supports mathematical co-processors, it is not copy protected, and is easy to install.

The program has been written using MS-QuickBASIC 4.0, which is one of the most advanced modular programming languages. The user is supplied with the original BASIC code, so that he can write new subprograms for his own special applications. If these new subprograms are submitted to Outokumpu Oy they can be also supplied to all other HSC-users with the new program versions, together with the name of the original author. The code for the GIBBS-program is not currently supplied.

HSC - DATABASE

The present HSC-database consists of over 2400 compounds, which means that most of the compounds whose thermodynamic data is available for higher temperatures can be found from this database. The basic thermodynamic data, i.e. enthalpy (H), entropy (S) and heat capacity (C) of elements and substances is saved in this database. The heat capacity values are given as the coefficients A, B, C and D of the heat capacity equation [1], where T is the temperature in K. The values have been saved in calories (1 cal = 4.184 J), because most of the original data is still in calories, joules can, however, also be used in the HSC-program.

$$C_p = A + B \cdot 10^{-3} \cdot T + C \cdot 10^5 \cdot T^{-2} + D \cdot 10^{-6} \cdot T^2 \qquad [1]$$

Thermodynamic data for solid and liquid phases is saved under the same formula name: for example, under the name Cu there is data for solid and liquid copper. Gas compounds, however, have their own records and names with the extension (g), for example Cu(g). If extension (l) is used the HSC-program automatically searches the records for the liquid phase, extrapolates H and S to 298.15 K and saves these with extension (l) in the OWNDB.HSC file, thereby permitting the **extrapolation of the liquid data for lower or higher temperatures.**

The HSC-database is saved in a random file which consists of records. The length of one record is 78 bytes. The following is saved in this space:

1. Chemical formula
2. Phase (s, l, g)
3. Reference
4. H = enthalpy of formation at 298.15 K,
   or the enthalpy of the phase transformation
5. S = entropy of formation at 298.15 K
   or the entropy of the phase transformation
6. A = 1. coefficient of the heat capacity equation
7. B = 2. coefficient of the heat capacity equation
8. C = 3. coefficient of the heat capacity equation
9. D = 4. coefficient of the heat capacity equation
10. T1 = lower temperature limit of the heat capacity equation
11. T2 = upper temperature limit of the heat capacity equation

Numerical values are saved in IEEE-format with approximately 7 digits' accuracy, using 4 characters (bytes) per single value. The number of records reserved for one element or compound depends on the number of temperature intervals of the heat capacity function.

So far the following references have been used for the HSC-database:

BKK: I. Barin, O. Knacke, and O. Kubascewski: Thermodynamical Properties of inorganic substances, Springer-Verlag, Berlin and New York, NY, 1973, Supplement 1977.

MILLS K. C. Mills: Thermodynamic Data for Inorganic Sulfides, Selenides and Tellurides, Butterworths, London 1974.

RUZINOV: L. P. Ruzinov and B. S. Guljanickij: Ravnovesnye prevrasoenija me-tallugiceskin reakeij, Moskva, 1975.

SAMSONOV: G. V. Samsonov: Fiziko-khimichekie svoitsva okislov, "Metallurgiya", Moscow, 1978.

There is data on further compounds available in other sources also, and this could possibly be added to the database at some future date. The whole database takes up about 350 kB of disk space.

BASIS OF CALCULATIONS

On the basis of the data in the HSC-database the enthalpy ($H°$) of a substance at temperature T and 1 atm pressure can be calculated using formula [2].

$$H°(T) = H_f(298) + \int_{298}^{T} c\,(T)\ dT + \sum H_t \qquad [2]$$

where H_f is the enthalpy of formation from the elements at 298.15 K, c is the heat capacity and H_t is the enthalpy of phase transformations. The entropy ($S°$) of a substance at temperature T can be calculated using formula [3].

$$S°(T) = S_f(298) + \int_{298}^{T} dH° \,/\, T \qquad [3]$$

where S_f is the entropy of formation at 298.15 K. On the basis of these enthalpy and entropy values the Gibbs free energy ($G°$) of formation at temperature T can be calculated from equation [4].

$$G° = H°(T) - T \cdot S°(T) \qquad [4]$$

For chemical reactions, the free energy change (ΔG_r) can be calculated as the difference between the free energies of products and reactants, which can be used to calculate the equilibrium constant (K) of the reaction, equation [5]. If $K > 1$, the reaction goes to the right, i.e. more products are generated, if $K < 1$, the reaction goes to the left, i.e. the products are not stable compounds.

$$\ln K = \Delta G_r / (-RT) \tag{5}$$

These formulas are used in different ways in all of the calculation options of the HSC-program. The calculations have in general been carried out with 7 digits of precision and in some cases with 15 digits of precision. The most important error source is the original thermodynamic data, and thus, if high accuracy is required, a survey of all the available thermodynamic data of the compounds concerned is recommended.

USING THE HSC - PROGRAM

The aim in the development work of the HSC-program has been that the using of the program must be as simple as possible, and the following means have been used towards that end: 1) the program is menu driven and 2) the user must give only the information which is essential for the specific problem, all other data which can be derived from that information or found from the HSC-database being given by the program.

The main menu of the program is shown in **Fig. 1**. Within the menu, it is possible to go to that option which is needed by choosing the corresponding letter. Either lower- or uppercase letters can be used. All valid answers are given in the menus.

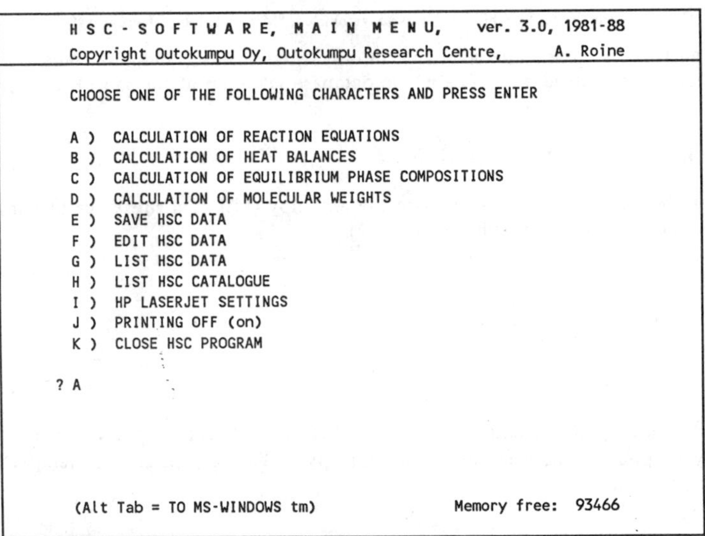

Fig. 1. The main menu of the HSC-program, options A ... K. The program shows this menu after it is started by choosing "HSC.PIF" in MS-Windows or writing "HSC" in the DOS-environment. All HSC-files must be in the \HSC-directory. The memory free value depends on the memory reserved in HSC.PIF and also on the RAM size of the computer.

Note ! *The responses of the user to the questions asked by the HSC-program have been underlined in all of the following examples !*

ENTHALPY, ENTROPY AND FREE ENERGY OF FORMATION

It is possible to calculate the enthalpy, entropy, or free energy values for elements or compounds by choosing option "A" from the HSC - MAIN MENU. The program then asks which temperature and energy units we want to use (kelvins or celsius and calories or joules). After these questions we can give the chemical formula of the compound, and the temperature or temperature interval in which we are interested.

An example of the input and output of the program is shown in **Fig. 2**. The accurary of the values is 7 digits, although the accuracy of the original thermodynamic data is in general much smaller.

```
TEMPERATURES IN KELVINS (=K) OR IN CELSIUS (=C) ?
k

RESULTS IN CALORIES (=C) OR IN JOULES (=J) ?
c

REACTION EQUATION OR CHEMICAL FORMULA ? (ENTER = MENU)
ZnO

TEMPERATURE(S) / K ?     (XXX-XXXX = Temperature interval !)
1473.15,1573.15

   ZnO

        T          Cp         H          S          G  REF.
        K  cal/(mol*K)  kcal/mol  cal/(mol*K)   kcal/mol
  1473.15      13.407   -68.754     29.365   -112.013  BKK
  1573.15      13.541   -67.407     30.250   -114.995  BKK


SAVE RESULTS USING NAME ? ( ENTER = NOT TO BE SAVED)
ZnO

REACTION EQUATION OR CHEMICAL FORMULA ? (ENTER = MENU, E = PREVIOUS EQUATION)
O2(g)

TEMPERATURE(S) / K ?     (XXX-XXXX = Temperature interval !)
273.15-1200

GIVE STEP FOR THE TEMPERATURE INTERVAL ? (ENTER = 100 DEGREES)
-


   O2(g)

        T          Cp         H          S          G  REF.
        K  cal/(mol*K)  kcal/mol  cal/(mol*K)   kcal/mol
   273.15       6.897    -0.174     48.396    -13.393  BKK
   300.00       7.016     0.013     49.048    -14.702  BKK
   400.00       7.310     0.731     51.111    -19.714  BKK
   500.00       7.500     1.472     52.764    -24.910  BKK
   600.00       7.649     2.229     54.145    -30.257  BKK
   700.00       7.778     3.001     55.334    -35.733  BKK
   800.00       7.898     3.785     56.380    -41.319  BKK
   900.00       8.011     4.580     57.317    -47.005  BKK
  1000.00       8.120     5.387     58.167    -52.780  BKK
  1100.00       8.227     6.204     58.946    -58.636  BKK
  1200.00       8.332     7.032     59.666    -64.567  BKK


SAVE RESULTS USING NAME ? ( ENTER = NOT TO BE SAVED)
-


REACTION EQUATION OR CHEMICAL FORMULA ? (ENTER = MENU, E = PREVIOUS EQUATION)
-
```

Fig. 2. Enthalpy, entropy and free energy of formation (option A). **Note:** *The responses of the user have been underlined.*

ENTHALPY AND FREE ENERGY CHANGE OF THE REACTION

The thermodynamic values for chemical reactions can be calculated with the same calculation option "A" which is used for individual elements and compounds, the only difference being that the user gives the reaction equation instead of the individual formula. The program automatically identifies the reaction equations from the individual formulas. An important feature of the HSC-program is that it checks the material balance of the reaction. If this is incorrect the equation can be edited. The range of the calculated equilibrium constant can be from -10^{308} to $+10^{308}$, and the range of the other variables from -10^{38} to $+10^{38}$.

An example of the oxidation of SO_2-gas is shown in **Fig. 3**. From industrial practice, we know that SO_2 does not react with water and oxygen to form sulphuric acid at lower temperatures, however on the basis of the equilibrium constant we can see that direct oxidation of the SO_2 to sulphuric acid is possible if an appropriate catalyst is used, because K > 1 at temperatures lower than 550 °C. After the calculations the results can be saved in such form that they can be transferred to **MS-Excel** or to **Lotus 123** for further calculations or graphics, **Fig. 4**.

VAPOR PRESSURES OF PURE ELEMENTS AND COMPOUNDS

Vapor pressures can be calculated by writing the reaction equation for the corresponding vaporization reaction (option "A" in the main menu). In this case, if the activity of the substance is 1, the vapor pressure in atm is equal to the equilibrium constant of the reaction. An example of this kind of calculation is shown in **Fig. 5**. Because the activity of the zinc is 1 the vapor pressure is equal to the equilibrium constant, as can be seen from equation [6].

$$K = p_{Zn} \, / \, a_{Zn} \tag{6}$$

If the gas phase consists of several gas components, reaction equations must be written for all of them. The total vapor pressure is then the sum of these partial pressures.

HEAT AND MATERIAL BALANCES

In many cases it is important to calculate the material and heat balances for theoretical or experimental processes. This can be done by choosing "B" from the main menu. After that the program makes the questions which are necessary, searches for the basic data and calculates the results. An example of the input information, which must be given is shown in **Fig. 6**, and the results in **Fig. 7**.

In all calculation options of the HSC-program, formulas can be written using normal notation, the only exception being formulas which start with a number, for example *2FeO*SiO2. It is also important to write the formulas in the same form which has been used in the HSC-database, otherwise the program will not be able to find the compound. The correct forms can be listed by choosing "H" from the main menu.

Option "B" can also be used to calculate adiabatic temperature of a flame, because the program extrapolates the final temperature of the products on the basis of the heat balance. An example of such calculations is given in **Fig. 8**. The heat losses and extra input heat must be set to zero in accordance with the definition of the adiabatic process.

MOLECULAR WEIGHTS OF THE COMPOUNDS

Atomic and molecular weights can easily be calculated by choosing option "D" from the main menu. Any kind of formula can be written and not only those which can be found from the HSC-database. This option also gives the contents of the elements in the compound as weight and atomic percents.

```
TEMPERATURES IN KELVINS (=K) OR IN CELSIUS (=C) ?
C

RESULTS IN CALORIES (=C) OR IN JOULES (=J) ?
C

REACTION EQUATION OR CHEMICAL FORMULA ? (ENTER = MENU)
SO2(g)+H2O(g)+1/2O2(g)=H2SO4(g)

TEMPERATURE(S) / C ?     (XXX-XXXX = Temperature interval !)
0-1000

GIVE STEP FOR THE TEMPERATURE INTERVAL ? (ENTER = 100 DEGREES)
-

     SO2(g)+H2O(g)+1/2O2(g)=H2SO4(g)

  TEMPERATURE                ENTHALPY  FREE ENERGY  EQUILIBRIUM CONSTANT
     T           T           CHANGE H   CHANGE G     OF THE REACTION K
     K           C           kcal/mol   kcal/mol
    273.15      0.00         -48.202    -31.926          3.518D+025
    373.15    100.00         -48.318    -25.938          1.558D+015
    473.15    200.00         -48.193    -19.952          1.646D+009
    573.15    300.00         -47.935    -14.007          2.195D+005
    673.15    400.00         -47.591     -8.115          4.313D+002
    773.15    500.00         -47.188     -2.278          4.406D+000
    873.15    600.00         -46.742      3.503          1.328D-001
    973.15    700.00         -46.267      9.232          8.444D-003
   1073.15    800.00         -45.772     14.910          9.188D-004
   1173.15    900.00         -45.267     20.542          1.489D-004
   1273.15   1000.00         -44.758     26.131          3.266D-005
  H2SO4(g) EXTRAPOLATED FROM  1000 K


SAVE RESULTS USING NAME ? ( ENTER = NOT TO BE SAVED)
H2SO4
```

Fig. 3. Enthalpy and free energy change of the reaction (option A).

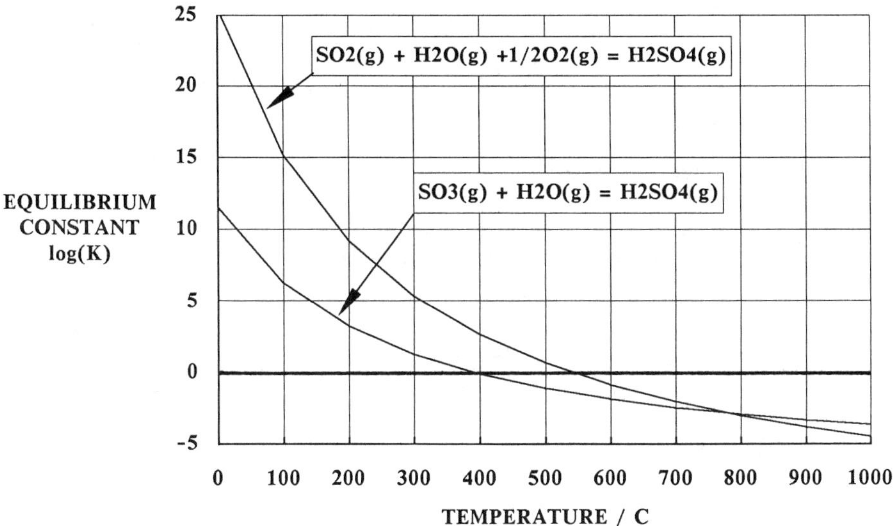

Fig. 4. Results of Fig. 3 represented graphically using MS-Excel.

```
TEMPERATURES IN KELVINS (=K) OR IN CELSIUS (=C) ?
C

RESULTS IN CALORIES (=C) OR IN JOULES (=J) ?
J

REACTION EQUATION OR CHEMICAL FORMULA ? (ENTER = MENU)
Zn=Zn(g)

TEMPERATURE(S) / C ?     (XXX-XXXX = Temperature interval !)
-200-1500

GIVE STEP FOR THE TEMPERATURE INTERVAL ? (ENTER = 100 DEGREES)
-


    Zn=Zn(g)

    TEMPERATURE                 ENTHALPY  FREE ENERGY  EQUILIBRIUM CONSTANT
         T          T           CHANGE H   CHANGE G      OF THE REACTION K
         K          C            kJ/mol     kJ/mol
       73.15     -200.00        131.769    122.312         4.497D-088
      173.15     -100.00        130.962    109.893         7.008D-034
      273.15        0.00        130.512     97.837         1.946D-019
      373.15      100.00        130.031     85.951         9.276D-013
      473.15      200.00        129.460     74.206         6.415D-009
      573.15      300.00        128.780     62.593         1.973D-006
      673.15      400.00        127.983     51.109         1.081D-004
      773.15      500.00        119.640     40.612         1.803D-003
      873.15      600.00        118.580     30.456         1.506D-002
      973.15      700.00        117.521     20.422         8.012D-002
     1073.15      800.00        116.461     10.497         3.083D-001
     1173.15      900.00        115.402      0.671         9.335D-001
     1273.15     1000.00        114.343     -9.065         2.355D+000
     1373.15     1100.00        113.283    -18.718         5.153D+000
     1473.15     1200.00        112.224    -28.293         1.008D+001
     1573.15     1300.00        111.164    -37.796         1.799D+001
     1673.15     1400.00        110.105    -47.232         2.983D+001
     1773.15     1500.00        109.046    -56.605         4.652D+001
     Zn EXTRAPOLATED FROM   1180 K


SAVE RESULTS USING NAME ? ( ENTER = NOT TO BE SAVED)
PZn
```

Fig. 5. Vapor pressures of the pure elements and compounds (Opt. A).
In this example the vapor pressure of zinc in atm is the same as the
equilibrium constant. Note: *The responses of the user have been
underlined.*

```
H S C - S O F T W A R E,   H E A T   B A L A N C E   M E N U   3.0
Copyright Outokumpu Oy, Research Centre, Pori, Finland, A. Roine

WHICH UNITS DO YOU WANT TO USE

   A )   kg FOR SOLIDS AND LIQUIDS AND Nm3 FOR GASES
   B )   kmol FOR ALL PHASES
   C )   BACK TO THE MAIN MENU


 ? b


DO YOU WANT TO GIVE INPUT FROM THE

   A )   KEYBOARD
   B )   FILE


 ? a



NOTE ! TYPE 0,0,0 WHEN YOU WANT TO MOVE TO THE NEXT STAGE
         EDIT USING THE CURSOR AND PRESS ENTER

INPUT SUBSTANCES OF THE PROCESS:

FORMULA,              TEMPERATURE/C, AMOUNT/kmol ?
Cu2S,25,1
FeS,25,2
SiO2,25,1
O2(g),25,4
0,0,0


EXTRA INPUT HEAT OF THE PROCESS / MJ ?
0


OUTPUT SUBSTANCES OF THE PROCESS:

FORMULA,              TEMPERATURE/C, AMOUNT/kmol ?
Cu,1100,2
*2FeO*SiO2,1400,1
SO2(g),1100,3
0,0,0


HEAT LOSS IN THE PROCESS / MJ ?
500


PROGRAM IS SEARCHING FOR DATA FROM THE HSC DATABASE, WAIT A MOMENT . . .
```

Fig. 6. Input data for the heat and material balance calculation (Option B), output of the program is given in the **Fig. 7**. This option can be used, for example, to check the material and heat balances of the experimental results. Note: *The responses of the user have been underlined.*

```
****** MASS BALANCE I ******************** MASS BALANCE II ************

          INPUT    OUTPUT    DIFF.      INPUT    OUTPUT     DIFF.
          kmol     kmol      kmol         kg       kg        kg

Cu        2.00     2.00      0.00       127.09   127.09     0.00
Fe        2.00     2.00      0.00       111.69   111.69     0.00
O        10.00    10.00      0.00       159.99   159.99     0.00
S         3.00     3.00      0.00        96.18    96.18     0.00
Si        1.00     1.00      0.00        28.09    28.09     0.00


**************************** HEAT BALANCE ****************************

INPUT:
=====
COMPONENT            TEMPERATURE      AMOUNT       kg      HEAT CONTENT
                        C            kmol        Nm3           MJ


Cu2S                  25.00          1.00       159.15      -79.50
FeS                   25.00          2.00       175.81     -200.83
SiO2                  25.00          1.00        60.08     -908.35
O2(g)                 25.00          4.00        89.65        0.00
EXTRA INPUT HEAT OF THE PROCESS                              0.00

OUTPUT:
======
COMPONENT            TEMPERATURE      AMOUNT       kg      HEAT CONTENT
                        C            kmol        Nm3           MJ


Cu                  1100.00          2.00       127.09       86.08
*2FeO*SiO2          1400.00          1.00       203.78    -1114.05
SO2(g)              1100.00          3.00        67.24     -726.42
HEAT LOSSES                                                 500.00
___________________________________________________________________

HEAT IS RELEASED / MJ                                       -65.71

CALCULATING FINAL TEMPERATURE OF THE PRODUCTS, PRESS ESC TO SKIP . . . .

FINAL TEMPERATURE OF THE PRODUCTS     1386.3 C


SAVE RESULTS USING NAME ? ( ENTER = NOT TO BE SAVED)
CuSMELT

DO YOU WANT TO CONTINUE, Y = YES , ENTER = MAIN MENU ?

—
```

Fig. 7. Results of the heat and material balance calculations (Opt. B).

```
******* MASS BALANCE I ******************** MASS BALANCE II ************

        INPUT     OUTPUT     DIFF.       INPUT     OUTPUT     DIFF.
        kmol      kmol       kmol          kg        kg        kg

C        4.00      4.00      0.00        48.04     48.04      0.00
H       10.00     10.00      0.00        10.08     10.08      0.00
N       49.00     49.00      0.00       686.33    686.33      0.00
O       13.00     13.00      0.00       207.99    207.99      0.00


*************************** HEAT BALANCE ****************************

INPUT:
=====
COMPONENT              TEMPERATURE       AMOUNT          kg    HEAT CONTENT
                           C             kmol           Nm3            MJ

C4H10(g)                 25.00           1.00         22.41       -124.68
O2(g)                    25.00           6.50        145.69          0.00
N2(g)                    25.00          24.50        549.13          0.00
N2(g)                    25.00          24.50        549.13          0.00
EXTRA INPUT HEAT OF THE PROCESS                                     0.00

OUTPUT:
======
COMPONENT              TEMPERATURE       AMOUNT          kg    HEAT CONTENT
                           C             kmol           Nm3            MJ

CO2(g)                 2000.00           4.00         89.65      -1143.46
H2O(g)                 2000.00           5.00        112.07       -776.36
N2(g)                  2000.00          24.50        549.13       1613.83
HEAT LOSSES                                                         0.00
______________________________________________________________________

HEAT IS RELEASED / MJ                                            -181.31

PROGRAM IS SEARCHING FOR DATA FROM THE HSC DATABASE, WAIT A MOMENT . . .

FINAL TEMPERATURE OF THE PRODUCTS     2124.2 C


SAVE RESULTS USING NAME ? ( ENTER = NOT TO BE SAVED)
BUTANE

DO YOU WANT TO CONTINUE, Y = YES , ENTER = MAIN MENU ?

  _
```

Fig. 8. Adiabatic temperature of the reaction. In this example the adiabatic temperature of butane flame (2124.2 °C) has been calculated using option "B" and assuming that there is no heat exchange with the surroundings.

EQUILIBRIUM COMPOSITIONS OF THE PHASES

The theoretical equilibrium compositions of the phases can be calculated by choosing option "C" from the main menu. This option makes an input file for the **GIBBS**-program, which calculates the equilibrium compositions of the phases. This program which was originally made by **Timo Talonen** and **Timo Syväjärvi** at the beginning of the 1970s has recently been translated to PC computers and modified to read the input files which are made by the HSC-program.

Equilibrium compositions are calculated using the free energy minimization method, which makes the program completely universal for all kinds of equilibrium calculations. The methods which are based on the equilibrium constants of different chemical reactions are not so easy to use because the independent reaction equations must first be derived.

The theoretical equilibrium calculations are very important for the development work of chemical and metallurgical processes, because they offer the possibility for simulating the effects of raw material amounts, activities, reaction temperatures, pressures, etc. on the compositions of the prevailing phases.

The HSC-program asks following information from the user:

1) **Formulas** of the components in each phase.

2) **Activity coefficients** for such components whose activity coefficients are not 1 (one is always the default setting).

3) **Input amounts** (in mol or kg) and **temperatures** of raw materials.

4) The raw material the amount of which will be increased stepwise and the size and number of the steps. This option makes it possible to calculate several equilibria in succession and to study the effects of the different variables on the equilibrium compositions.

5) **Reaction temperature** and **pressure**.

6) **Name for the input file.**

All other information is taken from the HSC-database or calculated by the HSC-program. The GIBBS-program generates the starting estimate for the equilibrium compositions, and starts to minimize the free energy by changing the composition of the phases. The number of iterations which are carried out depends on the precision criteria installed in the program. In the current version of the GIBBS-program the criteria is adjusted so that the final **error is only 0.1 to 0.5 %** for those components whose mole amounts are low and even smaller for those components whose mole amounts are high in the equilibrium state. This is a comparatively small error in comparison with errors of other similar programs, and, in addition, these other programs are usually much more difficult to use.

An example of an equilibrium calculation is shown in **Fig. 9**. The effect of the oxygen feed on this system is presented graphically in **Fig. 10** the points of which are calculated by increasing the input amount of the oxygen stepwise.

The errors can easily be estimated by comparing some of the equilibrium constants which can be calculated from the results of the GIBBS-program to those equilibrium constants which can be calculated by option "A". For example, the equilibrium constant of the reaction $CO(g)+1/2O_2(g)=CO_2(g)$ at 1300°C is 7.127E+4 when calculated using option "A" and 7.115E+4 when calculated from the equilibrium results, Fig. 9, the error thus being only 0.17 %.

Although the equilibrium calculations are very easy to carry out with the HSC- and GIBBS-programs some experience and knowledge of the basic principles of thermodynamics is, however, needed, since the probability of making serious errors in basic assumptions will otherwise be quite high.

There are several aspects which should be taken into account, because these might have considerable effects on the results and could, on the other hand, save a great deal of work. For example:

1) Before any calculations are made, the prevailing elements and compounds must be considered carefully in order to find all those components and phases which might be stable in the system.

2) Phase diagrams, as well as solubility and activity data could well be very useful when evaluating stable phase combinations.

3) If the activity coefficient of some component is set to be higher than 1 then it could be wise to establish its own phase for that component, in order to make it possible for the said component to precipitate when the activity reaches a value of one.

4) Raw materials must be given in their actual state (s, l, g), in order to get correct enthalpy and entropy values for the reactions.

5) Kinetic aspects must always taken into account when evaluating the final results.

FUTURE OF THE HSC - PROGRAM

The development work on the HSC-program will be continued at Outokumpu Oy on the basis of the current problems. Other HSC-users can, however, also join this development work by submitting their own numerical or graphical subprograms to Outokumpu Oy. In this way these new subprograms or modules can also be delivered to all other HSC-users with the new program versions together with the names of the original authors.

```
GIBBS-PROGRAM VER. 6.22, FOR MULTIPHASE EQUILIBRIUM COMPOSITION CALCULATIONS
Copyright (C) Outokumpu Oy, Outokumpu Research Centre, Pori, Finland 1974-88
T. Talonen, T. Syvajarvi and A. Roine

ZINC3

Temperature              1573.15 K
Pressure                 1.00E+00 bar
Volume                   1.96E+01 m3 ( NPT )
Reaction enthalphy       1.87E+04 kJ
Reaction entropy         1.45E+05 J/K
Iterations                    33   (Limit 80)

PHASE 1:          INPUT AMOUNT EQUIL AMOUNT EQUIL PRESS ACTIVITY  ACTIVITY
                  mol          mol          bar    COEFFICI
CO(g)             0.0000E+00   4.7228E+02   5.400E-01  1.000  5.400E-01
CO2(g)            0.0000E+00   1.8992E+02   2.170E-01  1.000  2.170E-01
Zn(g)             0.0000E+00   1.8135E+02   2.070E-01  1.000  2.070E-01
H20(g)            0.0000E+00   1.2713E+01   1.450E-02  1.000  1.450E-02
H2(g)             0.0000E+00   1.0714E+01   1.220E-02  1.000  1.220E-02
N2(g)             6.3730E+00   6.3730E+00   7.290E-03  1.000  7.290E-03
Pb(g)             0.0000E+00   9.1523E-01   1.050E-03  1.000  1.050E-03
PbS(g)            0.0000E+00   3.8255E-01   4.370E-04  1.000  4.370E-04
H2S(g)            0.0000E+00   7.2824E-02   8.330E-05  1.000  8.330E-05
S02(g)            0.0000E+00   1.1488E-02   1.310E-05  1.000  1.310E-05
ZnS(g)            0.0000E+00   8.5403E-03   9.760E-06  1.000  9.760E-06
S2(g)             0.0000E+00   6.1217E-03   7.000E-06  1.000  7.000E-06
SO(g)             0.0000E+00   2.0164E-03   2.310E-06  1.000  2.310E-06
ZnO(g)            0.0000E+00   5.6240E-05   6.430E-08  1.000  6.430E-08
02(g)             3.1010E+02   2.7900E-08   3.190E-11  1.000  3.190E-11
S03(g)            0.0000E+00   1.8107E-09   2.070E-12  1.000  2.070E-12
H2S04(g)          0.0000E+00   1.1006E-15   1.290E-18  1.000  1.490E-21
Total:            316.473      874.748      1.000

PHASE 2:                                   MOLE FRACT
FeO(l)            0.0000E+00   3.8106E+01   5.460E-01  1.000  5.460E-01
Si02(l)           2.0000E+00   1.0014E+01   1.440E-01  1.000  1.440E-01
CaO*Si02(l)       4.2000E+00   7.5000E+00   1.080E-01  1.000  1.080E-01
Al203*Si02        3.5000E+00   5.9000E+00   8.460E-02  1.000  8.460E-02
FeS(l)            0.0000E+00   5.6103E+00   8.040E-02  1.000  8.040E-02
ZnO               0.0000E+00   2.2365E+00   3.210E-02  3.000  9.620E-02
*2FeO*Si02(l)     0.0000E+00   3.8564E-01   5.530E-03  1.000  5.530E-03
ZnO*Fe203         2.1400E+01   6.3639E-03   9.120E-05  1.000  9.120E-05
PbO(l)            0.0000E+00   2.2218E-03   3.180E-05  1.000  3.180E-05
ZnSO4             3.1000E+00   7.9306E-13   1.140E-14  1.000  1.140E-14
PbSO4             1.3000E+00   1.5426E-13   2.210E-15  1.000  2.210E-15
Total:            35.500       69.761       1.000

PHASE 3:                                   MOLE FRACT
FeS               1.7000E+00   0.0000E+00   0.000E+00  1.000  0.000E+00
Al203*Si02        1.4000E+00   0.0000E+00   0.000E+00  1.000  0.000E+00
CaO*Si02          2.3000E+00   0.0000E+00   0.000E+00  1.000  0.000E+00
C                 6.6220E+02   0.0000E+00   0.000E+00  1.000  0.000E+00
H20               2.3500E+01   0.0000E+00   0.000E+00  1.000  0.000E+00
Total:            691.100      .000         0.000

PHASE 4:                                   MOLE FRACT
ZnO               1.5910E+02   0.0000E+00   0.000E+00  1.000  0.000E+00
Total:            159.100      .000         0.000

PHASE 5:                                   MOLE FRACT
CaO*Al203         1.0000E+00   0.0000E+00   0.000E+00  1.000  0.000E+00
Si02              1.0400E+01   0.0000E+00   0.000E+00  1.000  0.000E+00
Total:            11.400       .000         0.000
```

Fig. 9. Results of the equilibrium composition calculations (option C). In this example zinc oxides have been reduced with the coal, silica flux and oxygen to form fajalite slag and zinc gas. Activity coefficient of the ZnO in the slag is assumed to be 3 (CO/CO2 = 2.45).

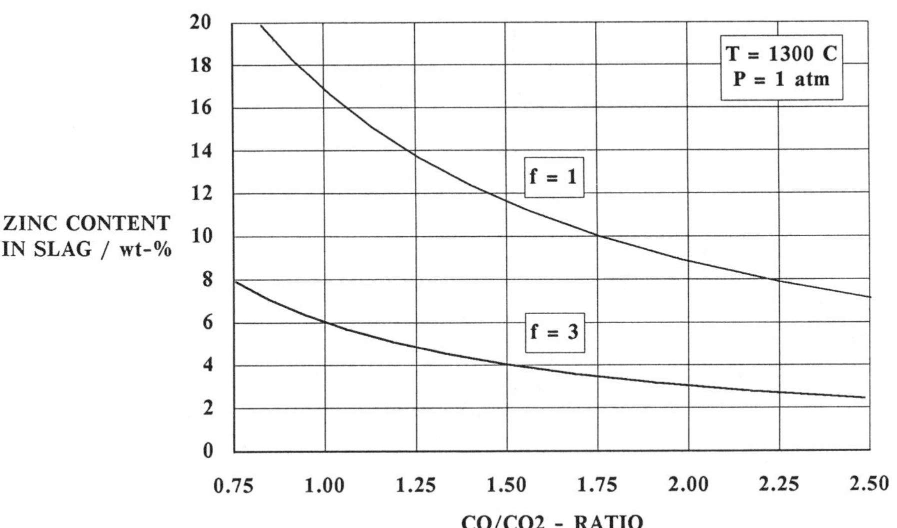

Fig. 10. Zinc content in the slag as a function of the CO/CO2-ratio in the gas assuming ZnO-activity coefficients to be 1 and 3 in the slag. The equilibrium compositions have been calculated by increasing the amount of oxygen stepwise in the system of **Fig. 9**. The figure was drawn using MS-Excel.

COMPLEX CHEMICAL EQUILIBRIA CALCULATIONS WITH THE THERMODATA SYSTEM

B.Cheynet

Thermodata, Domaine Universitaire de Grenoble, BP 66,
38402 Saint Martin d'Hères Cédex - France.

ABSTRACT

Since 1974, THERMODATA is engaged in the development of a
thermodynamic databank for inorganic and metallurgical systems and
its applications to practical problems. The on-line service enables
the user to undertake complex calculations of chemical equilibria,
using data automatically retrieved from the databases and to present
the results in a form appropriate to the application. The
capabilities of the system are discussed with emphasis on the
calculation methods used.

KEYWORDS

Databank ; metallurgy ; inorganic chemistry ; thermodynamics ; Gibbs
energy minimization ; complex chemical equilibrium ; thermodynamic
data.

INTRODUCTION

THERMODATA is an Integrated Information System providing complete
information on the physical chemistry of metals, alloys, inorganic
compounds and gaseous phases.

The main purposes are :

- to create, maintain and improve a bank of thermochemical data and
  to distribute information to users all over the world through
  networks,

- to solve any problem by developing computer software especially
  designed for the needs of industry and research engineering.

Four databases are integrated in the system :

THERMDOC   : bibliographical database,

```
THERMOCOMP  : thermodynamical   properties   of inorganic elements   and
              compounds,
THERMALLOY  : thermodynamical   properties   of metallic   alloys   and
              multicomponent systems,
THERMOSALT  : thermodynamical properties of molten salt mixtures.
```

The architecture of the system looks like an Expert-System. It holds five principal components :

- user interface,
- working codes,
- knowledge base,
- data-bases,
- expert interaction module.

User's request is transfered through the user interface to the working codes which combine knowledge and data to produce a coherent answer to a particular problem.

The working codes and knowledge data base have a modular architecture which allows a total independency of the different parts, linked by interfaces. So, they can integrate at any time new calculation procedures, models or data.

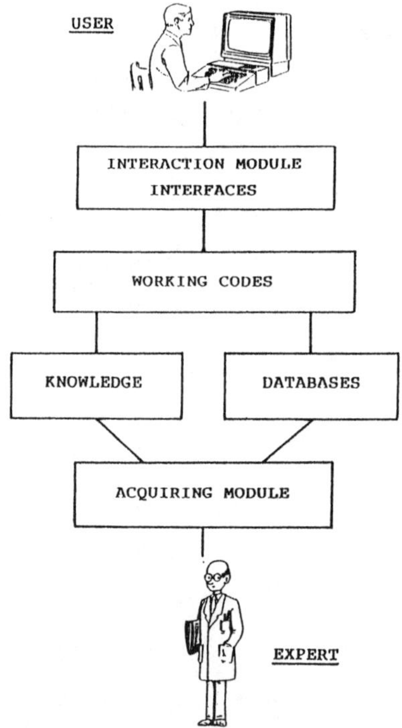

Fig. 1. Architecture of the Thermodata Integrated System.

In the particular case of complex chemical equilibria calculation, two codes have been integrated : Supercalc and Equicomp.

Supercalc is a general method for calculating the state of equilibrium of a multiphase multicomponent system, using a "hill climbing" minimization procedure. We principally use this code for the calculations of phase equilibria in multicomponent alloy systems.

Equicomp is a complex equilibrium determination software. The determination of the parameters is treated by a "direct search" minimization procedure. We use this code for gaz/condensed equilibria and very large multiphase multicomponent systems (25 elements, 500 compounds).

SUPERCALC

This method, developped and presented by Chevalier (1981) and Barbier and co-workers (1983) is based on the Gibbs free energy minimization of the system.

A multicomponent system, containing Φ phases at a given temperature and submitted to an unchanging pressure, can be defined by

$$P_1 n + P_2 n + \ldots P_f n + \ldots + P_\Phi n$$

where P_f is the fraction of n initial moles in phase f and satisfies the relation

$$\sum_{f=1}^{f=\phi} P_f = 1 \tag{1}$$

Taking the phases separately, one can determine the Gibbs free energy of the system, ΔG, which is equal to the summation of the Gibbs free energy of each phase, referred to the pure components in the same structural state. It can be expressed as

$$\Delta G = \sum_{f=1}^{\phi} P_f \Delta G^f = \sum_{f=1}^{\phi-1} P_f (\Delta G^f - \Delta G^\phi) + \Delta G^\phi \tag{2}$$

In order to define the Gibbs free energy of the system, the Gibbs free energy of each phase, ΔG^f, should be known as a function of temperature and composition. Usually, ΔG^f can be written at constant pressure

$$\Delta G^f = \Delta G^f (x_1^f, x_2^f, \ldots x_i^f, \ldots x_{N-1}^f, x_N^f, T) \tag{3}$$

Or in a condensed form

$$\Delta G^f = \Delta G^f (x_i^f, T) \tag{4}$$

with i = 1, ... N.

Since x_i^f is the atomic fraction of i in phase f, we have

$$\sum_{i=1}^{N} x_i^f = 1 \tag{5}$$

In eqn. (3), the Gibbs free energy of phase f can be related to the pure components in a given structural state. In eqn. (4), the atomic fractions x_i^f may be constants, dependent variables or independent variables, according to the nature of the phase. The following inequality is proved by the number of independent variables, N_ν^f, of phase f

$$0 \leqslant N_\nu^f \leqslant N - 1 \tag{6}$$

The dependent variables are expressed by linear combinations of the independent variables and the constants of the phases. On the assumption that the atomic fraction in phase f is

$$x_i^f = \frac{n_i^f}{nP^f} \tag{7}$$

and that

$$n_i = \sum_{i=1}^{\phi} n_i^f \tag{8}$$

the atomic fraction of components i of the initial melt is related to the different atomic fractions of the same component in the different phases by the equation

$$x_i = \sum_{f=1}^{\phi} P_f x_i^f = x_i^\phi + \sum_{f=1}^{\phi-1} P_f (x_i^f - x_i^\phi) \tag{9}$$

with i = 1, ..., N. These relations are called the generalized lever rule.

The compositions of the phases at the equilibrium are obtained by minimizing the Gibbs free energy of the system [expressed by eqn. (2)] in relation to the independent variables and on the assumption that the lever rule holds, using a hill-climbing technique described by Nelder and Mead (1965).

These authors describe the "hill-climbing" procedure in their paper :

We consider, initially, the minimization of a function of n variables, without constraints. P_0, P_1, ... P_n are the $(n + 1)$ points in n-dimensional space defining the current "simplex". [The simplex will not, of course, be regular in general]. We write y_i for the function value at P_i, and define.

h as the suffix such that $y_h = \max_i (y_i)$ [h for "high"]

and

l as the suffix such that $y_l = \min_i (y_i)$ [l for "low"].

Further we define $\overline{P}$ as the centroid of the points with $i \neq h$, and write $[P_iP_j]$ for the distance from P_i to P_j. At each stage in the process P_h is replaced by a new point ; three operations are used - reflection, contraction, and expansion. These are defined as follows : the reflection of P_h is denoted by P^*, and its co-ordinates are defined by the relation

$$P^* = (1 + \alpha)\,\overline{P} - \alpha P_h$$

where α is a positive constant, the reflection coefficient. Thus P^* is on the line joining P_h and $\overline{P}$, on the far side of $\overline{P}$ from P_h with $[P^*\overline{P}] = \alpha\,[P_h\overline{P}]$. If y^* lies between y_h and y_l, then P_h is replaced by P^* and we start again with the new simplex.

If $y^* < y_l$, i.e. if reflection has produced a new minimum, then we expand P^* to P^{**} by the relation

$$P^{**} = \gamma P^* - (1 - \gamma)\overline{P}.$$

The expansion coefficient γ, which is greater than unity, is the ratio of the distance $[P^{**}\overline{P}]$ to $[P^*\overline{P}]$. If $y^{**} < y_l$ we replace P_h by P^{**} and restart the processs ; but if $y^{**} > y_l$ then we have a failed expansion, and we replace P_h by P^* before restarting.

If on reflecting P to P^* we find that $y^* > y_l$ for all $i \neq h$, i.e. that replacing P by P^* leaves y^* the maximum, then we define a new P_h to be either the old P_h or P^*, whichever has the lower y value, and form

$$P^{**} = \beta P_h + (1 - \beta)\overline{P}.$$

The contraction coefficient β lies between 0 and 1 and is the ratio of the distance $[P^{**}\overline{P}]$ to $[P\overline{P}]$. We then accept P^{**} for P_h and restart, unless $y^{**} > \min(y_h, y^*)$, i.e. the contracted point is worse than the better of P_h and P^*. For such a failed contraction we replace all the P_i's by $(P_i + P_l)/2$ and restart the process.

A failed expansion may be thought of as resulting from a lucky foray into a valley (P*) but at an angle to the valley so that P** is well up on the opposite slope. A failed contraction is much rarer, but can occur when a valley is curved and one point of the simplex is much farther from the valley bottom than the others ; contraction may then cause the reflected point to move away from the valley bottom instead of towards it. Further contractions are then useless. The action proposed contracts the simplex towards the lowest point, and will eventually bring all points into the valley. The coefficients α, β, γ give the factor by which the volume of the simplex is changed by the operations of reflection, contraction or expansion respectively. Specific criterion is used for halting the procedure.

The reliability of this method depends straight on adopted thermodynamic data quality. For this reason we establish for each one of the systems a set of self-consistent thermodynamic data after a compilation of all thermodynamic properties as well as phase diagrams experimental informations available.

```
****** THERMODATA ****** Banque Multicomposants ******871007**
Systeme : AG-SN
             Phase : liquid
                   T= 1300.00K unite: J
               Ref AG:Agliq (stable)
               Ref SN:Snliq (stable)


                      Grandeurs partielles molaires AG
     x      Delta Gp    Delta Hp    Delta Sp       Act
    AG          AG          AG          AG          AG
  1.000           0           0     0.0000      1.0000
  0.900       -1489       -1190     0.2308      0.8713
  0.800       -3660       -3380     0.2154      0.7128
  0.700       -6303       -5163     0.8769      0.5581
  0.600       -9226       -5887     2.5692      0.4259
  0.500      -12299       -5299     5.3846      0.3205
  0.400      -15501       -3862     8.9538      0.2383
  0.300      -19014       -1875    13.1846      0.1722
  0.200      -23453         287    18.2615      0.1142
  0.100      -30827        2312    25.4923      0.0577
  0.000      -Infini        4408     Infini      0.0000


                      Grandeurs partielles molaires SN
     x      Delta Gp    Delta Hp    Delta Sp       Act
    SN          SN          SN          SN          SN
  0.000      -Infini      -40501     Infini      0.0000
  0.100      -38301      -15841    17.2769      0.0289
  0.200      -25654       -2854    17.5385      0.0932
  0.300      -17632        2707    15.6462      0.1957
  0.400      -12159        4041    12.4615      0.3247
  0.500       -8380        3459     9.1077      0.4606
  0.600       -5749        2251     6.1538      0.5875
  0.700       -3856        1185     3.8769      0.6999
  0.800       -2390         449     2.1846      0.8016
  0.900       -1137         104     0.9538      0.9002
  1.000           0           0     0.0000      1.0000
```

Fig. 2. Thermalloy numerical output.

We modelize the different phases - solutions and compounds - and we proceed to parameters optimization by using the software developped by Lukas and co-workers (1973).

Coherence of the obtained data permits their direct application to Supercalc calculation software.

In the THERMODATA Integrated System, the minimization procedure is stored in the working-codes memory, the models used for the representation of the Gibbs energies of the solutions phases and compounds in the knowledge base and the values of the model parameters are stored in the Thermalloy data base.

The user can obtain numerical or graphical outputs of thermochemical properties for a metallic fixed system, Fig. 2, phase diagrams, Fig. 3, and he can calculate the equilibrium state of a multicomponent system for fixed temperature and composition. The minimization procedure is linked with some advanced softwares developed to resolve specific problems like the crystallisation paths of a multicomponent alloy, Fig. 4.

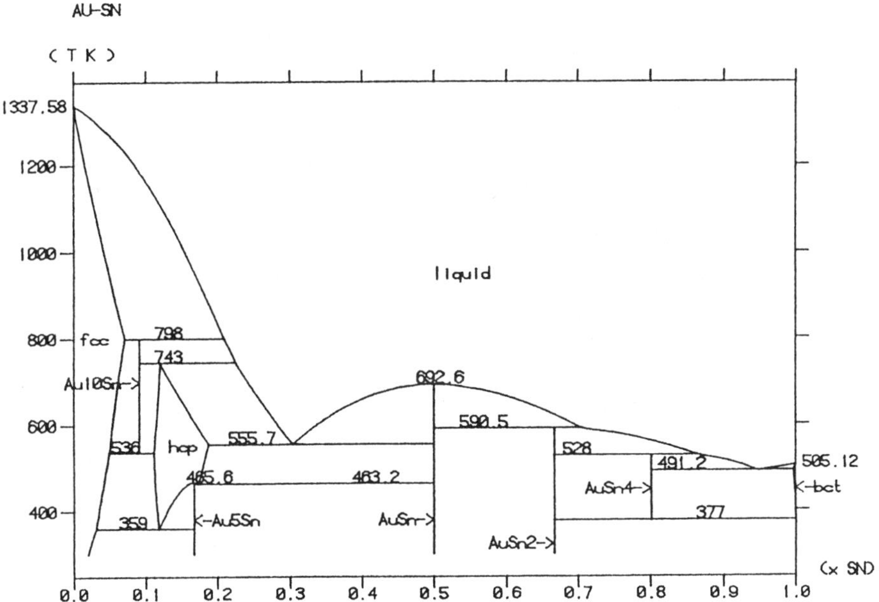

Fig. 3. Au-Sn calculated phase diagram.

TEMPERATURE RANGE	NUMBER AND NATURE OF EQUILIBRIUM PHASES	COMPOSITION OF EQUILIBRIUM PHASES AT TRANSITION TEMPERATURE		
			LIQUID	FCC
1800 → 1733	1 : LIQUID	x_{Co}	0.7	
		x_{Cr}	0.2	
		x_{Fe}	0.1	
1733 ↕ 1717	2 : LIQUID + FCC	x_{Co}	0.7	0.75430
		x_{Cr}	0.2	0.14453
		x_{Fe}	0.1	0.10117
		x_{Co}	0.63502	0.7
		x_{Cr}	0.28564	0.2
		x_{Fe}	0.07934	0.1
1717 → 1700	1 : FCC	x_{Co}	0.7	
		x_{Cr}	0.2	
		x_{Fe}	0.2	

Fig. 4. Crystallisation path in a ternary alloy Co-Cr-Fe.

EQUICOMP

This method is based on a general optimization technique which has
been applied to the chemical equilibrium problems. The minimum of
the Gibbs energy is found subject to the mass balance conditions and
the non-negative condition.

The total Gibbs Energy of a thermochemical system of NG gaseous and
NC condensed species is given by the following equation

$$\frac{G}{RT} = \sum_{i=1}^{NG} n_i \left(\frac{\mu^\circ_i}{RT} + \ln P + \ln x_i\right) + \sum_{i=1}^{NC} n_i \left(\frac{\mu^\circ_i}{RT} + \ln a_i\right)$$

where

n_i = number of moles of component i,

μ°_i = standard chemical potential of component i,

P = total pressure,

x_i = mole fraction of component i,

a_i = activity of component i.

The minimum of G corresponds to equilibrium state. We use for the minimization the "Direct Search" method, as Hooke and Jeeves (1961) have called their method. This method, first developped by the authors for solving numerical and statistical problems, has been adapted including constraints.

The objective is to minimize the value of the function G(ni) of the set of parameters ni by searching the appropriate ni values.

The application of direct search to a problem requires a space of points N which represent possible solutions, together with a means of saying that N1 is a "better" solution than N2 (written N1 > N2) for any two points in the space. There is presumably a single point N*, the solution, with the property N* > N for all N ≠ N*.

The method consists of making moves from a base point, which involve changing the ni in rotation by small amounts Δni and testing to see if any reduction has been made in G. The move is a success if ΔG < 0, and is a failure otherwise. The fact that no further progress can be made indicate that the solution has been found.

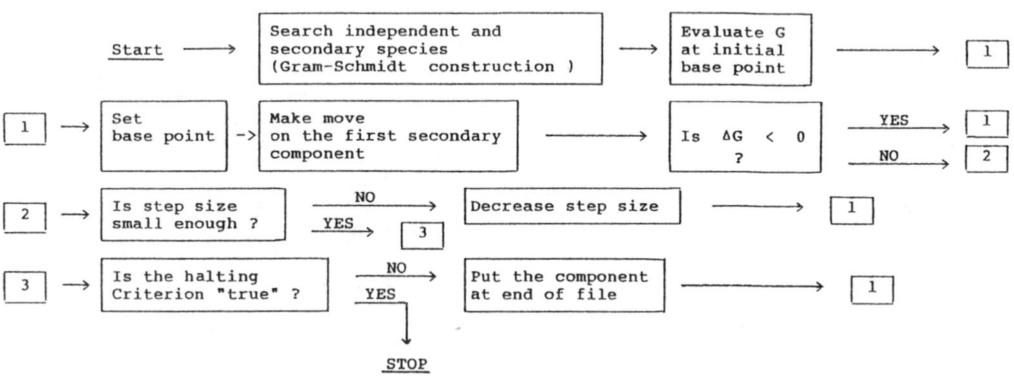

Fig. 5. Flow diagram for Equicomp minimization.

<u>Very Large System</u>

An example of the application of Equicomp to very large complex equilibrium (500 species) is the thermochemical analysis of volatilization, transport and evolution of high temperature volcanic gases.

The volcanic activity is driven by the degassing of the crust and
hearth mantle. These gases contribute to the main geochemical cycles
in the atmosphere, lithosphere and hydrosphere. Initially dissolved
in the magma, driven to the surface by slow convections and tectonic
events, these gases are at first in equilibrium with the silica melt
at a temperature of about 1200°C. When emitted, they cool in the
ground, react with rocks or mix with water or the atmosphere. For
years volcanologists have been trying to reconstruct the initial
composition and to explain the zoning of fumarolic incrustations and
sublimates, but the interpretation is rather difficult, due to the
complexity of the chemical system.

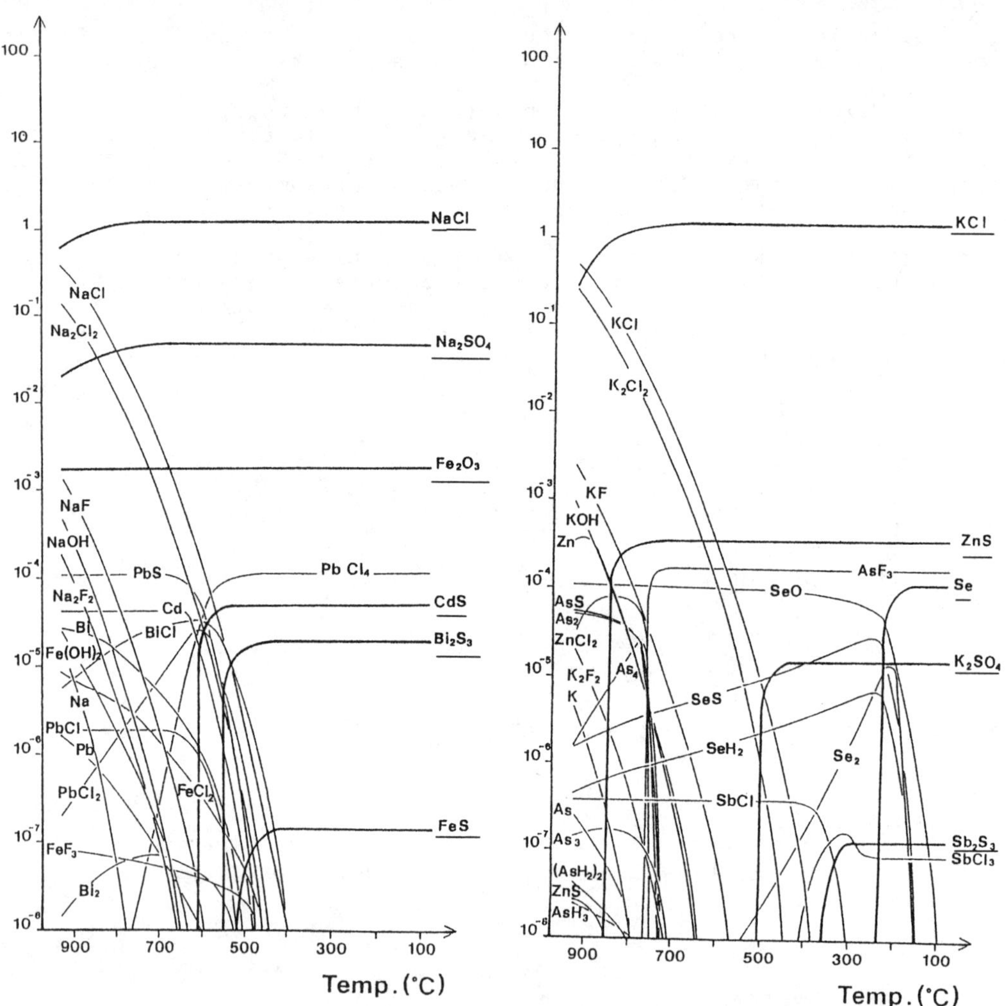

Fig. 6. Thermodynamic modelling of volcanic gases during cooling.
Hornito, Etna 1983 – (thick lines indicate condensed phases).

We approached this problem with a thermodynamic modelling. We first deduce the mass balance of the volcanic gas from the analysis and sampling using a field gas chromatograph, sampling bottles and condensors. During collection oxygen fugaciy and temperature is carefully mesured. Then wen calculate the equilibrium state of the system at high temperature and repeat such a calculation in 50°C step down to 100°C. Condensed phases which appear do not contribute to subsequent equilibrium. With decreasing temperature, various sublimates are observed, Fig. 6, (Le Guern, 1985 ; Quisefit and co-workers, 1988). In general, the calculations agree well with the temperature occurence of each sublimate phase and allow a better understanding of the chemistry in volcanic plumes.

Process Optimization

Refractory metal silicides (Mo, Ta, Ti, W) are good candidates to replace poly-Si from gates and interconnections of integrated circuits in VLSI technology. These materials are often obtained by the Chemical Vapo Deposition method. It is very useful to predict the operating conditions of CVD by means of a thermodynamic analysis.

For this reason, we modelize the different phases, solutions and compounds, and we proceed to parameters optimization (Chevalier and Vahlas, 1988). A one lattice substitutional model has been used to describe the different solution phases in these systems; the excess Gibbs energy has been developped using a Redlich-Kister polynomial expression. The temperature dependancy of all the solution phases and the stoichiometric compounds parameters has been represented by a classical six terms expression. All the data are stored in the Thermalloy database. We finally compare calculated phase diagrams with experimental ones, Fig. 7. Coherence of the obtained data permits their direct application to complex equilibria calculation Equicomp software.
Vahlas (1987), Million-Brodaz (1987) and Bernard (1987), built up computed chemical equilibria CVD phase diagrams, to put in evidence the influence, on $MeSi_2$ CVD, of parameters : Temperature, Total and partial reactants pressures, input gaz composition. Their results and conclusion are :

- WCl_6 is thermodynamically more appropriate than WF_6 as tungsten source, Fig. 8,

- Replacing SiH_4 by SiH_2Cl_2 extends in all cases pure $MeSi_2$ deposition domain, Fig. 9,

- Among WSi_2, $TaSi_2$, $TiSi_2$, the most suitable to deposit by CVD is $TiSi_2$, Fig. 10.

This example is an excelent demonstration of the contribution of an "a priori" thermodynamic approach with our integrated system. It get together the choice of $MexSiy$ data, based on Me-Si phase diagram optimization, thermodynamic data critical inventory of all species which can be involved in the equilibria and the power of a complex chemical equilibrium calculation software.

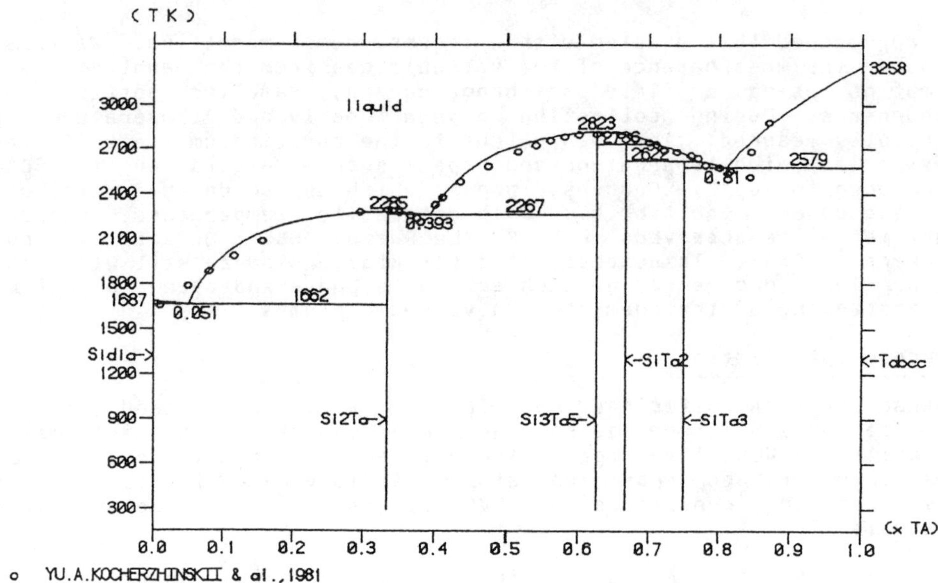

Fig. 7. Optimized Si-Ta phase diagram.

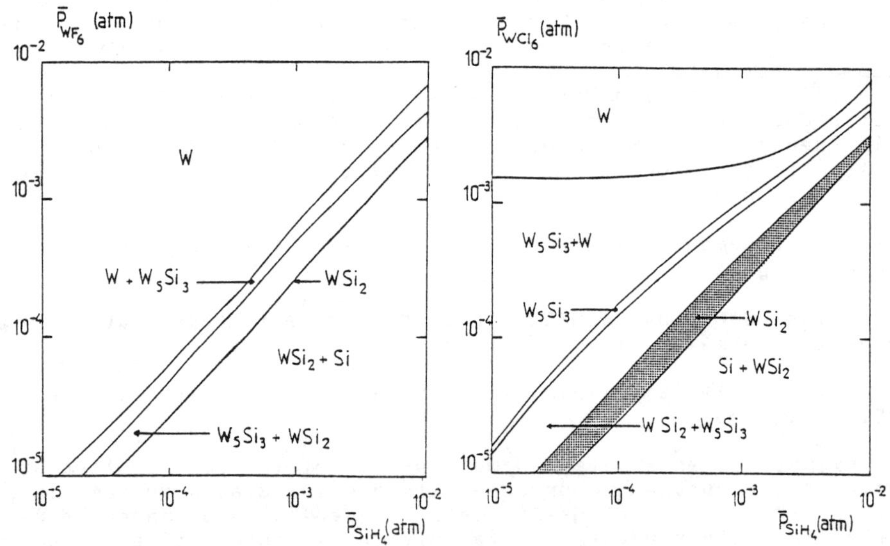

Fig. 8. Metal gas vector influence.
T=1000K, P(Ar) = 0.9 atm, P(WX6 + SiH4 + H2) = 0.1 atm.

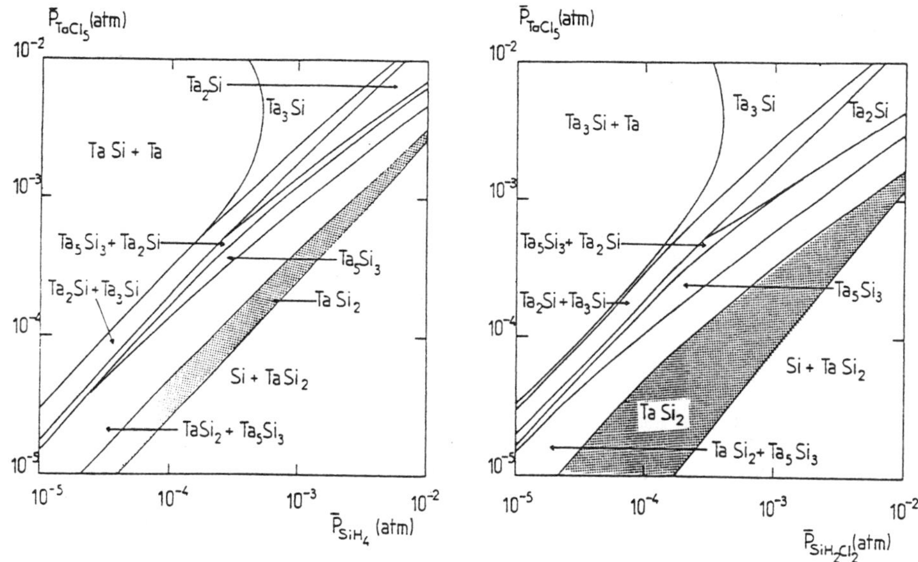

Fig. 9. Silicon gas vector influence.
T=1000K, P(Ar) = 0.9 atm, P(TaCl5 + SiH4 + H2) = 0.1 atm.

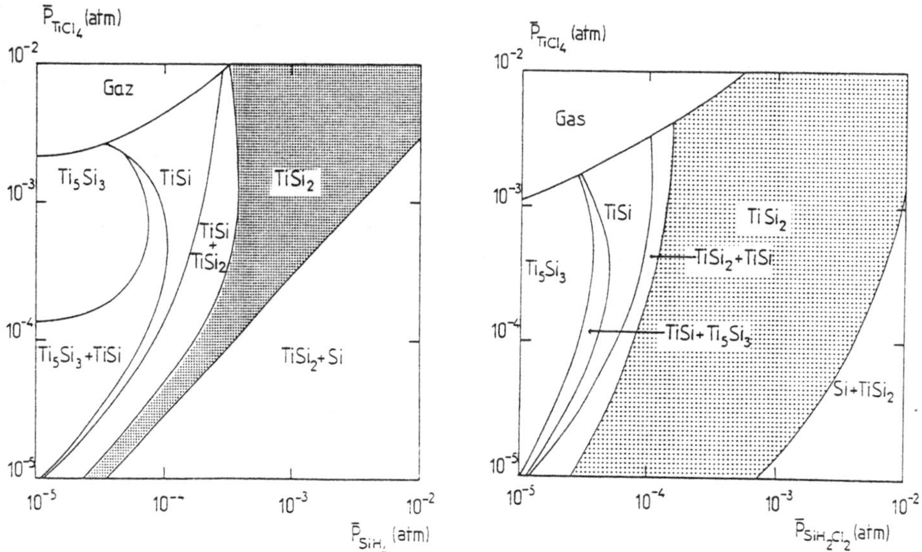

Fig. 10. TiSi2 is the most suitable.
T=1000K, P(Ar) = 0.9 atm, P(TiCl4 + SiH4 + H2) = 0.1 atm.

CONCLUSION

Every thermochimist can retain that a complex chemical calculation software linked to a critically assessed thermodynamic database is today a very powerfull tool which allows :

- to have a better understanding of very complicated chemical processes,

- to select the best material between a family of candidates potentially equivalent,

- to determine, the favourable working parameters for a reactor,

- to optimize every thermochemical process.

REFERENCES

Barbier, J.N., P.Y. Chevalier and I. Ansara (1983). Thermochimica Acta, 70, 173-188.
Bernard, C., C. Vahlas, J.F. Million-Brodaz and R. Madar (1987). Proc. Tenth Int. Conf. on CVD., Honolulu, Hawai, 700-710.
Chevalier, P.Y. (1981). Thèse Sci. Phys. INPG, France.
Chevalier, P.Y. and C.Vahlas (1988). Calphad XVII, 10-15 July, Berkeley, california, USA.

Hooke, R. and T.A. Jeeves (1961). J.Ass. Comput. Mach., 8, 212-229.
Kocherzhinskii, Yu.A., O.G. Kulik and E.A. Shishkin (1981). Dokl. Chem., 261, (1-3), 414-465.
Le Guern, F. (1985). IAVCEI, 1985 Scientific Assembly, 16-21 Sept., Naxos, (Sicily).
Lukas, H.L., E.T. Henig and B. Zimmerman (1973). Calphad, V1,3, 225.
Million-Brodaz, J.F., C. Vahlas, C. Bernard, J. Torres and R. Madar (1987). Proc. 6e Europ. Conf. on CVD., R.Porat Ed., Jerusalem, Israel, 280-297.
Nelder, J.A. and R. Mead (1965). Comput. J., 7, 308-313.
Quisefit, J.P., J.P. Toutain, B. Cheynet and A. Creusot (1988). C.R.Acad. Sci. Paris, t. 306, série II, 387-392.
Vahlas, C., G. Baret, J.F. Million-Brodaz, C. Bernard and R. Madar (1987). Proc. 6e Europ. Conf. on CVD., R. Porat Ed., Jerusalem, Israel, 255-263.

The Thermodynamics Workbench: A Simulation Support System for Research and Education

K. J. Meltsner* and G. Kalonji**
*GE Corporate Research and Development Center, Schenectady, New York
**Dept. of Materials Science and Engineering, Mass. Inst. of Technology, Cambridge, Massachusetts

ABSTRACT

A restatement of traditional thermodynamics has been developed to make it more amenable to computer solution. The Thermodynamics Workbench is a computer program designed for modeling thermodynamic problems. It incorporates and demonstrates this approach to thermodynamics and is designed to simplify the creation and use of thermodynamic models. The Workbench is unique in its ability to handle thermodynamic problems in which a generalized approach to equilibrium is required, and its power comes from its unique representation of thermodynamics.

KEYWORDS

Thermodynamics; artificial intelligence; knowledge representation; modeling; frame-based systems.

INTRODUCTION

The Thermodynamics Workbench is a computer program designed to perform thermodynamic calculations for both teaching and research. It consists of a number of modules written in Common LISP and has been used to solve problems ranging from phase equilibria in metallurgical systems to problems in solid thermodynamics. The Workbench is general in design, in that given appropriate libraries, it can be used for problems in areas other than thermodynamics or materials science. It consists of an information management system, a user-interface package, a symbolic and numeric math package, and a set of thermodynamics libraries. It is designed to be used either as an interactive program or as a part of a larger program.

The Workbench has been designed to use a computer science metaphor known as *object-oriented programming*. In this approach, data and subprograms are "encapsulated" into *objects* so that the programmer does not need to know about their internal design. These objects are then manipulated by a small set of commands. The command set usually consists of requests to furnish information about the object and orders to change that information (Abelson and Sussman, 1985). Each Workbench object corresponds to a thermodynamic subsystem or connection between subsystems.

The Workbench incorporates a generalized approach to thermodynamics. Thermodynamic information can be entered in equation form with no limits as to the form of the equation. Problems involving energies of interaction between subsystems (e.g., coherent spinodal decomposition) can be handled by using a solution method which includes a free energy term for that interaction in the system's thermodynamic potential. Specialized methods may also be used when greater efficiency is desired. In addition, the Workbench provides a simple user interface with context-sensitive help, and one version provides a graphically-oriented user interface.

Previous Systems for Thermodynamics and Modeling

The first systems for thermodynamic modeling were dedicated programs written to calculate specific types of equilibria (Kaufman, 1979; Nash, 1986). These were programs written to solve a single type of problem; a different program was needed for each phase diagram feature. As computers and the thermodynamic (calculation of phase diagram or "CALPHAD") approach both became more accepted, interactive thermodynamics programs soon followed, and now there are several general-purpose thermodynamics programs (Nash, 1986; Sundman, Jansson and Andersson 1985).

Basic Concepts

<u>Computer science concepts.</u> When we designed the Workbench, we decided it should provide facilities to describe models as simply and as generally as possible. After trying several other representation schemes, we chose to use a *frame* system (Minsky, 1974) combined with a *value dependency system* (Haase, 1986) to represent thermodynamic knowledge. The combination of the two methods has several important features:

- default values can be specified and equations used to find desired quantities as needed

- information can be stored to avoid recalculation

- methods and default values can be inherited from a parent object, allowing specialized or more efficient methods and values to be used when possible, but not requiring each object to have full definitions for every quantity

- previously calculated information is discarded when a model's parameters are changed

The frame system stores equations and values for the models, while the value dependency system stores the justifications for all derived equations and values, in order to assure their validity.

Much of the Workbench's power and flexibility comes from its ability to automatically create procedures from equations, eliminating programming for most models. Internally, each quantity's equation is represented as a list of quantities it can be derived from and an algebraic expression for the quantity's value. Any quantity may have more than one equation. When the Workbench needs to find a quantity's value, it finds the equation and constructs a procedure from it. The procedure is smart enough to check if any of the expression's variables is not known, and if so, it tries the next equation until it finds one which has all of its required quantities defined. If no equation works, the procedure returns "unknown value" to the frame system. The ability to handle multiple equations allows the Workbench to define quantities in terms of multiple sets of parameters, since the value dependency system is used to maintain consistency among the derived and user-provided quantities.

<u>Thermodynamic concepts.</u> Workbench thermodynamics is a variation of traditional thermodynamics (Gibbs, 1948). The program defines and manipulates all its models in terms of the variational approach. The criterion for equilibrium is the extremum of the appropriate thermodynamic potential with respect to the transfer of the unrestricted extensive quantities. The algorithms used currently will find a stable or metastable equilibrium whenever possible. While this may not be the simplest way to find equilibria, it is a robust and general criterion, applicable to all thermodynamic systems.

Borrowing from computer science terminology, each thermodynamic subsystem is called an *object*. Objects are interconnected with special objects called *walls*, which can regulate the flow of extensive quantities. Each one is an *instance* of a thermodynamic model, a set of procedures or equations describing the behavior of the object. Most models are specialized versions of more general models, requiring that some procedures be *inherited* from the *parent* object. All thermodynamic objects come from a single general model and inherit basics such as methods to determine total entropy or enthalpy.

The choice of an appropriate thermodynamic potential is determined by the global system conditions. Equilibria can be found with such conditions as irreversible transfer of volume, constant temperature or constant chemical potential.

As a general rule*, the Workbench finds an equilibrium for connected objects by finding the extremum of an appropriate thermodynamic potential function. In practice, this involves finding the minimum of the system's energy or an appropriate Legendre transformation of the energy with respect to the conservative transfer of the quantities unrestricted by the connecting walls. For N unrestricted extensive quantities and M objects, the system state variable is described by $N \cdot (M\text{-}1)$ parameters. The number of parameters can be further reduced by conditions such as constant temperature or pressure.

The system's thermodynamic potential is then minimized using standard optimization techniques, such as Powell's method for finding minima when derivative information is not available (Acton, 1970). When a minimum is found, the objects' states are updated. All extensive quantity transfers are conservative, and there is no chance for the program to find a physically unreachable state.

Minimization with respect to extensive quantities has an advantage over other methods (Gaye and Lupis, 1970): it reduces a problem with nonlinear constraints to one with linear bounds (negative amounts of the extensive quantities are not allowed). The bounds can be handled using a simple penalty function (Scales, 1985) or a restricted line search algorithm. This reduces the difficulty of minimization by at least a factor of two or three. The calculation of equilibria as a function of extensive quantities also provides valuable information about the phase fractions without additional computation.

<u>Applications of Computer Science to Thermodynamics</u>

<u>Symbolic mathematics.</u> The Workbench incorporates a small symbolic math system. The symbolic math system is used to manipulate equations entered by users and find derivatives of those equations. While these capabilities are limited, they do allow the Workbench to provide symbolic answers for many problems when parameters are unknown.

```
Workbench-II> show b2
B2:
    OMEGA: 9000
    TEMPERATURE: 450
Workbench-II> show all equation b2 molar-gibbs-mixing-excess
Information about the equation for B2's MOLAR-GIBBS-MIXING-EXCESS:
Value:
  with:
   X-B = X-B
   X-A = X-A
   OMEGA = OMEGA
```

---

* Its method is determined by the wall's type. Solution methods, like other properties, are inherited from parent models.

```
((OMEGA * X-A) * X-B)
```

Here we have a regular solution model named "b2." We have defined values for the temperature and the regular solution constant (Ω). We have not defined its composition (x-a is the mole fraction of a). We can look at the equation for the excess molar Gibbs free energy of mixing and for the mole fractions. The Workbench, if numeric values are not available, can try to derive an algebraic expression for any quantity defined using an equation:

```
Workbench-II> show all b2 x-a
Information about B2's X-A:
Value:
((NTOTAL - NB) / NTOTAL)
Workbench-II> show all b2 x-b
Information about B2's X-B:
Value:
(NB / NTOTAL)
```

If no value can be found for a quantity, the Workbench uses the equation as the quantity's symbolic value, and then substitutes any known values into the equation and simplifies it:

```
Workbench-II> show all b2 molar-gibbs-mixing-excess
Information about B2's MOLAR-GIBBS-MIXING-EXCESS:
Value:
(9000.0 * NB * (NTOTAL + (- NB)) * NTOTAL ^ -2)
```

Frames for object-oriented programming. We implement all thermodynamic objects as individual frames so they can inherit equations from more general models and use the value-dependency system to calculate and store derived quantities. A simple subsystem, such as a single phase, usually consists of a frame to hold parameters and derived values. The inheritance system then provides the appropriate equations and procedures to calculate the latter from the former.

For example, if an equation for the regular solution coefficient in a parent frame named "regular" is defined, it can be inherited by its children frames and applied to locally stored values:

```
Workbench-II> set regular omega equation
   Equation> (alpha * 8.314 * temperature)
Workbench-II> show all b2 omega
Information about B2's OMEGA:
Value: (3741.3 * ALPHA)
Workbench-II> set b2 alpha 2.1
Setting B2 ALPHA to 2.1
Workbench-II> show all b2 omega
Information about B2's OMEGA:
Value: 7856.729
Depends upon:
B2   ALPHA
B2   TEMPERATURE
REGULAR   OMEGA   EQUATION
```

The value for omega in "b2" is shown to be derived from the values for alpha and temperature (locally stored), and the equation for it from the "regular" frame.

DESIGN CONSIDERATIONS

Finding values

<u>Inheriting values.</u> The most important condition for inheriting a value is that it not be derived from any other values. This is to force the Workbench to derive any values that are used from the values that are accessible from the original frame.

If an underived value cannot be found in a parent frame, the Workbench then checks for a method to find it. If one is found, the Workbench invokes the method with the name of the original frame and slot. If a value is found by the procedure, it is stored in the "value" facet of the original slot. If a method cannot be found, the program checks to see if there is an equation or code fragment available, and constructs a method if there is, and then uses it.

In practice, this should all be transparent to the user. For example, if we define a "junk-1" quantity in the "regular" frame and an equation for "junk-2" which is derived from it, we can use those values in "regular's" children frames:

```
Workbench-II> show all regular junk-1
Information about REGULAR's JUNK-1:
Value: 10
Depends upon:
BINARY  JUNK-1
Workbench-II> show all b2 junk-1
Information about B2's JUNK-1:
Value: 10
Depends upon:
BINARY  JUNK-1
```

The value for "junk-1" can be inherited into a child frame named "binary," and the Workbench will show where that frame's value for "junk-1" came from.

```
Workbench-II> set binary junk-2 equation
   Equation> (junk-1 * 2)
Workbench-II> show all binary junk-2
Information about BINARY's JUNK-2:
Value: 20
Depends upon:
BINARY  JUNK-1
BINARY  JUNK-2  EQUATION
```

Both the value and the equation are inherited and used to derive "junk-2's" value in the child-frame.

```
Workbench-II> set b2 junk-1 15
Setting B2 JUNK-1 to 15
Workbench-II> show all b2 junk-2
Information about B2's JUNK-2:
Value: 30
Depends upon:
B2  JUNK-1
BINARY  JUNK-2  EQUATION
```

If a new local value is set, the inherited one is removed. When the user wants a value for "junk-2," the Workbench can then use the new value to derive its value.

<u>Maintaining consistency through dependency information.</u> The only way that inheritance can work well is if the Workbench keeps track of which values and equations depend on each other. The Workbench maintains every value or equation as an *item* composed of three pieces of information:

{value depends-upon dependents}

where:
>   **value** is the item's value (a numeric value or an equation for the quantity)
>   **depends-upon** is a list of the items upon which the current item depends
>   **dependents** is a list of the items which depend upon the current item

<u>Finding dependency information.</u> The Workbench maintains a *stack* of all values that it is currently searching for. For example, if we are looking for the value of Gibbs free energy in a binary alloy frame, and to find it we need the total number of moles in the system, the stack might look like this.

```
value of binary's total moles
value of binary's Gibbs free energy
```

If we need to find the number of moles of one of the species, we add another layer onto the stack:

```
value of binary's moles of A
value of binary's total moles
value of binary's Gibbs free energy
```

When we find the value of a layer, we peel off the top layer and add the name of the item we were looking for as a "depend-upon" item for all the other items:

```
value of binary's total moles
          depends-upon (value binary's moles of A)
value of binary's Gibbs free energy
          depends-upon (value binary's moles of A)
```

And when we find the value of the total number of moles, we can do the same with it:

```
value of binary's Gibbs free energy
          depends-upon (value binary's moles of A, total moles)
```

This handles the "depends-upon" information. The reciprocal information, the "dependents" list, is generated at the same time.

Manipulating the values in the Workbench

The process of setting a value has several steps. First, the program checks to see if a value can be found for the quantity without setting it. If one can, the program reconciles the value it found with the request to change the slot's value.   It may remove an old value and its dependents, or if the value is derived from other values, it asks the user which ones should be removed. If one can't be found, the program sets the value and returns the quantity's new value.

The Workbench also has the capability of removing values previously set or inherited without setting new ones. Values are removed both automatically and by user command. There are several reasons to explicitly remove a value: the user wants to change the model's basis, a more general value can be inherited, or the lack of a value will force a method to execute the next time the value is asked for, which in turn will cause some sort of side effect. When a value inherited from a parent frame is removed, the value in the parent frame is not altered. This can lead to some confusion if the user tries to find the value and the program inherits the same value that was just removed.

The value dependency system automatically removes values whenever the values upon which they depend are removed or altered. This is to ensure all of the values in the model remain consistent with each other. For example, if an equation is removed, any subroutine generated from it or any value which was calculated from that subroutine will also be removed. After a value and all of its dependents are removed, the program also removes it from the dependents lists of all the quantities upon which it depends. This is to ensure that if a new value is set later, those quantities will not erroneously list the value as a dependent.

Providing multiple bases for a model

The Workbench provides the ability to define a model in terms of several sets of parameters. This is needed so that models can allow users to set the value of any reasonable quantity. The price paid for this flexibility is that a model can be said to have several bases, and this can cause confusion when the model is used. The Workbench will not automatically adjust previously set quantities to be consistent when a quantity is set, but it will ask the user to resolve any inconsistency.

The purpose for allowing multiple definitions of a quantity is to mimic the flexibility of algebra, in which a set of equations may be solved for any unknown variable if the system of equations is sufficiently specified. The basic problem is the difference between the use of the equals ("=") sign in algebra and in most programming languages. The equals sign means *equality* in algebra; the quantities on one side are known to be equal to the quantities on the other side. In programming, the equals sign means *assignment*. The quantity on the left hand side is set to the value of the expression on the right hand side. This simplifies matters; it is not necessary to handle solving for the unknown quantity in terms of the known ones, or to handle conflicts between the values of the quantities.

The Workbench provides the ability to specify multiple definitions for a quantity, checks for over-specification of a model when values are set, and catches recursion when a problem is underspecified. To support these features, the current version of the Workbench requires the model's designer to specify all the possible algebraic relations between quantities. This is generally not onerous; most models can inherit their basic definitions from more general models, and more complex relations are usually used in a single direction.

Recursion as a result of under-specification. Recursion* at its simplest will occur when one quantity which is unspecified is defined in terms of another, and the second quantity is unspecified and defined in terms of the first. The chain of dependent quantities can be much longer, but the Workbench will catch circular dependencies of any length. For example, if we define two quantities in terms of one another and they are both unspecified, the Workbench could get caught in an infinite loop if it did not catch circular dependencies:

```
Workbench-II>  set b2 junk-1 equation
   Equation>  (junk-2 * 2)
Workbench-II>  set b2 junk-2 equation
   Equation>  (junk-1 / 2)
```

---

* The Workbench does not use the mathematical definition of recursion in this case; that is, the definition of a function in terms of itself. Instead, it uses the more common computer science definition of what could be termed *mutual recursion*, or a quantity which is defined in terms of another quantity which in turn is defined as function of the first.

If we defined a value for "junk-1," we would have absolutely no trouble in finding a value for "junk-2." In the Workbench, this sort of recursion only occurs when a problem is under-specified.

<u>Enforcing consistency with multiple bases.</u> The Workbench uses dependency information to find a value's origins, since if a value does not depend on any other value, it has to have been set. It checks any attempt to set a value to see if it is consistent with previously set information by calculating the value from the values already in the frame system and checking it against the value to be set. If they aren't equal, the Workbench generates a list of the primitive (set) values that were necessary to generate it, and asks the user to which value should be removed to make the system consistent.

If a frame has previously inherited a value from a parent frame the Workbench will suggest that the inherited value be removed since the program normally restricts the search for values to the current frame when it attempts to decide if a model is over-specified. If the dependent value had not previously been inherited, it would not be found when the model's consistency was checked.

A good example of a model with multiple bases is a regular solution model defined with its chemical composition in terms of both mole fractions and mole numbers:

```
Workbench-II>  set b2 na .5
Setting B2 NA to 0.5
Workbench-II>  set b2 ntotal 2
Setting B2 NTOTAL to 2
Workbench-II>  b2 x-a
B2's X-A is 0.25
Workbench-II>  set b2 x-a .5
Current value is: 0.25
B2 X-A depends on:
1    B2    NTOTAL    2
2    B2    NA    0.5
Remove which, or cancel request with [0]>2
Setting B2 X-A to 0.5
Workbench-II>  show all b2 na
Information about B2's NA:
Value: 1.0
Depends upon:
B2    NTOTAL
B2    X-A
BINARY    X-B    EQUATION
BINARY    NB    EQUATION
BINARY    NA    EQUATION
```

We have a binary model named "b2." In it, we can define values for the number of moles of A and the total number of moles. If we want to set the mole fraction of A, the program asks us which one of the previously set values needs to be removed to maintain the model's self-consistency. When we check the number of moles of A, we can see that it has been derived from the mole fraction of A and total number of moles using a number of previously defined equations.

<u>Changing a basis.</u> The program can change a model's basis. For example, a wall's minimization routines must operate upon extensive quantities. To do this, it finds the current basis as defined by the user, change the model's basis so that is suitable for the wall, and when the wall is finished, return it to its original basis, but with the new values the wall has calculated. To do this safely, the model designer must provide a list of the variables that the new basis must be able to provide values for (e.g., with a minimization wall, this would be the system's thermodynamic potential).

Ephemeral evaluations

The Workbench incorporates the ability to perform calculations without maintaining dependency information. This feature is included so that certain operations, such as preparing data for graphing or finding an equilibrium, can be made faster. The program can run two or three times faster when it is not necessary to maintain complete dependency information.

The main drawback to using ephemeral evaluation is that the programmer must explicitly remove any calculated values when a model's parameters are changed. It is also necessary to program defensively and use Common LISP's ability to ensure that a "clean-up" procedure be executed even after an error occurs. The ephemeral evaluation mode must be turned off before control of the Workbench returns to the user, or the model will behave peculiarly. If the model's basis does not include the ephemeral quantities, it is necessary to switch the model's basis to include them. Different procedures which use ephemeral evaluation handle this in different ways. The minimization wall automatically changes the model's basis and saves the old one before attempting any ephemeral evaluations. After the minimization, the wall restores the old basis, but with the new values from the minimization. The graphics routine sets the independent quantity for the graph to the first of its values so that the frame system will ask the user to resolve any inconsistency or over-specification. After finding the values of the dependent quantities, it then makes the frame's evaluation ephemeral with respect to the independent quantity, and when it is finished, it restores the original basis.

User-interface issues

Since the Workbench is intended for both students and researchers, it was necessary to make it as simple as possible to use. While the current user interface is crude compared to many commercial programs, it is usable and easy to learn.

<u>"Natural-language" interface.</u> We have chosen to use a "natural-language" interface based on augmented transition trees (Winston and Horn, 1984) because we believe natural-language interfaces allow the user more flexibility (Samad, 1986) when a program must run on a "dumb" terminal. While menu-oriented systems may be simpler, it is difficult to express complex operations with them. With a natural language interface, multiple operations (select objects, display quantities, etc.) can be linked together into one or two sentences. For example:

```
show temperature and pressure gas
```

The `show` command displays the temperature and pressure of the model named "gas."

```
connect hot-object and cold-object via wall1
set hot-object temperature 298
wall1 go
wall1 hot-object and cold-object temperature and pressure
```

In this example, two objects are connected with a wall, one of the object has its temperature changed, and the temperature of the composite system and the subsystems are found.

The chief problem with this style of interface is that it requires exact typing. Unfortunately, the interface tends to be unforgiving of certain types of simple errors, such as extra parentheses or brackets.

<u>Help facilities.</u> The Workbench includes a help system. The LISP commands used to define Workbench commands also automatically generate documentation for them, allowing the Workbench to provide comments to the user as to the next expected input when errors are made.

```
Workbench-II> dump new ?
Input should be:
file name:
        file name
Workbench-II> ?
Input should be:
command:
        CONNECT
        SET frame name:
          thermo. object name

        MAKE
        EXIT
        KILL frame name(s):

        REMOVE frame name:
          thermo. object name
            [...]
```

The Workbench keeps track of what input is expected at any point. The first item before the colon is the sort of input the Workbench expects, and the item or items after it are specific examples.

The Workbench also uses the same information to tell the user what is wrong when there is an error:

```
Workbench-II> dump dffdasdsf
                    ^
Looking for: frame name
```

Graphics. The Unix version of the Workbench uses "gp," an external graphics program, to generate all of its graphical output. The program is invoked by the Workbench and creates the graph using a text file written for it by the Workbench.   Since "gp" is device-independent, the Workbench uses it to provide screen and hardcopy output. The data files can also be edited and used with other programs.

Implementing a Macintosh II Version. Recently, an Apple Macintosh II version of the Workbench has been created. This version allows the user to use the standard Macintosh menus and dialog boxes, in addition to the normal Workbench interface (Fig. 1). It also provides the ability to automatically "paste" calculation results into the Macintosh Clipboard so that they can be plotted using a scientific graphics program. The Macintosh II version of the Workbench currently runs in under 3 megabytes of memory using Coral Common LISP. We have found that its performance is comparable to or faster than the Digital Equipment Corp. (DEC) VAXStation II version using DEC VAXLisp.

USING THE WORKBENCH

Designing a Model

Inheriting from parent frames. When a user wants to enter a new model, it can be a simple process. Since most metallurgical models can be accommodated as variants on the regular solution model, the first step is to enter an expression for Ω, the regular solution coefficient. Let us consider the case of a model for an Al-Zn α (bcc) phase:

Ω's equation for Al-Zn α is:

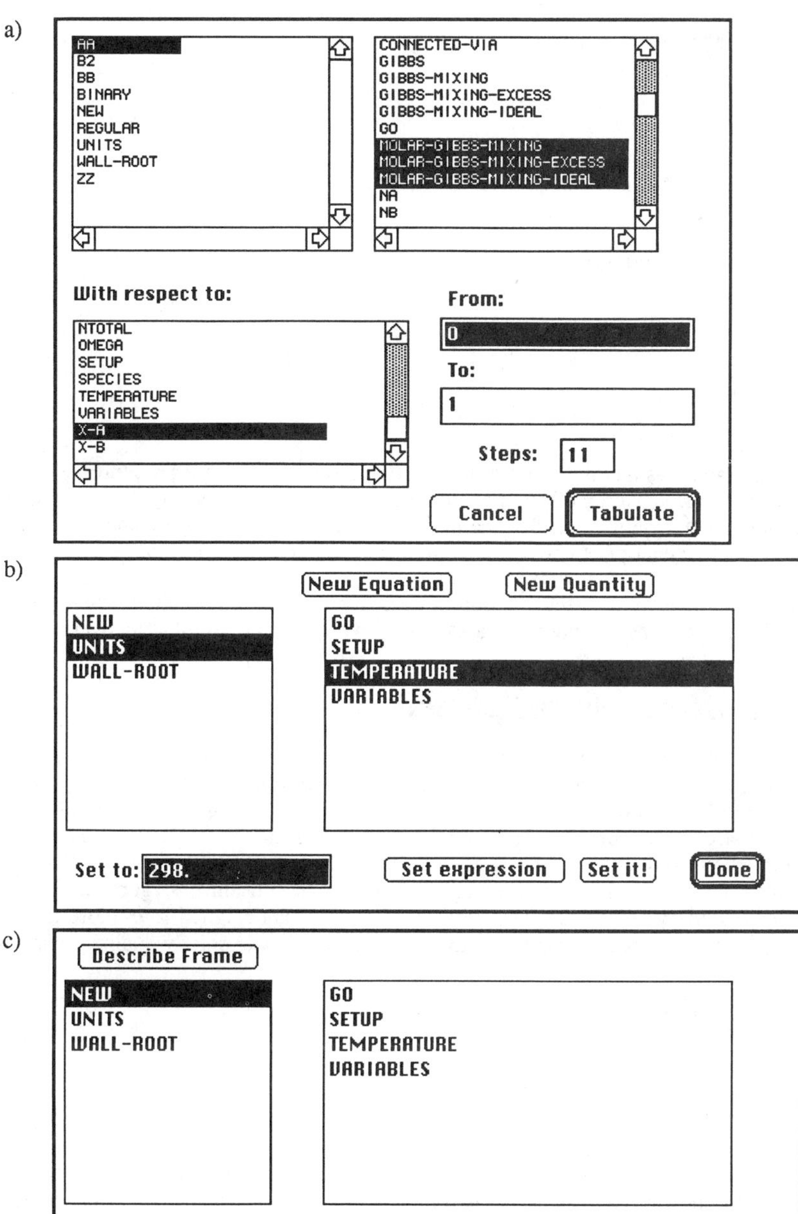

Fig. 1: Several dialog boxes from the Macintosh version of the Workbench.
a) Tabulate properties dialog box.
b) Set property values dialog box.
c) Show property values and equations dialog box.

$$(19200\ X_{Al} + 9600\ X_{Zn})\ (1 - \frac{T}{4000})$$

And in the Workbench, it would look like this:

```
Workbench-II> show all equation aa omega
Information about the equation for AA's OMEGA:
Value:
  with:
   TEMPERATURE = TEMPERATURE
   X-B = X-B
   X-A = X-A
((13200 * X-A + 9600 * X-B) * (1 - TEMPERATURE / 4000))
Dependents are:
AA  OMEGA  IF-NEEDED
AA  OMEGA
```

We can define the model as a child of a regular solution model and inherit its expressions for mole fractions, excess Gibbs free energy of mixing, etc. More complex models may need the inclusion of special terms, such as magnetic contributions to the free energy. In this case, the equation for the phase's molar Gibbs free energy can be redefined to include a magnetic contribution.

Sharing model elements simplifies the creation of complex models and highlights elements that are truly different. It shows a class of models as a hierarchy, which can be useful as an overview.

Teaching thermodynamics with the Workbench

We have used the Workbench (in several versions) to help teach an introductory materials thermodynamics course. As a result, the program has evolved to be more effective as a teaching aid. In fact, most of the driving force to improve the Workbench has been a direct result of student comments. Response to the current version of the Workbench has been good. Students used the Workbench with three assignments in a graduate thermodynamics class. The first assignment's covered piezoelectricity and symmetry in crystals. The second and third assignments explored the thermodynamics of binary solutions. The third assignment was designed to explore the thermo–dynamics of coherent phase transformations, a topic covered in class in chiefly qualitative terms.

DISCUSSION

Workbench Thermodynamics

The Workbench provides an environment in which any sort of thermodynamic problem can simulated and studied. Its generality is a direct result of its representation scheme. Thermodynamics as embodied in the Workbench is a synthesis of two important concepts: traditional Gibbsian thermodynamics and object-oriented programming. The combination is a natural one; the internal details of a subsystem do not matter as long as the subsystem can provide information about changes in the value of a thermodynamic potential when extensive quantities are transferred in or out of it. This modularity allows us to represent the subsystems as separate objects if we wish. We can then use the information about the thermodynamic potential, along with information about the extensive quantities present in a subsystem, to find an equilibrium state for a set of connected subsystems. It is not necessary to know any details of the subsystems' representations, and this is the source of the

Workbench's power. Different models can be selected for each subsystem and the Workbench will still be able to find an equilibrium state.

A general approach is more useful. For example, it can handle the thermodynamics of precipitation. When we consider the precipitation of a second phase, we often find it is a gross simplification to assume there is no interaction between the phases (Cahn, 1984). If we want to simulate a more realistic system, the Workbench allows us to enter information about the interaction between its subsystems. These interactions can include strain energies arising from coherent precipitation or interfacial energies arising from incoherent precipitation. As far as we know, no other general-purpose thermodynamics program handles these energies correctly. For example, Thermo-Calc's model representation method (Sundman, Jansson and Andersson, 1985) requires any interaction energies to be expressed as a contribution to one phase's excess Gibbs free energy of mixing (Grujicic and Olson, 1988). The interaction energy cannot even be calculated separately if desired; only the Gibbs free energy of mixing can be calculated. In addition, it is not possible to model an alloy in which the interaction energy depends on the composition or amount of more than one phase. With the Workbench, such a model can be entered with a single equation.A more serious deficiency is that most thermodynamics programs depend on the "common tangent" assumption: chemical activities in a multi-phase region are constant. This assumption is simply not true when effects such as coherency are taken into account (Cahn and Larche, 1984). The Workbench does not make such an assumption and it handles all problems correctly by minimizing the system's thermodynamic potential.

The Workbench's generality is useful in other ways. Thermo-Calc cannot find a miscibility gap unless it is given a starting composition outside the gap; the Workbench can determine whether a phase will separate at any composition. The user can also specify an interaction energy for such a separation so that coherent spinodal decomposition can be simulated as easily as the incoherent case. The Workbench user does pay a price for this generality. Equilibria take longer to find than with other programs. However, speed is not a factor if the problem requires a generalized approach.

A general approach to thermodynamics is also necessary if the Workbench is to be extended further. By replacing the minimization procedure with one which transfers quantities based on potential differences in the connected subsystems, kinetic problems such as diffusion can be simulated.

Suggestions for future work

There are two areas in which significant work needs to be done. First, the Workbench needs a library of kinetic walls. The Workbench's basic approach to thermodynamics is well-suited to handle quasi-thermodynamic problems such as many diffusion and nucleation problems. The correct way to handle them in the Workbench would be to devise walls which transfer extensive quantities in a manner governed by the kinetics of the connection between its objects. This would be a simple and powerful way to handle these problems. This would be an unlimited area for exploration; it would be especially interesting to work on methods to use different types of kinetics simultaneously.

Second, the Workbench needs a real symbolic math system. We plan to couple the Workbench with a commercial symbolic math program. This link should be transparent, and will provide the user with complete symbolic calculus and algebra capabilities. We plan to use this to provide automatic definition of differential quantities and direct solution of simple equilibria.

The other planned expansion is the addition of a phase diagram display library. A simple method to derive binary and pseudo-binary diagrams will use the fact that a binary phase diagram is a projection of the minimum Gibbs free energy surface of a system of phases. The problem of constructing such a projection is identical to the computer science problem of eliminating hidden surfaces in a solid model from view. In our case, each region can be thought of as a solid curve in three dimensions: Gibbs free energy, composition and temperature. The only regions which would be visible when one looked down the Gibbs free energy axis would be the stable ones. If the diagram were to be rotated and sectioned, a series of two-dimensional plots of Gibbs free energy as a function of temperature or composition would be obtained.   Such diagrams would be invaluable in teaching alloy thermodynamics and phase diagram construction. The researcher could also make use of computer graphics techniques which can shade the curves as to their slopes (which would be functions of entropy and chemical potential) or make them transparent so metastable regions can be examined.

CONCLUSIONS

- The Workbench includes a novel approach to thermodynamics which both simplifies the solution of problems and allows a wider class of problems to be attacked.

- The Workbench's frame-based information management system, value-dependency maintenance system, symbolic math package, and numerical math package allows it to use a powerful and generalized approach to solve a variety of thermodynamics problems.

- The frame system provides support for the creation of models that is not available from other thermodynamics programs.

ACKNOWLEDGEMENTS

MIT's Project Athena provided funding for the development of an earlier version of the Workbench. Development has also been supported by funding from the National Science Foundation's Presidential Young Investigator Award program. Mr. Craig Counterman provided "gp," a data plotting program.

REFERENCES

Abelson, H. and Sussman, G. J. (1985). *Structure and Interpretation of Computer Programs*. The MIT Press, Cambridge.

Acton, F. S. (1970). *Numerical Methods that Work*. Harper & Row, New York.

Cahn, J. W. and F. Larche (1984). "A Simple Model for Coherent Equilibrium," *Acta Metall.*, **32**, 1915-1923.

Gaye, H. and Lupis, C. H. P. (1970). *Scrip. Metall.*, **4**, 685-692.

Gibbs, J. W. (1948). *The Collected Works of J. Willard Gibbs*, Vol. I, Yale Univ. Press, New Haven.

Grujicic M. and Olson, G. B. (1988). "Ferrite—M_2C Coherent Phase Equilibrium in AF1410 Steel."

Haase, K. W. Jr. (1986). *ARLO: Another Representation Language Offer*. Technical Report 901, MIT Artificial Intelligence Laboratory, Cambridge. 14-16.

Kaufman, L. (1979). *CALPHAD*, **3**, 27-44.

Minsky, M. (1974). *A Framework for Representing Knowledge*. Memo No. 306, MIT A. I. Laboratory, Cambridge.

Nash, P. (1986). Approaches to the Computer Representation of Phase Diagrams. In L. H. Bennet (ed.), *Computer Modeling of Phase Diagrams*, The Metallurgical Society, Warrendale. 331-342.

Samad, T. (1986). *A Natural Language Interface for Computer-Aided Design*. Kluwer Academic Publishers, Boston.

Scales, L. E. (1985). *Introduction to Non-linear Optimization*. MacMillan Publishers Ltd., London.

Sundman, B., Jansson, B. and Andersson, J.-O. (1985). *CALPHAD*, **9**, 153-190.

Winston, P. H. and Horn, B. K. P. (1984). *LISP*, 2nd ed., Addison-Wesley, Reading, MA.

COMPUTATIONS USING MTDATA OF METAL - MATTE - SLAG - GAS
EQUILIBRIA

A. T. Dinsdale*, S. M. Hodson**, T. I. Barry*, J. R. Taylor***

*Division of Materials Applications,
**Division of Information Technology and Computing
National Physical Laboratory, Teddington, Middlesex TW11 OLW, UK

***Johnson Matthey Technology Centre, Blount's Court,
Sonning Common, Reading, Berkshire RG4 9NH, UK

ABSTRACT

The principle of calculating phase equilibria in multicomponent, multiphase systems
from a knowledge of the Gibbs energy of the subsystems is well established. This
paper demonstrates the practical realisation of this principle for the
pyrometallurgical extraction of values from copper-nickel-iron sulphide ores. The
data for the matte, slag and other phases have been assembled and equilibrium
calculations made using MTDATA, the NPL databank for Metallurgical Thermochemistry,
incorporating MULTIPHASE, a reliable and efficient calculation procedure which uses
true Gibbs energy minimisation. A description of MULTIPHASE is given.

KEYWORDS

Chemical equilibrium calculation; Thermodynamic modelling; MULTIPHASE; Gibbs energy
minimisation; Matte slag gas alloy equilibria; Cu-Fe-Ni-S; Pyrometallurgy.

INTRODUCTION

Problems in metallurgy and materials science generally involve many components.
This is particularly true of problems in pyrometallurgy, where not only the
components of the ore but also the 'reagents' used for treating them need to be
considered. A complete description of these complex processes places great demands
on the thermodynamic models for the multicomponent, multiphase system, the
availability of critically assessed data and computational procedures. In addition,
facilities are required for easy control of the calculation and data by the user,
and for the production of graphical and tabular output.

MTDATA

The aim of the present paper is to illustrate these points by reference to
calculations on systems of practical importance undertaken using MTDATA, the NPL
metallurgical thermochemistry databank (Barry and colleagues, 1987), which is used
for data assessment and for the calculation of phase equilibria for a broad range
of applications in metallurgy, materials science and inorganic chemistry.

For these purposes MTDATA incorporates thermodynamic data, facilities for data management and a set of calculation modules. The most important of these is MULTIPHASE, which, as its name suggests, is used mainly for calculating chemical equilibria in multicomponent, multiphase problems, although it is equally at home with a simpler problem such as the speciation of sulphur in the gas phase.

MULTIPHASE is able to calculate and plot equilibria in systems with up to 20 components under a variety of constraints which include temperature, pressure or volume, composition (mole or weight) and potentials of products. The results of stepped calculations can be plotted selectively (for example the composition of an individual phase or the distribution of a particular component) as a function of the originally stepped variable or dependent variables.

As described more fully below and by Hodson (1988), MULTIPHASE operates in either or both of two stages. The first stage is a true Gibbs energy minimisation of the integral Gibbs energy of the system (Helmholtz energy for constant volume calculations). The amounts of substances calculated to be present at less than about 10^{-6} moles are set to zero. The second stage is needed only when the detailed knowledge of speciation or the distribution of components present in very small amounts is required. It normally uses the results of a first stage or a previous stage 2 calculation to provide an initial guess. A particularly useful feature of stage 1 is that, although it can use an initial guess, this is not a requirement and the procedure is inherently highly reliable.

Inevitably multicomponent calculations relevant to industrial problems entail the need for data for all the subsystems involved and the ability to use a variety of models to describe the behaviour of alloy phases, slags, salts, mattes, minerals, and aqueous solutions as a function of temperature, pressure and composition. For the purpose of assessment, validation and management of data for these subsystems MTDATA incorporates other modules, which can also be used for problem solving.

THERMOTAB   is used for tabulating and plotting data for individual substances and chemical equations and for amending and storing data. The equations can be auto-balanced.

COPLOT   calculates and plots predominance area diagrams (also called phase stability or Pourbaix diagrams) for multicomponent systems not involving non-ideal solutions.

ACCESS   allows the selective retrieval of data for systems of up to 20 components for chosen phases. The resultant datafiles can be used by MULTIPHASE, TERNARY and COPLOT.

BINARY   calculates and plots binary temperature - composition diagrams and lists invariant points. The diagrams may be annotated.

G_PLOT   is used for tabulating and plotting thermodynamic functions for binary systems.

TERNARY   provides an interactive, automatic procedure for calculating and plotting phase boundaries and tielines in ternary systems in mole or weight fraction. Experimental data and labels may be added.

FITANDPLOT   is a flexible procedure for fitting data to chosen functions. Facilities include automatic error checking of input data for typing errors, control of fitting accuracy and range splitting, as well as import and export of data from and to THERMOTAB.

The operation of MTDATA is controlled through an easily learned user interface, in

which the normal sequence is: "command, parameter, value !". Figure 1 below shows
the sequence of commands used in setting up the calculations for Fig. 2b.

define datafile **'MATSLAGGAS03'** !
Recover data from the datafile prepared in advance

list system **components phases** !
List the components and their status, the phases and their status

classify absent phase(1) **p(2) p(3) p(5) p(6) p(7) p(8) p(9) p(10) free
component(5)** !
*Since the calculation is made for a high temperature, it is efficient to
remove all the solid phases from consideration. Oxygen, the fifth
component is classified as free, allowing its amount to be determined by
MULTIPHASE based on the initial guess set below. Classify is also used to
ask MULTIPHASE to look for miscibility gaps and to fix reference states.*

set temperature 1573 !
The temperature is in kelvins, the default pressure is 101325 Pa.

set n(1) 1 n(2) 6.52 n(3) 1 n(4) 1.3 n(5) 15 n(6) 3.3 n(7) 1.18 !
*In this example the amounts of components are set up in molar terms.
Amounts could have been entered in terms of mass or specified for
substances rather than for components. The amount set for the free
component, i.e. n(5) for oxygen, is used as an initial guess.*

set l_p(SO2<**gas**>**) -1** !
The partial pressure of sulphur dioxide is fixed.

step n(8) 0.2 20 0.5 !
*The amount of nitrogen (component 8) is varied from 0.2 to 20 moles in
steps of 0.5 mol.*

list system **settings** !
The settings can be checked.

compute initial last_active_set print graph !
*Various compute options are available. In this example the last_active
set of variables is used to initiate the next calculation and output is
to be graphical rather than tabular.*

      Fig. 1. Setting up MULTIPHASE for the calculation of Fig.
            2b. The components are respectively: Ni, Fe,
            Cu, S, O, Si, Ca, N. Bold print indicates actual
            entries. Abbreviations may be used for keywords
            providing they are unique. Comments are in italics.

THE PYROMETALLURGICAL TREATMENT OF NICKEL-IRON-COPPER MATTES

A basic principle of many processes for extracting copper and nickel from sulphide
ores is to overlay the molten ore or matte with a lime-silica-iron oxide slag and
to blow with air or oxygen. Oxidation of iron occurs, which as a result partitions
to the slag phase, and sulphur is carried off as sulphur dioxide. The further
processing depends very much on the composition of the ore, including the amounts
of impurities, some of which may themselves be of value. The following report does
not relate to any particular process and is intended only to illustrate the
calculation procedures. Not all factors have been taken into account in the

thermodynamic modelling. For example, the solution of Cu, Ni and S in the slag is not considered.

The initial compositions of the mattes and slags used in the calculations below are given in Table 1. The mass of the slag is made three times that of the matte so that its composition does not change markedly. In practice the slag would be replenished either from time to time or continuously. In continuous processing the slag could be made to flow countercurrent to the matte. Two different matte compositions are used to illustrate different aspects of processing.

TABLE 1   Initial Compositions of the Matte and Slag phases

Component	matte A wt%	moles	matte B wt%	moles	Component	slag wt%	moles
Ni	26.7	1.0	7.9	0.3	CaO	10.0	1.18
Fe	25.4	1.0	25.0	1.0	'FeO'	60.0	5.52
Cu	28.9	1.0	48.4	1.7	SiO_2	30.0	3.33
S	18.9	1.3	18.6	1.3			
Total mass/g	219.7		223.0			662.9	

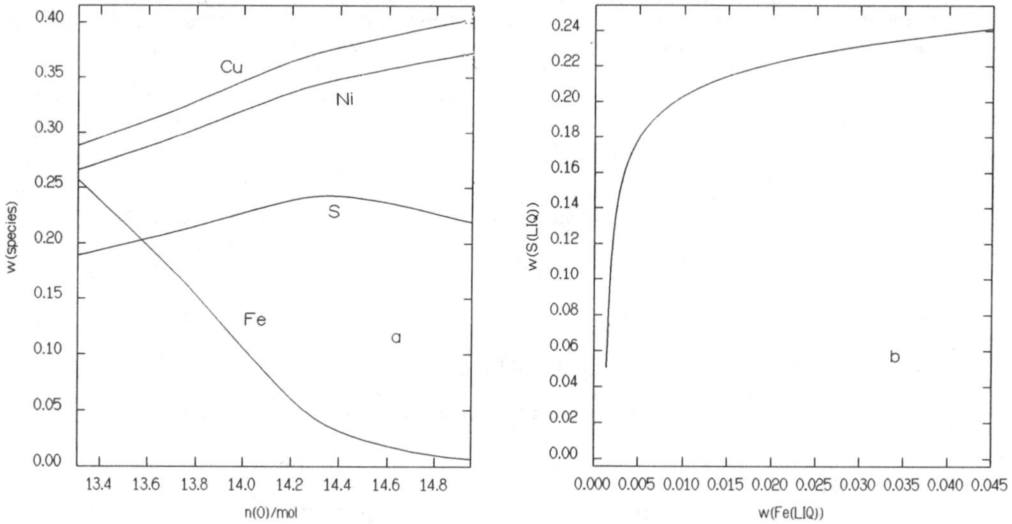

Fig. 2. (a) Composition of matte A as a function of oxygen content of the system at 1573 K. The starting amount of oxygen before blowing is 13.36 mol.
(b) Interdependence of the weight fraction of iron and sulphur in the liquid for a particular nickel-copper ratio, slag composition and sulphur dioxide pressure.

In agreement with practical experience, calculations show that initially the added oxygen is consumed mainly in the process of oxidation of iron, which is transferred to the slag. As a result the proportion of sulphur in the matte actually increases. In Fig. 2a oxygen is added progressively until a substantial pressure of sulphur dioxide develops. Most of the iron is removed from the matte during this phase.

In the second phase the partial pressure of sulphur dioxide is considered to be
maintained at 0.1 atm. This calculation could have been undertaken in a number of
ways. That chosen was to allow the oxygen amount to be determined by MULTIPHASE
when the volume of the off-gas was progressively expanded by blowing in nitrogen.
Figure 1 shows the sequence of commands that were used to set up the calculation. A
great variety of graphs are obtainable from the results of stepped calculations and
they need not necessarily be presented in terms of the originally stepped variable.
Figure 2b shows the weight fraction of iron plotted as a function of the weight
fraction of sulphur for the conditions set up in Fig. 1. The relationship between
these two variables is a function of the composition of the slag and the relative
proportions of nickel and copper in the matte, the temperature and the partial
pressure of sulphur dioxide. A knowledge of this relationship is important to the
control of the blowing process in order to prepare the matte for subsequent
processing. Good agreement between calculated and experimental results has been
demonstrated by Dinsdale, Hodson and Taylor (1988). The strongly non-ideal
behaviour is reflected in both the calculated and experimental results.

Liquid-Liquid Immiscibility in the Matte

A feature of copper-rich, nickel-poor mattes is that they tend to develop
liquid-liquid immiscibility. Thus at 1573 K matte B exists as two liquids. As the
amount of iron is reduced, the two liquids become miscible but immiscibility again
develops as the amount of sulphur declines and a sulphur-poor phase separates.

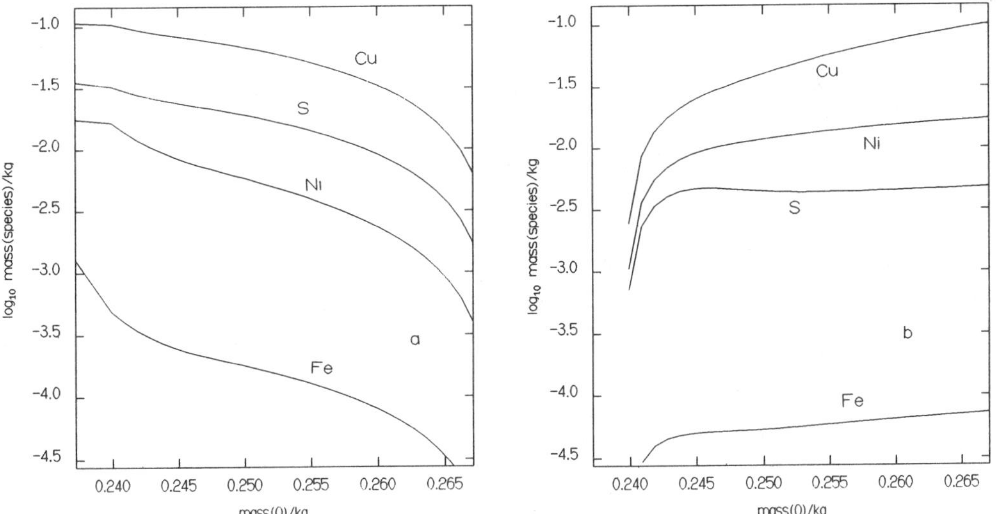

Fig. 3. Two-liquid formation in matte B as a function of
        oxygen content of the system at 1573 K.
        Compositions of (a) the sulphur-rich and (b) the
        sulphur-poor liquid.

The proportions of the two phases and the distribution of components between them
in this second regime are illustrated in Fig. 3. The proportion of iron in the
matte has been reduced to low levels and thus the composition lies nearly within
the Cu-Ni-S ternary system of Fig. 4. In this diagram, the position of the
miscibility gap, which is not perfectly in accord with the experimental data
reviewed by Chang, Neumann and Choudray (1979), suggests that for the compositions
of Fig. 3 phase separation should begin at about 30 wt% sulphur. Because the

tielines in Fig. 4 run at an angle to lines of constant copper-nickel ratio, enrichment of nickel occurs in the sulphur-poor phase. If a differential flow is established between the phases it is possible to reduce the level of nickel in the sulphur-rich phase (which will follow the upper edge of the miscibility gap) to low levels. This possibility is exploited in the Noranda process, the thermodynamics of which has been explored by Nagamori and Mackey (1978).

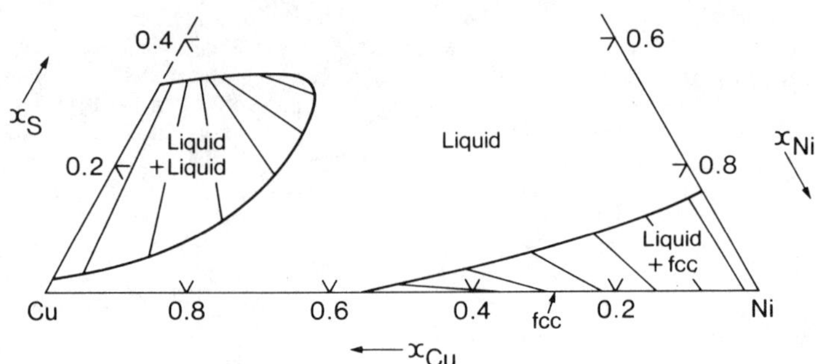

Fig. 4. Calculated phase diagram for the Cu-Ni-S system for
1573 K.

MULTIPHASE offers several options of tabular output; an example of these is given in Fig. 5 below, which shows the results (edited to save space) of a single step in the calculations displayed graphically in Fig. 3.

```
STAGE 1 Results
Temperature =  1573.00 K
Fixed gas pressure =  1.013250E+05 Pa
Calculated gas volume =  7.172848E-01 m3
```

	Phase	Spec	Amount	Mole fraction
		-ies	mole	Partial pressure
				Notional activity
Phase LIQUID				
Ni(LIQ)	4	1	0.19876	0.20751
Fe(LIQ)	4	2	0.00095	0.00100
Cu(LIQ)	4	3	0.61835	0.64556
S(LIQ)	4	4	0.13979	0.14594
Phase total is			0.95786	1.00000
Phase LIQUID				
Ni(LIQ)	4	1	0.10124	0.05658
Fe(LIQ)	4	2	0.00323	0.00181
Cu(LIQ)	4	3	1.08153	0.60446
S(LIQ)	4	4	0.60326	0.33716
Phase total is			1.78926	1.00000
Phase SLAG				
SiO.50	11	1	1.44802	0.16179
Fe2/30	11	2	0.04811	0.00538
CaO	11	3	0.00017	0.00002
FeO	11	4	1.90744	0.21312

```
SiO.50Fe2/30        11    5      0.20923    0.02338
SiO.50CaO           11    6      1.10093    0.12301
SiO.50FeO           11    7      3.84182    0.42925
Fe2/30CaO           11    8      0.02883    0.00322
Fe2/30FeO           11    9      0.31542    0.03524
CaOFeO              11   10      0.05007    0.00559
Phase total is                  8.95004    1.00000

Phase GAS
Cu<g>               12    1      0.00012    0.00002
N2<g>               12    4      4.99999    0.89976
OS<g>               12    8      0.00114    0.00021
O2S<g>              12   10      0.55571    0.10000
S2<g>               12   12      0.00003    0.00001
Phase total is                  5.55703    1.00000
```

```
Component      Chem. Pot.     Activity        Moles
Ni          -1.023565E+05   3.991462E-04   3.000000E-01
Fe          -1.566855E+05   6.267199E-06   6.520000E+00
Cu          -9.278681E+04   8.296679E-04   1.700000E+00
S           -2.181046E+05   5.722415E-08   1.300000E+00
O           -2.814964E+05   4.493439E-10   1.560896E+01
Si          -5.128231E+05   9.356250E-18   3.300000E+00
Ca          -5.594041E+05   2.656610E-19   1.180000E+00
N           -1.716695E+05   1.993016E-06   1.000000E+01
Gibbs Energy = -9.956548E+06 J
```

Fig. 5   Edited results showing the composition of the slag
         and liquid phases for a single set of conditions.
         Ferric iron in the slag is identified as Fe2/30.

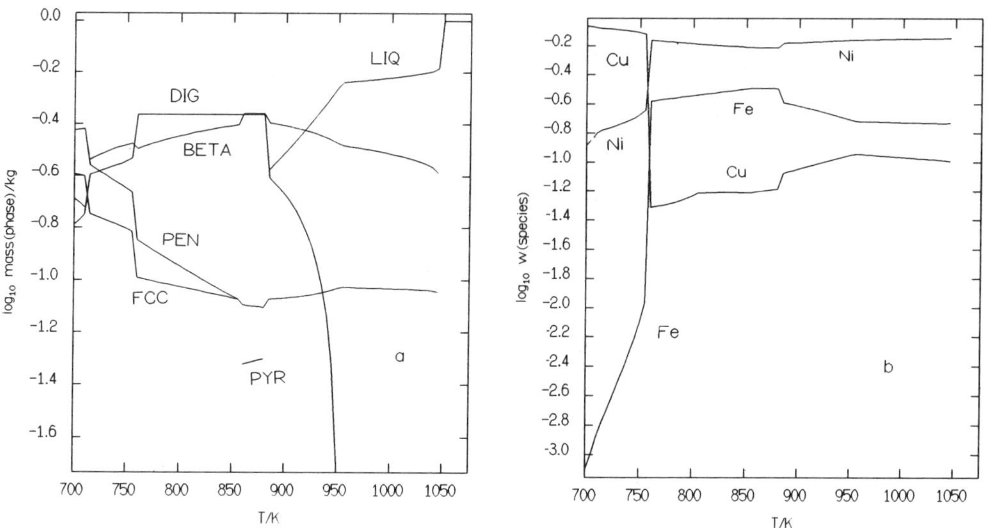

Fig. 6. Phase equilibria as a function of temperature in
        matte A   depleted   in sulphur and   iron.   The
        composition is Ni 0.325, Fe 0.1, Cu 0.352, S
        0.223 parts by weight. (a) mass of phases, cf
        Table 2, (b) composition of the fcc phase.

Solidification and recrystallisation

Further separation of the components occurs during crystallisation of the matte on
cooling. It may sometimes be advantageous not to reduce the iron to very low
levels; among other things this avoids loss of values to the slag. Figure 6a shows
the predicted sequence of phases formed during cooling of a matte of original
composition A in which the amount of iron has been reduced to 10 wt% and the
sulphur to 22.3wt%. The compositions of the fcc phase are given in Fig. 6b and the
distribution of components between the phases is given in Fig. 7.

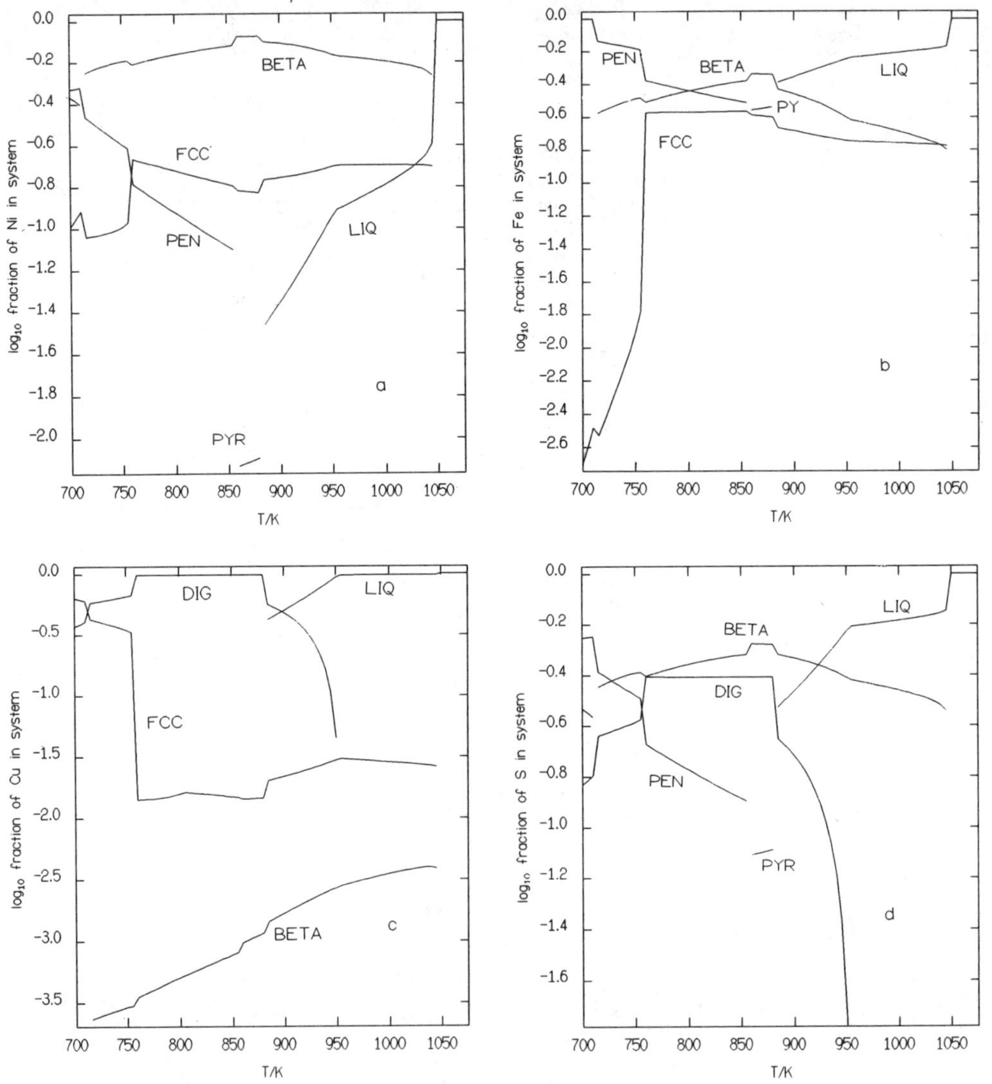

Fig. 7. Distribution of the components of the matte between
        the phases as listed in Table 2. The total mass of
        each phase is given in Fig. 6a.

Solidification sets in at 1050 K with the formation of the beta and metallic fcc phases. Nickel is strongly segregated to the beta phase, which approximates to Ni_3S_2 in composition. Plots similar to Fig. 6b show that the iron maintains its concentration in the diminishing amount of liquid, whilst the concentration of copper is enhanced. At 950 K digenite precipitates, depleting the liquid in copper and sulphur and leading to the disappearance of the liquid at 900 K. At this temperature iron is partitioned to the beta and transient pyrrhotite phase, which has only a small temperature range of stability before the emergence of pentlandite. Not obvious in these plots but evident in the copper-nickel phase diagram is a tendency to immiscibility in the fcc phase at low temperature. This contributes to a further transformation at 750 K, in which the amount of copper combined with sulphur as digenite sharply declines, whereas that of nickel and iron combined with sulphur in pentlandite increases.

The calculations relate only to a single composition. Nevertheless they demonstrate that the crystallisation processes in a multiphase matte can be very complex and that calculations based on thermodynamic data can contribute greatly to the selection and optimisation of processing methods.

THERMODYNAMIC MODELS AND DATA

The phases and thermodynamic models are given in Table 2.

TABLE 2   Phases and Thermodynamic Models

gas	Ni-Fe-Cu-O-S-N	ideal gas
slag	$CaO-FeO-Fe_2O_3-SiO_2$	cellular model
liquid	Ni-Fe-Cu-S	variable sublattice
fcc	Ni-Fe-Cu	Redlich Kister
bcc	Ni-Fe-Cu	Redlich Kister
beta('Ni_3S_2')	Ni-Fe-Cu-S	Redlich Kister
pyrrhotite	(Ni,Fe)S	Redlich Kister
pentlandite	$(Ni,Fe)S_{0.889}$	Redlich Kister
digenite	Cu_2S	Stoichiometric
heazlewoodite	Ni_3S_2	Stoichiometric
millerite	NiS	Stoichiometric
Ni7S6	Ni_7S_6	Stoichiometric

The data for the sulphide phases were derived exclusively from the assessments of Dinsdale (1982, 1984) and Fernandez Guillermet and colleagues (1981). The metal/matte liquid is described by a two-sublattice model with variable site fractions similar to that of Hillert and colleagues (1985). Both binary and ternary experimental data were taken into consideration in the assessment. Most of the several crystalline phases are modelled as solid solutions. The data for the slag phase are based with some modification on the work of Gaye and Welfringer (1984) and Taylor and Dinsdale (1988), who used the cellular model of Kapoor and Frohberg (1971). Other models used elsewhere for slags include the sublattice model and the quasichemical model of Pelton and Blander (1986). Data for the gaseous species were obtained from the SGTE database (Barry, 1985; Ansara and Sundman, 1987). All the data were converted to the G-Hser formalism, in which the Gibbs energy data for all substances are referred to the properties of the pure elements in their stable states at 298.15 K and 101325Pa, rather than to the elements in a specified phase at current temperature.

MULTIPHASE

Mathematical Form

The equilibrium state of a metallurgical or chemical system may be determined by solving, for a given temperature and pressure, the following problem:

$$\text{minimise}_{n} \; G = \sum_{j=1}^{N} n_j * \mu_j \tag{1}$$

such that
$$\sum_{j=1}^{N} a_{ij} * n_j = r_i \qquad\qquad i=1,2,\ldots M \leq N$$

which may be written as
$$A * n = r$$

and
$$n_j \geq 0$$

where

G   is the Gibbs energy which is to be minimised by varying the values of the n_j, $j = 1,2,\ldots N$

n_j is the amount in moles of species j present in the system; each chemical substance with a different phase designation being considered as a distinct chemical species,

N   is the number of species in the system,

μ_j is the chemical potential of species j, which may be a function of some or all of the species amounts in the same phase; it also depends on temperature and pressure, (the form of μ_j depends on the model of the phase under consideration and is constructed such that $dG/dn_j = \mu_j$),

a_{ij} is the number of units of component i per species j,

M   is the number of components in the system,

r_i is the number of moles of component i in the system.

The expression G is a nonlinear function of the species amounts and contains parameters dependent on temperature and pressure which have been determined from simple systems and experimental evidence. These parameters are considered as known data as far as MULTIPHASE is concerned. A similar problem is to be solved if it is required to set the volume of the system constant instead of the pressure.

Note that the problem (1) involves minimising a nonlinear objective function G subject to linear constraints including bounds on the variables. To solve this class of problem the following method may be used which forms the basis for Stage 1 of MULTIPHASE.

Optimisation Method

At the beginning of each iteration of this method a certain number (S, say) of constraints will be considered active, ie currently satisfied as equalities. In problem (1) these will comprise the M component abundance equations (which are equality constraints and so must always be satisfied) and S-M of the n_j set to zero. The latter correspond to species in absent phases. The S active constraints $(N \geq S \geq M)$ may be written as

$$C * n = \begin{bmatrix} r \\ 0 \end{bmatrix} \qquad \text{where } C = \begin{bmatrix} A \\ I_{S-M} \end{bmatrix}$$

and I_{S-M} is a subset of the rows of the identity matrix.

An orthogonal factorisation of the matrix C^T is now formed. At the kth iteration this gives

$$Q^T * \begin{bmatrix} U \\ 0 \end{bmatrix} = C^T = [A^T \quad I^T_{S-M}]$$

where $Q^T * Q = I$ (the identity matrix) and U is upper triangular.

Let Q^T be partitioned such that $Q^T = [Q^T_1 \quad Z]$, where Q^T_1 is composed of the first S columns of the $N*N$ matrix Q^T. Then the columns of Z satisfy the relation $C * Z = 0$ and so define the null space of the active constraints.

If the approximation to the solution at the start of iteration k is n^k, a better approximation n^{k+1} is given by

$$n^{k+1} = n^k + w * p$$

where w is a scalar and the vector p is calculated from

$$Z^T * H * Z * p_z = - Z^T * g \qquad (2)$$

$$p = Z * p_z$$

The vector g is the first derivative vector of G at n^k and H is the second derivative (Hessian) matrix calculated at the same point. The scalar w is chosen so that the new point is at a minimum along p. After minimising on the current subspace (providing no bound has been violated) the projection of g on the space Z is zero, ie

$$Z^T * g^{k+1} = 0 \text{ at } n^{k+1} \qquad (3)$$

This method is analogous to Newton's method for unconstrained optimisation with the gradient replaced by the gradient projected on the active set of constraints and the second derivative matrix replaced by its projection on the active set.

To solve equations (2) the matrix $Z^T * H * Z$ may be reduced to its Cholesky factors $L * D * L^T$, where L is unit lower triangular and D is diagonal, and the vector p_z is calculated by forward and backward substitution.

Instead of doing a full Cholesky factorisation at each iteration the program performs quasi-Newton updates to the Cholesky factors.

When the minimum on the current set of active constraints has been found the vector of Lagrange multipliers l is calculated from

$$C^T * l = g^{k+1}$$

ie

$$Q^T * \begin{bmatrix} U \\ 0 \end{bmatrix} * l = g^{k+1}$$

ie

$$U * l = Q_1 * g^{k+1}$$

These multipliers are used to determine whether any of the current active constraints of the form $n_j = 0$ should be released from its bound and allowed to take a positive value. Iterations are continued until the Lagrange multipliers are all positive, at which point a minimum has been found to the original constrained problem. At each iteration the factors Q and U are updated if constraints have been added to or deleted from the active set C.

It is a feature of the method that a lower value of the objective function is found at the end of each iteration while maintaining feasibility with respect to the constraints. This is one reason that the method is so reliable. More details of the method may be found in Gill, Murray and Wright (1981). Software to solve optimisation problems of various types may be found in the NPL Numerical Optimisation Software Library (Hodson, 1983).

MULTIPHASE, Stage 1

The form of the function μ_j involves a term in $\ln(n_j)$ (except for the μ_j associated with stoichiometric species). This can give numerical problems on the computer if n_j is small, so in Stage 1 of MULTIPHASE the bounds

$$n_j \geq 0$$

are replaced by

$$n_j \geq 10^{-6}$$

for other than stoichiometric species. This of course limits the accuracy attainable in n_j but does allow the determination of whether a phase is present in a reliable way. If more accuracy is required a second stage is performed as described later.

Example, Alloy System

The method above may be used to solve an alloy system. For simplicity an alloy system consisting of two metals and two phases is considered but the method is applicable to multicomponent, multiphase systems. The model given here is of a Redlich-Kister type.

Let n_1^L denote the amount of metal 1 in the first (say liquid) phase,
    n_1^S the amount of metal 1 in the second (say solid) phase,
    n_2^L the amount of metal 2 in the liquid phase,
    n_2^S the amount of metal 2 in the solid phase.

Typically the Gibbs energy of the system may be given by

$$
\begin{aligned}
G = \ & (n_1^L + n_2^L) * (a_1^L * x_1^L + a_2^L * x_2^L + R * T * x_1^L * \ln(x_1^L) + R * T * x_2^L * \ln(x_2^L) \\
& \qquad + x_1^L * x_2^L * (c_{12}^{L} + d_{12}^{L} * (x_1^L - x_2^L) + \dots + g_{12}^{L} * (x_1^L - x_2^L)^4)) \\
& + (n_1^S + n_2^S) * (a_1^S * x_1^S + a_2^S * x_2^S + R * T * x_1^S * \ln(x_1^S) + R * T * x_2^S * \ln(x_2^S) \\
& \qquad + x_1^S * x_2^S * (c_{12}^{S} + d_{12}^{S} * (x_1^S - x_2^S) + \dots + g_{12}^{S} * (x_1^S - x_2^S)^4))
\end{aligned}
$$

where $x_1^L = n_1^L / (n_1^L + n_2^L)$, $x_2^L = n_2^L / (n_1^L + n_2^L)$
and $x_1^S = n_1^S / (n_1^S + n_2^S)$, $x_2^S = n_2^S / (n_1^S + n_2^S)$.

The constants $a, c, \dots g$ are known functions of temperature, T, but the minimisation is carried out at a given temperature. R is the gas constant.

The component abundance equations are in this case

$$n_1^L + n_1^S = r_1 \qquad \text{and} \qquad n_2^L + n_2^S = r_2$$

ie the total quantity of each metal is known but not its phase distribution.

For a ternary system with three components terms of the form $h_{jkl} * x_j * x_k * x_l$ may be included in the expression for G.

<u>Example, Ideal Gas and Stoichiometric Substance Mixture</u>

A mixture consisting of an ideal gas and several stoichiometric substances is now considered. (Each stoichiometric substance is considered as a phase containing one species.) In this case the problem becomes

$$\text{minimise } G = \sum_{j=1}^{NG} n_j * (d_j + R*T*\ln(x_j)) + \sum_{j=NG+1}^{N} n_j * d_j$$

where $x_j = n_j / (n_1 + n_2 + \ldots + n_{NG})$

(which satisfy the identity $\sum_{j=1}^{NG} x_j = 1$)

and such that $\sum_{j=1}^{N} a_{ij} * n_j = r_i$ $i=1,2,\ldots,M$

and $\qquad n_j \geq 0$

The values of d_j, a_{ij}, r_i, the gas constant R and the temperature T are assumed given, as is the pressure P; the d_j are functions of T and P, NG is the number of species in the gaseous phase, NC the number of stoichiometric phases and N = NG + NC.

If all the gaseous species are expected to be present in significant amounts (ie such that x_j are all greater than 10^{-6}, say) the method described earlier may be used. However, it is sometimes required to determine very accurately concentrations of species that may be present in extremely small quantities. The model for the gas phase will allow this if the special form of the gradient is exploited, as now described; this is the method used in Stage 2 of MULTIPHASE.

<u>MULTIPHASE, Stage 2</u>

The gradient vector of G for the system above is of the form

$$g_j = d_j + RT*\ln(x_j) \qquad j=1,2,\ldots,NG$$

$$g_j = d_j \qquad j=NG+1,\ldots,N$$

For very small values of x_j, $\ln(x_j)$ will clearly be large and negative.

Each major iteration of the optimisation method described earlier found a zero of the projected gradient Z^T*g, with the restriction that the current active set $C * n = \begin{bmatrix} r \\ 0 \end{bmatrix}$ remain active.

If the transformation
$$y_j = \ln(x_j) \qquad \text{for } j=1,2,\ldots,NG$$
is made and
$$t = n_1 + n_2 + \ldots + n_{NG}$$
then the equations (3)
$$Z^T*g = 0$$

(which are in fact nonlinear algebraic equations in the n_j) become linear functions of y_j only, but the component abundance equations become nonlinear in y_j and t. The net effect of the transformations is to write $n_j = t*\exp(y_j)$ for the gaseous species.

At each stage of the transformed optimisation problem then, the requirement is to solve

$$a_{1,i} * t * \exp(y_1) + \ldots + a_{NG,i} * t * \exp(y_{NG}) \tag{4}$$

$$+ a_{NG+1,i} * n_{NG+1} + \ldots + a_{NG+NC,i} * n_{NG+NC} = r_i \qquad i=1,2,\ldots,M$$

$$\exp(y_1) + \ldots + \exp(y_{NG}) = 1 \tag{5}$$

$$\hat{Z}^T * y = -(1/RT) * Z^T * d \tag{6}$$

(where $\hat{Z}^T$ is composed of the first NG columns of the matrix Z)

and $n_j = 0$ for some condensed species.

It is also required that $n_j \geq 0$ for the remaining condensed species.

The set of equations may be simplified if a different matrix Z is chosen from that described earlier; the property $C * Z = 0$ is retained but the orthogonality condition is relaxed (ie $Z^T Z \neq I$) so that a certain structure may be imposed.

Suppose that the columns of C be permuted to obtain a matrix $\tilde{C}$ in such a way that if it is partitioned into

$$\tilde{C} = \begin{bmatrix} \tilde{A}_1 & \tilde{A}_2 & \tilde{A}_3 \\ I & 0 & 0 \end{bmatrix}$$

and if the vector n is permuted correspondingly so that $\tilde{n} = [\tilde{n}_1 \quad \tilde{n}_2 \quad \tilde{n}_3]$ where $\tilde{n}_1$ are the amounts of species currently considered absent, then the $M * M$ matrix $\tilde{A}_2$ is chosen to be of full rank M. It is also desirable that $\tilde{n}_2$ should correspond to dominant species. Note that if a Stage 1 calculation has been performed first then the dominant species are known. It is now possible to calculate a matrix $\tilde{Z}$, partitioned in a corresponding manner, so that

$$\tilde{Z}^T = [0 \quad \tilde{Z}_2^T \quad -I]$$

where $\tilde{Z}^T$ is an $(N-S) * N$ matrix and where $\tilde{Z}_2^T$ is an $(N-S) * M$ matrix. If the species corresponding to the vector $\tilde{n}_2$ are called basic and those corresponding to $\tilde{n}_3$ are called nonbasic then the equations (6) define each nonbasic variable from the vector $\tilde{y}_3$ in terms of the set of basic variables in $\tilde{y}_2$. The nonbasic variables $\tilde{y}_3$ can now be eliminated from the set of equations (4) and (5) and a set of M+1 nonlinear equations is derived for the basic variables $\tilde{y}_2$ together with the variable t. This set is solved iteratively satisfying a linearised set at each iteration.

Initially it will not be known which stoichiometric phases have $n_j = 0$, so a guess is made, a matrix C set up, as described earlier, and an initial matrix Z determined. A sequence of optimisation problems is then solved until the correct C and hence the correct Z is determined, and the solution to the original problem obtained. At each iteration, if some n_j which were previously nonzero become zero, they are added to the current active set, C, a new Z determined and the equations solved again. Next, the Lagrange multipliers corresponding to the original problem (1) need to be calculated by solving

$$\begin{bmatrix} \tilde{A}_1^T & I \\ \tilde{A}_2^T & 0 \\ \tilde{A}_3^T & 0 \end{bmatrix} l = \tilde{g} \; .$$

This is readily solved using a QU decomposition of $\tilde{A}_2$ obtained earlier. If the multipliers corresponding to condensed species bounds are not all positive one of the condensed species is released from its zero bound, a new Z determined and the process repeated.

There are potential difficulties in this approach in that, if the choice of active set -is wrong, (ie incorrect phases have been chosen,) and thus Z is incorrect, there may be no solution to the equations maintaining nonnegativity of all n_j, in which case the calculated Lagrange multiplier may not be meaningful. Also, unlike the first method, a lower value of the Gibbs energy is not necessarily found at each iteration and this may affect convergence. Therefore an initial minimisation using Stage 1 is performed with the nonnegativity constraints for the gaseous species replaced by bounds of the form

$$n_j \geq 10^{-6}$$

This gives a good choice of active set and good starting values for Stage 2.

CONCLUDING REMARKS

The methods described here are generalised in MULTIPHASE to solve multiphase, multicomponent systems containing mixtures of gases, aqueous solutions, stoichiometric species and solution phases with interactions between species. Phase models available in MULTIPHASE include Redlich-Kister extended to multiple sublattices with fixed site ratios or to sublattices with variable site ratios, and, an extended Kapoor-Frohberg model for liquid slags. A magnetic component is added if needed. New models may be added as required.

Other computation methods are described in the proceedings of this conference. There will always be a need for improvements in these methods and, if the needs of industry for thermodynamic analysis of processes are to be met, a major requirement is for data, and in particular data critically assessed on the basis of well established models.

ACKNOWLEDGEMENTS

The authors gratefully acknowledge the work of the many contributors to MTDATA since its inception in 1970 and in particular their indebtedness to R. H. Davies, the manager of MTDATA. Thanks are also due to J. A. Gisby, N. J. Pugh, D. C. Cornwall, who assisted R. H. Davies in devising and implementing many recent improvements to MTDATA, and to T. G. Chart for his advice and help in the critical assessment of data. This work was sponsored in part by Matthey Rustenberg Refiners.

REFERENCES

Ansara, I. and B. Sundman (1987). The Scientific Group Thermodata Europe. In P. S. Glaeser (Ed), *Computer handling and dissemination of data*, 154-158. Elsevier Science Publishers, Amsterdam.
Barry, T. I., A. T. Dinsdale, R. H. Davies, J. Gisby, N. J. Pugh, S. M. Hodson, and M. Lacy (1987). MTDATA Handbook: *Documentation for the NPL Metallurgical and Thermochemical Databank*. National Physical Laboratory, Teddington.
Barry, T. I. (1985). High temperature inorganic chemistry and metallurgy. In T. I. Barry (Ed), *Chemical Thermodynamics in Industry: Models and Computation*, Chap. 1, 1-39. Blackwells Scientific Publications, Oxford.

Chang, Y. A.,   J. P. Neumann and U. V. Choudray (1979).   Phase Diagrams   and Thermodynamic Properties of Ternary Copper-Sulphur-Metal Systems, *INCRA Monograph VII, The Metallurgy of Copper.* INCRA, New York.

Dinsdale, A. T., T. G. Chart, T. I. Barry and J. R. Taylor (1982). Phase equilibria and thermodynamic data for the Cu-S system. *High Temperatures - High Pressures, 14,* 633-640.

Dinsdale, A. T. (1984). The generation and application of thermodynamic data. *Thesis,* Brunel University.

Dinsdale, A. T., S. M. Hodson and J. R. Taylor (1988). Application of MTDATA to the modelling of slag, matte, metal, gas phase equilibria. *Proceedings 3rd Internat. Conf. Molten slags and fluxes, Strathclyde.* Institute of Metals, London.

Fernandez Guillermet A., M. Hillert, B Jansson, B. Sundman (1981). An assessment of the Fe-S system using a two-sublattice model for the liquid phase. *Metall. Trans. 12B,* 745-754.

Gaye, H. and J. Welfringer (1984). Modelling of the thermodynamic properties of complex metallurgical slags. In H. A. Fine and D. R. Gaskell (Eds), *Proceedings 2nd Internat. Symp., Metal Slags and Fluxes, Lake Tahoe,* 357-375. Metall. Soc. AIME. New York.

Gill, P. E., W. Murray, M. H. Wright (1981). *Practical Optimisation,* Academic Press, New York.

Hillert, M., B. Jansson, B. Sundman and J. Agren (1985). A two-sublattice model for molten solutions with different tendency to ionization. *Metall. Trans., 16A,* 261-266.

Hodson, S. M. (1983). *Brief guide to the NPL Numerical Optimisation Software Library,* National Physical Laboratory, Teddington.

Hodson, S. M. (1988). A mathematical description of MULTIPHASE: A computer program for the calculation of complex phase equilibria in chemical and metallurgical systems. *NPL Report DITC(A) series.*

Kapoor, M. L., and G. M. Frohberg (1971). Theoretical treatment of activities in silicate melts. *Proceedings Symp. Chemical Metallurgy of Iron and Steel, Sheffield.* 17-22, The Iron and Steel Institute, London.

Nagamori, N. A. and P. J. Mackey (1978). Thermodynamics of copper matte converting, Part 1: Fundamentals of the Noranda Process. *Metall. Trans., 9B,* 255-265.

Pelton, A. D. and M. Blander (1986). Thermodynamic analysis of ordered liquid solutions by a modified quasichemical approach - Application to silicate slags. *Metall. Trans., 17B,* 805-815.

Taylor, J. R. and A. T. Dinsdale (1988). Thermodynamic and phase diagram data for the system $CaO-SiO_2$ system. *CALPHAD,* submitted for publication.

METALLURGICAL APPLICATIONS OF THERMO-CALC

Bo Sundman

Division of Physical Metallurgy
Royal Institute of Technology (KTH)
S-100 44 Stockholm, Sweden

ABSTRACT

A general presentation is first made of Thermo-Calc and some of its databases. The assessment procedure necessary to generate a database is discussed. An example of a calculation using a slag database is given.

KEYWORDS

Thermodynamics; databases; Slags; Metallurgy; Software;

INTRODUCTION

Thermo-Calc is a software system (Sundman, Jansson and Andersson, 1985) can be used for equilibrium and phase diagram calculations for very general kinds of thermochemical systems. It was originally developed for calculations in steels and alloys and has been used for that purpose in almost ten years. The last few years there has been a considerable interest to use Thermo-Calc also for other types of calculations e.g. slags or molten salts but there has been no database for such systems until a database from IRSID, the French steel research institute, became available (Gaye and Columbet, 1984). This database consists of assessed data for the system Al2O3-CaO-FeO-Fe2O3-MgO-MnO-SiO2. IRSID uses a "cell model" (Kapoor and Frohberg, 1974) for the liquid oxide and this had to be implemented into Thermo-Calc also. An enhancement made at KTH was to merge the IRSID database, which only covers the oxides, with a database for dilute solution of about 20 elements in liquid iron (Sigworth and Elliot, 1974) in order to make it possible to apply the Thermo-Calc software for realistic slag calculations including equilibria between the slag and the liquid steel directly.

SOFTWARE

Thermo-Calc consists of modules as shown in Fig. 1. Each module has a specific purpose, in one of the modules the user selects database and defines his system i.e. which components he is interested in. Then there are two equilibrium calculation modules, the oldest POLY-1 is dedicated to alloy systems whereas the new, POLY-3, can handle more general equilibrium calculation e.g. gas equilibria and equilibria with slags. In both calculation modules there is a postprocessor for tabular or graphical output of the results. There is also a module for tabulation data or chemical reactions and one module for interactive manipulation of thermodynamic data.

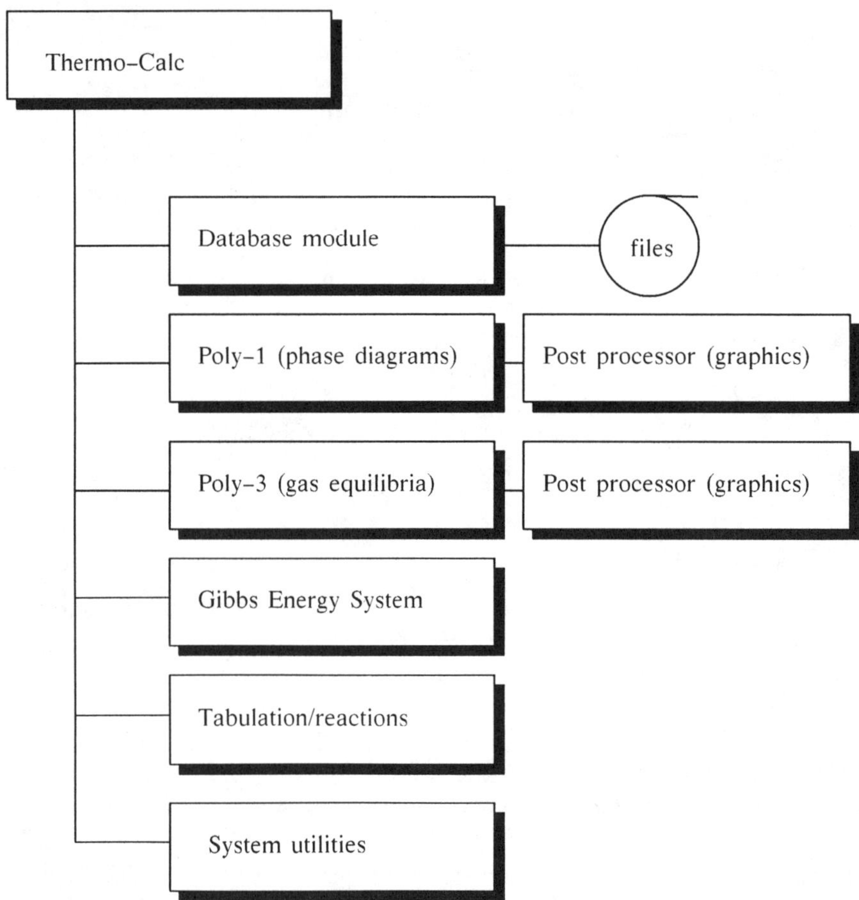

Fig. 1. The modular structure of Thermo-Calc.

All modules have the same user interface where the user has a "conversation" with the program by giving it commands and answering the questions the program may ask. Each command is usually very elementary and the user may execute them in almost any order. By combining the elementary commands in the POLY modules the user can create a very general set of conditions for an equilibrium calculation.

The results of a calculation can be listed or presented graphically in the postprocessor. It is possible to define arbitrary functions of state variables in order to generate the most useful presentation.

DATABASES

Once one has access to a general thermodynamic tool like Thermo–Calc the important question is how good databases there are for a particular problem. The important fact with a thermochemical database is that one may make extrapolations. A database with just a few assessed systems, the IRSID database has just twenty binaries, can be used to extrapolate into a multicomponent system and predict properties where there are no experimental information. In Thermo–Calc the following databases are most useful

- the SGTE solution database with about 150 assessed systems,
- the SGTE substance database with about 2000 substances and 1000 gaseous species,
- the SGTE aqueous database for dilute aqueous solutions,
- the Fe–base database with about 25 assessed ternary systems in the iron rich corner,
- the IRSID/Sigworth–Elliot database for slag/liquid iron equilibria

SGTE stands for Scientific Group Thermodata Europe (Ansara and Sundman, 1986) and is a consortium of seven research organisations and universities in England, France, Germany and Sweden. SGTE has a long term project to provide a general purpose thermochemical database with assessed data.

The SGTE databases are extended each year. The other databases are updated only occasionally. The Fe–base database is the best available for calculations of steels in the solid range.

ASSESSMENT PROCEDURE AND MODELS

In is a highly qualified work to provide databases for thermochemical calculation. One cannot compare with traditional bibliographical or numerical databases because it is not sufficient to collect the available data data for a system. That is actually only the beginning as the experimental data must be interpreted in terms of thermochemical models and a number of model parameters determined which can describe the experimental data as accurate as possible. Both the selection of model and which data to trust when there are inconsistencies can be very difficult.

The advantage of assessing the data in terms of models is that it makes it very easy to extrapolate. Thus one may compute equilibria in multicomponent system using assessments of only binary or ternary systems. In multicomponent system the amount of experimental data is usually very scarce and the possibility to predict properties of multicomponent systems is a very important feature of a system like Thermo–Calc.

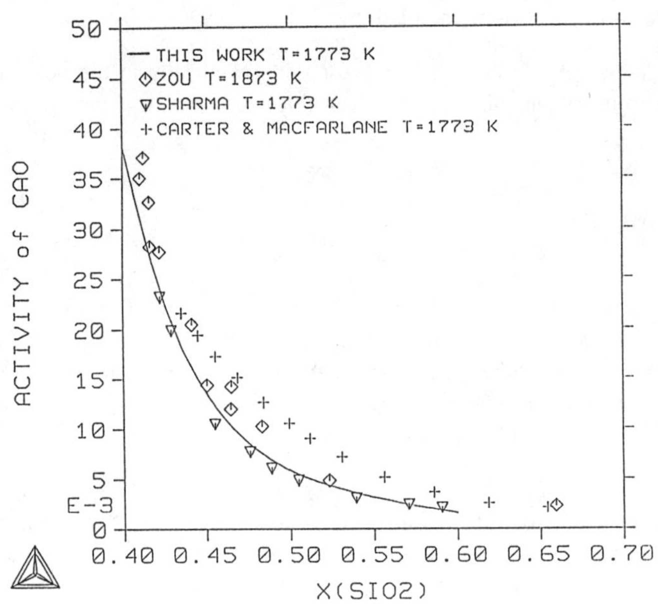

Fig. 3. Calculated activities in CaO–SiO2 compared with experiments.

Other groups are also working on databases for oxide systems, most important perhaps the FACT group in Montreal. They have developed a new model for the liquid including also a qusichemical entropy term. One should also mention prof. Saxena at Brooklyn College, New York who work with geological systems and a special interest for high pressure properties.

EXAMPLE OF APPLICATION

The number of real case studies using the slag database is still limited but the experiences so far are very promising. Most Thermo–Calc customers handles their calculations on–line from their own terminals and only occasionally they present what they have managed to do. The following example should give an idea of the abilities of Thermo–Calc and the IRSID/Sigworth–Elliot database. In order to show how the solution is found most of the conversation between the user and the computer is presented.

In the example the user wanted to increase the Chromium content of a steel from 20 to 25 per cent. He then had problems with clogging in the continuous casting of the material. The solid oxide formed was approximately Cr2O3. In the calculation below a possible solution was found by increasing the Manganese content.

In the examples a typewriter font is used and input from the user is witten with a larger size. Comments inserted in order to make the intention of the user more clear is written with the current font.

The IRSID database does not give a good description of the solid oxides but treats them all with fixed stoichiometry. This is not satisfactory for more accurate calculations and therefore we have at KTH started a project assessing some oxide systems using more sophisticated models i.e. the compound energy model (Hillert, Jansson and Sundman, 1988) and a new ionic model for the liquid (Hillert and co-workers, 1985). The advantage of the new model for the liquid is that it can have more than one anion whereas the cell model is limited to a single anion Thus we will be able to incorporate sulphur or any other anionic species. A second advantage is that the model gives a uniform description of the liquid treating the metallic and oxide liquids as a single phase with a miscibility gap. This project has just started and so far Fe-O, CaO-SiO2, Al2O3-CaO and Ca-Fe-O are in various stages of completion. In Figs. 2 and 3 some results from the assessment of CaO-SiO2 (Hillert, Sundman and Xizhen, 1987) are shown. The data for the solid oxides are assessed from 298.15 up to the melting point and this makes it possible to compute enthalpies and heat capacities. The current project at KTH is to assess most ternaries in Al-Ca-Fe-Mg-Si-O system within two years. Parallel projects will lead to an assessment of some sulphur system starting with Cu-Fe-S.

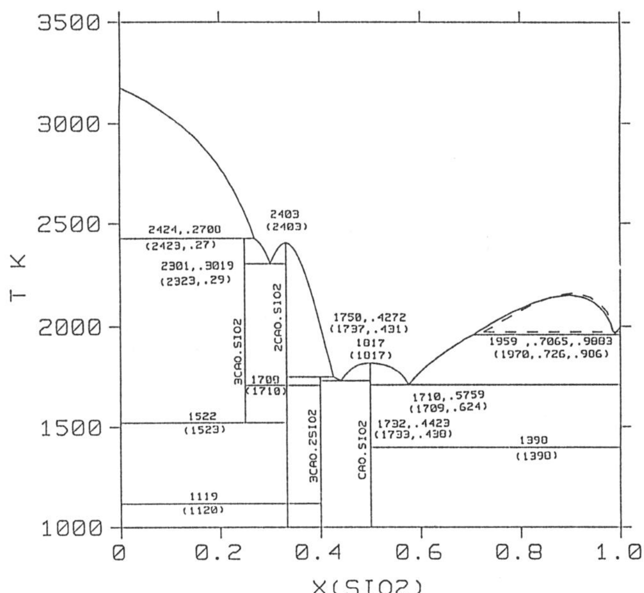

Fig. 2. The calculated phase diagram for CaO–SiO2.
Experimental temperatures in parenthesis.

The program is started and presents itself.

```
$ tc

THERMO CALC version D
Copyright: Division of Physical Metallurgy, KTH, S 100 44 Stockholm, Sweden
Double precision version linked at  1-JUN-1988 11:10:22
```

The user goes to the database module to define his system. The commands to Thermo-Calc can be abbreviated and the full command is given on the following line. The user knows all commands and does not use the extensive help facilities of Thermo-Calc much. A less advanced user may press a ? as answer at any point in order to obtain more help.

```
SYS: GO D
   ... the command in full is GOTO_MODULE

Current database: SGTE DATA BASE (Extended Thermo-Calc version)

TDB_SSOL: SW SLAG
   ... the command in full is SWITCH_DATABASE
Current database: Dilute Fe liquid and slag and oxides with Fe-Ca-Si-Mn-Mg-Al
```

The user first defines his system

```
TDB_SLAG: DEF-SYS MN CA SI CR FE O
   ... the command in full is DEFINE_SYSTEM
TDB_SLAG: GET
   ... the command in full is GET_DATA
-OK-
```

There are no data for CR2O3 in the slag database. These data are appended from the SGTE substance database (called SSUB).

```
TDB_SLAG: APPEND SSUB
   ... the command in full is SWITCH_DATABASE
TDB_SSUB: DEF-SPECIE
   ... the command in full is DEFINE_SPECIES
SPECIES:CR2O3
TDB_SSUB: L-S C
   ... the command in full is LIST_SYSTEM
CR2O3_S     :CR2O3:
CR2O3_L     :CR2O3:
TDB_SSUB: GET
   ... the command in full is GET_DATA
-OK-
TDB_SSUB: GO P-3
   ... the command in full is GOTO_MODULE

POLY_3 version 1.0, Test release December 1986
```

```
POLY_3: ?
  ... the command in full is HELP
BACK                        GOTO_MODULE              SAVE_WORKSPACES
CHANGE_STATUS_COMPONENT      HELP                     SELECT_EQUILIBRIUM
CHANGE_STATUS_PHASE          INFORMATION              SET_ALL_START_VALUES
CHANGE_STATUS_SPECIES        LIST_AXIS_VARIABLE       SET_AXIS_VARIABLE
COMPUTE_EQUILIBRIUM          LIST_CONDITIONS          SET_CONDITION
DEFINE_COMPONENTS            LIST_EQUILIBRIUM         SET_REFERENCE_STATE
DELETE_INITIAL_EQUILIB       LIST_INITIAL_EQUILIBRIA  SET_START_VALUE
ENTER_CONSTANT               LIST_STATUS              SHOW_VALUE
ENTER_EQUILIBRIUM            LIST_SYMBOLS             STEP_FOR_INITIAL_EQUILIB
ENTER_FUNCTION               LOAD_INITIAL_EQUILIBRIUM STORE_INITIAL_EQUILIB
ENTER_TABLE                  READ_WORKSPACES
EVALUATE_FUNCTION            RECOVER_START_VALUES
EXIT                         REINITIATE_MODULE
```

Some basic information on POLY-3 is obtained with the INFORMATION command. State variables are very important as they are used in order to set conditions. Note that many questions has a default answer given within //.

```
POLY_3: INFO
  ... the command in full is INFORMATION
WHICH SUBJECT /PURPOSE/: ?

INFORMATION
   This gives information on the following subjects:

   Purpose       State varibles

WHICH SUBJECT: STATE

State variables
        In many cases POLY-3 asks for or lists values of state
variables. Examples of state variables are temperature, mole fraction,
enthalpy etc. In POLY-3 a general notation method has been designed
for a predifined set of state variables. This list is not exhaustive
but the remaining state variables can be obtained by combinations of
the predefined ones.

MNEOMONIC                   MEANING

Intensive properties
T                           temperature
P                           pressure
AC(component)               activity
MU(component)               chemical potential

Normalizing of extensive properties:
For all extensive properties a suffix can be added to the mneomonic
name to indicate a normalized extensive property.

        Extensive property Z for the whole system:
        Z           extensive property Z in the whole system.
        ZM          extensive property Z per mols of atoms in the system
        ZW          extensive property Z per mass (gram) in the system
        ZV          extensive property Z per volume (m3) in the system
```

```
Extensive property Z for a phase
Z       extensive property Z per mole of formula unit of the phase
ZM      extensive property Z per mole of atoms of the phase.
ZW      extensive property Z per mass (gram) of the phase.
ZV      extensive property Z per volume (m3) of the phase.
```

```
Energetic Extensive properties (note: suffixes M, W and V can be applied)
S                           entropy of the system
S(phase)                    entropy of a phase
V                           volume of the system
V(phase)                    volume of a phase
G                           Gibbs energy of the system
G(phase)                    Gibbs energy of a phase
H                           entalphy of the system
H(phase)                    entalphy of a phase
F                           Helmholtz energy of the system
F(phase)                    Helmholtz energy of a phase
```

```
Amount of components (note: suffixes M, W and V can be applied)
N(component)                Number of moles in the system
N(phase,component)          Number of moles in a phase.
B(component)                Mass in the system
B(phase,component)          Mass in a phase
```
Remark: instead of the mneomonics NM and BW, X and W can be used to
designate mole fraction and mass fraction.

```
Total amount in the system (note: suffixes M, W and V can be applied)
N                           Number of moles in the system
B                           Mass in the system
```

```
Amount of a phase in the system (note: suffixes M, W and V can be applied)
NP(phase)                   number of moles of a phase
BP(phase)                   mass of a phase
VP(phase)                   volume of a phase
```
Remark: the normalizing properties are for the total system

```
Constitution
Y(phase,species#sublattice)     site fraction
```
Note: This quantity is dependent upon the model choosen for the phase.

Derivatives of state variables can evaluated using a dot "."
between two state variables, for example HM.T is heat capacity.

WHICH SUBJECT:

POLY_3: LI-ST P
   ... the command in full is LIST_STATUS
*** STATUS FOR ALL PHASES

GAS	ENTERED	CAO	ENTERED	FEOLIQ	ENTERED
FE_LIQUID	ENTERED	CAO_SIO2	ENTERED	MN2O2_SIO2	ENTERED
SLAG	ENTERED	CR2O3_L	ENTERED	MNO	ENTERED
CA2O2_SIO2	ENTERED	CR2O3_S	ENTERED	MNO_SIO2	ENTERED
CA3O3_SI2O4	ENTERED	FE2O2_SIO2	ENTERED	SIO2	ENTERED
CA3O3_SIO2	ENTERED	FE2O3	ENTERED	WUSTITE	ENTERED

Now the conditions are set. The pressure is one atm in Pascal. The amounts of
Fe, Ca, Si and O are given as number of moles. It is actually not very important

what values are used because they will mainly determine the ratio slag/fe–liquid
and that is not essential in the present case.

```
POLY-3: S-C T=1800,P=101325,N(FE)=100,N(CA)=2,N(SI)=2,N(O)=6
    ... the command in full is SET_CONDITION
```

The amounts of Cr and Mn is given as weight fractions in the Fe–liquid phase

```
POLY_3: S-C W(FE-L,CR)=.2,W(FE-L,MN)=.004
    ... the command in full is SET_CONDITION
POLY_3: L-C
    ... the command in full is LIST_CONDITIONS
T=1800, P=101325, N(FE)=100, N(CA)=2, N(SI)=2, N(O)=6, W(FE_LIQUID,CR)=2E-1,
   W(FE_LIQUID,MN)=4E-3
DEGREES OF FREEDOM 0

POLY_3: S-A-S
    ... the command in full is SET_ALL_START_VALUES
Automatic start values for phase constitions? /Y/:

POLY_3: C-E
    ... the command in full is COMPUTE_EQUILIBRIUM
 14 ITS,  CPU TIME USED  13 SECONDS

POLY_3: L-E
    ... the command in full is LIST_EQUILIBRIUM
Output file: /TERMINAL/:
Options /VWCS/:

OUTPUT FROM POLY-3

Conditions:
T=1800, P=101325, N(FE)=100, N(CA)=2, N(SI)=2, N(O)=6, W(FE_LIQUID,CR)=2E-1,
   W(FE_LIQUID,MN)=4E-3
DEGREES OF FREEDOM 0

Temperature 1800.00,  Pressure  1.013250E+05
Number of moles of components  1.37914E+02,  Mass  7.27102E+00
Total Gibbs energy -1.81972E+07,  Enthalpy  6.52481E+06,  Volume   .00000E+00

Component                 Moles      Fraction   Activity   Potential   Ref.state
CA                        2.0000E+00 1.1025E-02 5.4984E-13 -4.2248E+05 SER
CR                        2.7038E+01 1.9335E-01 5.6006E-04 -1.1206E+05 SER
FE                        1.0000E+02 7.6808E-01 6.1357E-04 -1.1069E+05 SER
MN                        8.7568E-01 6.6164E-03 2.3360E-06 -1.9407E+05 SER
O                         6.0000E+00 1.3203E-02 5.2842E-13 -4.2308E+05 SER
SI                        2.0000E+00 7.7253E-03 1.2513E-08 -2.7233E+05 SER

FE_LIQUID#1                   Status ENTERED
Number of moles 1.2806E+02,   Driving force   .0000E+00
FE 7.94189E-01   MN  4.00000E-03   O   5.13011E-04
CR 2.00000E-01   SI  1.22241E-03   CA  7.56512E-05

SLAG#1                        Status ENTERED
Number of moles 9.8551E+00,   Driving force   .0000E+00
O  3.82424E-01   SI 1.96936E-01   FE 8.29597E-03
CA 3.29599E-01   MN 8.27449E-02   CR   .00000E+00
```

We have now calculated the equilibrium with the original Chromium content of 20 weight per cent. Now increase this to 25..

```
POLY_3: S-C W(FE-L,CR)
  ... the command in full is SET_CONDITION
Value /.2/: .25

POLY_3: C-E
  ... the command in full is COMPUTE_EQUILIBRIUM
 10 ITS,  CPU TIME USED  9 SECONDS
POLY_3: L-E
  ... the command in full is LIST_EQUILIBRIUM
Output file: /TERMINAL/:
Options /VWCS/:

OUTPUT FROM POLY-3
Conditions:
T=1800, P=101325, N(FE)=100, N(CA)=2, N(SI)=2, N(O)=6,
   W(FE_LIQUID,CR)=2.5E-1, W(FE_LIQUID,MN)=4E-3
DEGREES OF FREEDOM 0

Temperature 1800.00,  Pressure  1.013250E+05
Number of moles of components 1.47048E+02,  Mass  7.74591E+00
Total Gibbs energy -1.92038E+07,  Enthalpy 7.17710E+06,  Volume   .00000E+00

Component                 Moles      Fraction   Activity   Potential   Ref.state
CA                        2.0000E+00 1.0349E-02 6.7872E-13 -4.1933E+05 SER ?
CR                        3.6189E+01 2.4293E-01 6.9334E-04 -1.0886E+05 SER ?
FE                        1.0000E+02 7.2099E-01 5.7367E-04 -1.1170E+05 SER ?
MN                        8.5914E-01 6.0935E-03 2.4540E-06 -1.9333E+05 SER ?
O                         6.0000E+00 1.2393E-02 4.7286E-13 -4.2474E+05 SER ?
SI                        2.0000E+00 7.2517E-03 1.4099E-08 -2.7054E+05 SER ?

FE_LIQUID#1                   Status ENTERED
Number of moles 1.3721E+02,   Driving force  .0000E+00
FE  7.44070E-01  MN  4.00000E-03  O   5.03470E-04
CR  2.50000E-01  SI  1.37351E-03  CA  5.26549E-05

SLAG#1                        Status ENTERED
Number of moles 9.5603E+00,   Driving force  .0000E+00
O   3.82827E-01  SI  1.96088E-01  FE  6.59095E-03
CA  3.41022E-01  MN  7.34726E-02  CR   .00000E+00

CR2O3_S#1                     Status ENTERED
Number of moles 2.7869E-01,   Driving force  .0000E+00
CR  6.84202E-01  CA   .00000E+00  FE   .00000E+00
O   3.15798E-01  MN   .00000E+00  SI   .00000E+00
```

The Cr2O3 oxide is now stable. To supress it one may now try to decrease the oxygen potential by adding more Manganese. To get the exact amount needed to avoid forming forming Cr2O3 set the status of Cr2O3 to FIX with zero amount and allow the Manganese content to vary.

```
POLY_3: C-S-P CR2O3_S
  ... the command in full is CHANGE_STATUS_PHASE
Status: /ENTERED/: FIX
```

```
Number of moles /0/:
POLY_3: L-C
  ... the command in full is LIST_CONDITIONS
T=1800, P=101325, N(FE)=100, N(CA)=2, N(SI)=2, N(O)=6,
   W(FE_LIQUID,CR)=2.5E-1, W(FE_LIQUID,MN)=4E-3
FIX PHASES
CR2O3_S
DEGREES OF FREEDOM -1
POLY_3: S-C  W(FE-L,MN)
  ... the command in full is SET_CONDITION
Value /.004/: ?
          The value that should be assigned to the state variable.
The value NONE (given in full!) means that the condition is cleared (reseted).
Value /.004/: NONE
POLY_3: L-C
  ... the command in full is LIST_CONDITIONS
T=1800, P=101325, N(FE)=100, N(CA)=2, N(SI)=2, N(O)=6, W(FE_LIQUID,CR)=2.5E-1
FIX PHASES
CR2O3_S
DEGREES OF FREEDOM 0

POLY_3: C-E
  ... the command in full is COMPUTE_EQUILIBRIUM
  6 ITS,  CPU TIME USED   5 SECONDS
POLY_3: L-E
  ... the command in full is LIST_EQUILIBRIUM
Output file: /TERMINAL/:
Options /VWCS/:
OUTPUT FROM POLY-3

Conditions:
T=1800, P=101325, N(FE)=100, N(CA)=2, N(SI)=2, N(O)=6, W(FE_LIQUID,CR)=2.5E-1
FIX PHASES
CR2O3_S
DEGREES OF FREEDOM 0

Temperature 1800.00,  Pressure  1.013250E+05
Number of moles of components 1.47271E+02,  Mass  7.75834E+00
Total Gibbs energy -1.92511E+07,  Enthalpy  7.19503E+06,  Volume   .00000E+00

Component                 Moles      Fraction   Activity   Potential   Ref.state
CA                        2.0000E+00 1.0332E-02 6.9417E-13 -4.1899E+05 SER
CR                        3.6134E+01 2.4217E-01 6.9402E-04 -1.0885E+05 SER
FE                        1.0000E+02 7.1983E-01 5.7264E-04 -1.1173E+05 SER
MN                        1.1378E+00 8.0573E-03 3.2332E-06 -1.8920E+05 SER
O                         6.0000E+00 1.2373E-02 4.7255E-13 -4.2475E+05 SER
SI                        2.0000E+00 7.2401E-03 1.2933E-08 -2.7184E+05 SER
```

```
FE_LIQUID#1                    Status ENTERED
Number of moles 1.3741E+02,    Driving force    .0000E+00
FE   7.42908E-01  MN  5.28552E-03  O   5.02570E-04
CR   2.50000E-01  SI  1.25152E-03  CA  5.21899E-05


SLAG#1                         Status ENTERED
Number of moles 9.8627E+00,    Driving force    .0000E+00
O    3.79282E-01  SI  1.92338E-01  FE  6.58071E-03
CA   3.28070E-01  MN  9.37286E-02  CR   .00000E+00


CR2O3_S#1                      Status FIX
Number of moles  .0000E+00,    Driving force    .0000E+00
CR   6.84202E-01  CA   .00000E+00  FE   .00000E+00
O    3.15798E-01  MN   .00000E+00  SI   .00000E+00
```

We can read the Manganese composition in the Fe–liquid above as 0.53 weight per cent. Changing to this value may prevent clogging.

ACKNOWLEDGEMENT

Financial support from Nordisk Industrifond and the Swedish Board for Technical Development (STU) is gratefully acknowledged.

REFERENCES

Ansara, I and Sundman, B. (1986). The Scientific Group Thermodata Europe, Proc Conf. CODATA, Ottawa, July 1986.

Hillert, M., Jansson, B., Sundman, B. and Ågren, J. (1985). A Two–Sublattice model for Molten Solutions with a Different Tendency for Ionization. Metall. Trans. A., 16A, p 261–266.

Hillert, M., Jansson, B. and Sundman, B. (1988). Application of the Compound–Energy Model to Oxide systems, Z. Metallkde, 79, p 81–87.

Hillert, M., Sundman, B. and Xizhen W. (1987). An Assessment of the CaO–SiO2 system, TRITA–MAC 355, Royal Institute of Technology, Stockholm, Sweden.

Gaye, H. and Coulombet, D. (1984). EC report PCM RE 1064, CECA No 7210.

Kapoor, M. L. and Frohberg, M. G. (1984). Archiv Eisenhuttenw, 45, p 663.

Sigworth, G. K. and Elliot, J. F. (1974). Metal Science, Vol 8, (1974), p 298.

Sundman, B., Jansson, B., and Andersson J–O (1985). The Thermo–Calc databank system. CALPHAD, 9, p 153–190.

HETEROGENEOUS EQUILIBRIUM CALCULATIONS
WITH MULTICOMPONENT SOLUTION MODELS
– SOLGASMIX AND THE F*A*C*T SYSTEM

W.T. Thompson**, G.Eriksson*, A.D. Pelton* and C.W. Bale*

* Ecole Polytechnique de Montreal,
Box 6079, Station A
Montreal, Quebec, Canada H3C 3A7

** Royal Military College of Canada,
Kingston, Ontario, Canada K7K 5L0

ABSTRACT

The introduction of solution databases into the F*A*C*T system
will be discussed from the point of view of use in mass-
constrained Gibbs energy minimization computations using an
advanced version of the SOLGASMIX program. Examples will be shown
illustrating various treatments of solution thermodynamics
including those applicable to metallic solutions, slags, molten
salts and aqueous solutions. Special features of the interactive
program will be emphasized.

KEYWORDS

Thermodynamics, solution models, database, computing, Gibbs
energy minimization, slag, aqueous equilibria

INTRODUCTION

For a number of years, a database system called the Facility for the Analysis of Chemical Thermodynamics (F*A*C*T) has operated from Montreal. This system is telephone accessible at a flat rate by Datapac and Telenet throughout North America and also overseas at somewhat higher cost by standard linkages to these telecommunication networks. F*A*C*T, and its French language counterpart F*A*I*T, were designed as a teaching and research aids for the academic environment as well as a services to industry. F*A*C*T / F*A*I*T now offers a full range of programs encompassing most of the requirements of metallurgy and materials specialists. These include heat balances, predominance diagrams, Pourbaix diagrams, binary and ternary phase diagram computations and a general Gibbs energy minimization program called EQUILIB which incorporates an advanced version of SOLGASMIX (Eriksson, 1975). In addition to a large database developed from many standard compilations (e.g. Barin, Knacke and Kubaschewski, 1973, 1977; Stull, Prophet and co-workers, 1971) which are automatically read as necessary by the programs, an important feature is the user-friendly interface. Given the periodic nature of recourse to thermochemical computations by most engineers and scientists, this aspect of F*A*C*T is very important. The necessity in a computational system to resort to a manual to comprehend operational requirements after a short hiatus from regular usage will discourage the majority of users from using even the most advanced and useful programs. This paper will focus on the user interface and in particular deal with the manner in which users identify complex solution phases in the F*A*C*T EQUILIB program.

IDEAL SOLUTIONS

An example of conventional usage of the EQUILIB program is shown in Fig. 1. The user simply defines the initial state of the system by entering the reactants in the desired proportions. The final state may be determined in a variety of ways. When, as in the present example, each reactant state is specified (298 K in each case, with hydrostatic pressure 1 atm and most stable phase assumed by default) the product equilibrium may be constrained by fixing the enthalpy change in addition to the final pressure. Following the specification of the overall composition (in this case by specifying "A"), any two other properties will suffice in EQUILIB so long as one is an intensive variable (temperature or pressure). The input is completed with a selection of species to be considered in the product equilibrium. Presently 500 are permitted. Solutions are indicated among the species selection by enclosing groups within slashes. Species trailing the last slash are treated as pure phases. The output from the program is shown in Fig.2. Notice that gaseous ions are easily introduced into the calculation following their appearance as possible product species in the listing in Fig.1. The caption beneath Fig. 2 will clarify the significance of the numerical output.

DILUTE NON-IDEAL SOLUTIONS

Although it is reasonable to treat gaseous solutions as ideal, particularly at high temperature and moderate pressure, ideality is unreasonable for a majority of applications when some degree of precision is needed. The oxidation of Cr from molten Cu-Cr alloys provides a simple case in point. The Cu-Cr system is of a simple eutectic variety with negligible solid solubility however the Cr-liquidus shows the characteristic S-shape characteristic of strong positive deviation from ideal melt behaviour. This results in a low chromium solubility in molten copper. Clearly, quantitative calculations involving the melt phase must acknowledge this effect. When the molten alloy is dilute in Cr, and Henry's Law is applicable, a feature of the options menu allows the user to introduce temperature dependent activity coefficients. The user simply enters the overall composition of the Cu-Cr-O system that is of interest as shown in Fig. 3. Subscripts are not entered in this case implying that extensive property changes are not to be computed. The product temperature and pressure are set at 1373 K and 1 atm. In this case, the option to list all possible product species from the main F*A*C*T database as well as from the user's private file is exercised by typing "list" as shown at the bottom of Fig 3. The listing is shown in Fig. 4. At the bottom of this figure the user has grouped species 2-11 in the ideal gas phase, 15 and 17 in the melt phase and wishes to consider pure solids 19-27 as possibly coexisting in the equilibrium phase assemblage. To introduce an activity coefficient for Cr dissolved in liquid Cu, option 2 is selected as shown in Fig. 5. The user simply enters the species number in the list shown in Fig. 4 and then enters the two constants in the equation shown near the bottom of Fig. 5. The standard state is implied when the species number is entered (17, pure liquid Cr), and since the logarithm of the activity coefficient is dimensionless, units are of no concern from the user's point of view. The output from the computation shown in Fig. 6. The activity coefficient expression is retained until such time as a new set of reactants is entered or a new species selection is made.

SOLUTION MODELS

Although the introduction of concentration independent activity coefficients is a useful extension to EQUILIB, its application is restricted for practical purposes to dilute solutions where Henry's Law is appropriate. In the general case, it is necessary to express the activity of a component not only as a function of temperature but also of concentration of that component. Furthermore, in systems with more than two components, the influence of the other components on the activity coefficient and other solution properties must also be described by appropriate mathematical models. Presently, five models are available. These include a dilute solution Wagner parameter style formalism, a dilute aqueous solution model incorporating Pitzer-Brewer coefficients, a sub-lattice model, a generalized empirical power series representation, and a quasi-chemical slag model.

DILUTE SOLUTION MODEL WITH WAGNER PARAMETERS

In this treatment the user may enter and store interaction parameters describing the concentration and temperature dependence of activity coefficients of species in dilute multicomponent systems dominated by a single solvent. The model takes the form (Pelton and Bale, 1986):

$$\ln\gamma_i/\gamma_i^0 = \ln\gamma_{solvent} + \epsilon_{i1}X_1 + \epsilon_{iw}X_2 + \epsilon_{i3}X_3 + \cdots + E_{iN}X_N \quad [1]$$

A feature of this model is the handling of the solvent activity coefficient with the equation:

$$\ln\gamma_{solvent} = -\frac{1}{2} \sum_{j=1}^{N} \sum_{k=1}^{N} \epsilon_{jk}X_jX_k \quad [2]$$

The inclusion of this often small term is needed to satisfy the Gibbs-Duhem equation which can be significant at high solute concentrations. A simple example of the use of the treatment is in the solidification of steel in an atmosphere of carbon monoxide gas. The input is shown in Fig. 7 for the case of a 0.2 weight percent steel. The selection of an option prior to entering the equation (refer to Fig. 5 option 5) has caused the numbers preceding the species to be interpreted as grams. The listing of species has included a user-entered hypothetical species for carbon dissolved in delta iron. The arrow in Fig. 8 is a reminder that this species has been user-supplied. The introduction of hypothetical species is effectively another way to deal with non ideal solution behaviour. At the bottom of the species list in Fig. 8, a special group (39-41) appears preceded by a short heading. For this group interaction parameters are available. By selecting species as indicated, the product equilibrium will be computed considering a binary Henrian solution of delta iron and carbon, a multicomponent melt phase treated by the interaction parameter formalism as well as a gas phase. The output is shown in Fig. 9. The masses of each condensed phases are shown as well as the composition in weight percent. The volume of the gas phase is reported in addition to the volumetric composition. Should it be desired, the user could enter interaction parameters for all solutes in the delta iron and include therefore in the computation the solubility, for example, of oxygen in this phase. Moreover, a third solution phase for austenite and other components such as Mn, Si, P, etc., could be introduced which would make possible the complete thermodynamic modelling of the solidification of steel.

MOLTEN OXIDE (SLAG) MODEL

The input shown in Fig. 10 is similar to that in Fig. 7 except that a slag phase is introduced and the hot metal contains Si and Mn in addition to just carbon. The carbon monoxide is formed by the introduction of oxygen. This input effectively deals with basic oxygen steelmaking reactions. The species list this time concludes with a 5 component molten oxide phase (153-157). No special invocation by the user is necessary to expose these data. This is automatically done when the input elements would make possible their usage. The species selection at the bottom of Fig. 10 includes an ideal gas phase, a non-ideal hot metal phase (treated with interaction parameters), a non-ideal slag phase, and a selection of possible pure condensed phases. The output shown in Fig. 11 provides the equilibrium when 100 grams of hot metal containing 1.1% Si, 0.8% Mn and 4.1% C in contact with 10 grams of a 35%lime -35%magnesia -30%silica slag is reacted with 5 grams of O2. phase. In performing the computation, the system internally takes care of matters such as standard state without questioning the user (Pelton and Blander, 1986). The mathematics and physical chemistry is described elsewhere along with convincing evidence to show that the model describes quite precisely available experimental data over the entire range of possible compositions including the region of immiscibility.

SUB-LATTICE MODEL

Certain types of solutions are properly treated only when the ideal term for the entropy of mixing recognizes restrictions on the placement of atoms, ions or complex ions. Molten salts are well known examples of solutions where species in the solution cannot mix entirely randomly. Anions and cations are regarded as being distributed on interpenetrating but distinct kinds of sites. The example in Fig. 12 shows the computed equilibrium between molten carbonate' fuel cell electrolyte and a typical fuel gas containing sulfur. The computation of dissolved sulfate level is made using the sublattice model (Pelton, 1988) available to the EQUILIB program and parameters which have been adjusted and stored to fit available information on this system.

PITZER-BREWER AQUEOUS SOLUTION MODEL

Data stored in the main F*A*C*T database for dilute aqueous species force the activity to approach the molality as the solution becomes progressively more dilute. Unfortunately, the level of concentration at which activity ceases to closely approximate the molality is much below the solubility of most salts. It is therefore necessary to use an empirical treatment to represent the known behaviour especially if precipitation computations are to be reliable. The Pitzer-Brewer treatment (Harvie, Moeller and Weave, 1984) is available to EQUILIB. An example of a computation is shown in Fig. 13. The user is considering the solubility of rock salt in dilute hydrochloric acid when contact with carbon dioxide yielding a total gas volume of 500 liters at 298 K. By basing the input on 55.5 moles of H2O as shown near the top of Fig 13, it is apparent that the initial state involves a 1 molal HCl to which is added 3 moles of NaCl. The computed equilibrium shown in Fig. 14 includes the redox potential and the pH which appear beneath the subscript for the aqueous phase.

CONCLUSION

The ability to easily introduce solution models into EQUILIB considerably extends the scope of this versatile program incorporating an advanced version of SOLGASMIX. Users may make use of a small but growing public database of important solutions or use a F*A*C*T program to develop their own models which are subsequently automatically presented whenever the combination of input elements warrants their possible introduction into the computation.

ACKNOWLEDGEMENTS

The authors acknowledge the support of the Natural Sciences and Engineering Research Council and the Computing Centres of Ecole Polytechnique and McGill University.

REFERENCES

1) Barin, I., Knacke, O. and Kubaschewski, O. (1973, 1977).
    Thermochemical Properties of Inorganic Substances.
    Springer-Verlag, Berlin

2) Eriksson, G. (1975).
    SOLGASMIX a computer program for calculation of equilibrium
    compositions in multiphase systems.
    Chem. Scr. 8, 100-103.

3) Harvie, C.E., Moeller, N. and Weave, J.H. (1984).
    The prediction of mineral solubilities in natural waters: The
    $Na-K-Mg-Ca-H-Cl-SO_4-OH-HCO_3-CO_3-CO_2-H_2O$ system to high ionic
    strengths at 25°C.
    Geochim. Cosmochim. Acta, 48, 723-751.

4) Pelton, A.D. (1988).
    A database and sublattice model for molten salt solutions.
    CALPHAD: Comput. Coupling Phase Diagrams Thermochem., 12,
    127-142.

5) Pelton, A.D. and Bale, C.W. (1986).
    A modified interaction parameter formalism for non-dilute
    solutions.
    Metall. Trans. A., 17A, 1211-1215.

6) Pelton, A.D. and Blander, M. (1986).
    Thermodynamic analysis of ordered liquid solutions by a
    modified quasichemical approach - Application to silicate
    slags.
    Metall. Trans. B., 17B, 805-815.

7) Stull, D.R., Prophet, H. and co-workers (1971). JANAF Thermochemical Tables.
    National Bureau of Standards, Washington.

```
****** ENTER REACTANTS ******  (23JUN88)
(OR PRESS "RETURN" FOR LAST ENTRY)

>         <1-A>  FE*S      +        <A>  02        =

ENTER NEW SUBSCRIPTS, TYPE "R", ENTER "HELP" OR PRESS "RETURN"
>R
                  (298)                    (298)

****************************************************************************
  <A>   T PROD   P PROD  DELTA H  DELTA G  DELTA V  DELTA S  DELTA U  DELTA A
        (K)      (ATM)    (J)      (J)      (L)      (J/K)    (J)      (J)
****************************************************************************
 ---S1 G  ------------------------------------------------------------------

>0.65     *       1        0

THERE MAY BE A DELAY WHILE THE LOCATION OF
ALL DATA ON COMPOUNDS CONTAINING THE REACTANT ELEMENTS
IS DETERMINED.

GASEOUS  SPECIES    1 -  19
LIQUID   SPECIES   20 -  27
SOLID    SPECIES   28 -  45

ENTER CODE NUMBERS, ENTER "LIST" TO DISPLAY, OR ENTER "HELP"
****(CHANGES TO EQUILIB REQUIRE THE USE OF SLASHES ABOUT GASEOUS SPECIES)***

>/1-19/22-45

PRESS "RETURN" WHEN READY FOR OUTPUT
ENTER AN OPTION NUMBER, OR ENTER "O" FOR OPTIONS MENU
```

Fig. 1. Standard input to the F*A*C*T EQUILIB program. User supplies lines that are marked with > at the extreme left. The relative number of moles of FeS and 02 are specified with variable "A" to facilitate a change in subsequent computations. The subscripts indicate the reactant temperatures, 298 K for both reactants in this case. Most stable phase for each at standard atmospheric pressure is assumed. The reaction yielding the equilibrium at pressure 1 atm. is adiabatic or isenthalpic. The species included in the computation include all gases in an ideal solution and 24 condensed phases (species 22-45).

```
>
        <1-A>    FE*S      +         <A>   02        =
                 (298)                     (298)

        0.52747                 (   0.59932        S02              T
                                +   0.25310        02               T
                                +   0.62274E-01    S0               T
                                +   0.58302E-01    O                T
                                +   0.25105E-01    Fe               T
                                +   0.15987E-02    S                T
                                +   0.24571E-03    S03              T
                                +   0.44504E-04    S2               T
                                +   0.99640E-05    S20              T
                                +   0.12811E-06    03               T
                                +   0.95587E-09    S3               T
                                +   0.17144E-10    O[+]
                                +   0.16453E-10    E[-]
                                +   0.55824E-12    O[-]
                                +   0.13205E-12    02[-]
                                +   0.33827E-13    S4               T
                                +   0.46770E-19    S5               T)
                                (  3006.5,  1.00       ,G)

                                +   0.33676        FeO
                                (  3006.5,  1.00       ,L1,   1.0000      )
```

WHERE 'A' ON THE REACTANT SIDE IS 0.650

THE CUTOFF CONCENTRATION HAS BEEN SPECIFIED TO 0.100E-19

DATA ON 25 PRODUCT SPECIES IDENTIFIED WITH 'T' HAVE BEEN EXTRAPOLATED

```
***************************************************************************************
    DELTA H     DELTA G     DELTA V     DELTA S     DELTA U     DELTA A    REACT V
     (J)         (J)         (L)         (J/K)        (J)         (J)        (L)
***************************************************************************************
---S1 G  ----------------------------------------------------------------------------
       0.0   -713263.2   0.114E+03      98.181    -11573.5   -724836.7   0.159E+02
```

Fig.2. The output from EQUILIB associated with Fig. 1.
Total moles of gas is 0.52747. Mole fraction of
each gas precedes the formulation. The adiabatic
product temperature is 3006.5 K. The moles of
the only condensed phase (FeO liquid) is 0.33676.

```
****** ENTER REACTANTS ****** (23JUN88)
(OR PRESS "RETURN" FOR LAST ENTRY)

>     0.6  CU    +     0.4   CR    +    0.15 O2 =

ENTER SUBSCRIPTS, ENTER "HELP", OR PRESS "RETURN"
>

**************************************************************************
   T PROD   P PROD   V GAS    DELTA H   DELTA G   DELTA S   DELTA U   DELTA A
    (K)      (ATM)    (L)    <---- MAY SPECIFY IF SUBSCRIPTS PROVIDED ----->
**************************************************************************
>   1373      1

THERE MAY BE A DELAY WHILE THE LOCATION OF
ALL DATA ON COMPOUNDS CONTAINING THE REACTANT ELEMENTS
IS DETERMINED.

GASEOUS  SPECIES   1 -  11
LIQUID   SPECIES  12 -  17
SOLID    SPECIES  18 -  27

ENTER CODE NUMBERS, ENTER "LIST" TO DISPLAY, OR ENTER "HELP"
****(CHANGES TO EQUILIB REQUIRE THE USE OF SLASHES ABOUT GASEOUS SPECIES)***
>LIST
```

Fig. 3. The input corresponding to the oxidation of a
Cu-Cr alloy with pure oxygen. The temperature
is set to 1373 K. The option to list all
possible Cu - Cr - O species in the databases
is exercised.

POSSIBLE PRODUCT COMPOUNDS FOUND ("<---" WILL IDENTIFY YOUR PRIVATE DATA)
UNNUMBERED SPECIES MEANS INCOMPLETE DATA SET

```
 1  CU              <---   G1  VAPOUR       298.0 K -  2900.0 K 1
 2  CU2                    G1  GAS          298.0 K -  6000.0 K 2
 3  CU                     G1  GAS         2848.0 K -  3400.0 K 1
 4  CR*O                   G1  GAS          298.0 K -  6000.0 K 2
 5  CR                     G1  GAS         2945.0 K -  3100.0 K 1
 6  O3                     G1  GAS          298.0 K -  2000.0 K 1
 7  O2[-]                  G1  GAS          298.0 K -  6000.0 K 2
 8  O2                     G1  GAS          298.0 K -  3000.0 K 1
 9  O[+]                   G1  GAS          298.1 K -  6000.0 K 2
10  O[-]                   G1  GAS          298.0 K -  6000.0 K 2
11  O                      G1  GAS          298.0 K -  3000.0 K 2

12  CU              <---   L1  LIQUID      1356.0 K -  2843.0 K 1
13  (O)             <---   L2  DIS.FE(L)   1800.0 K -  2200.0 K 1
14  CU2O                   L1  LIQUID      1509.0 K -  2000.0 K 1
15  CU                     L1  LIQUID      1357.0 K -  2848.0 K 1
16  CR*O3                  L1  LIQUID       470.0 K -   600.0 K 1
17  CR                     L1  LIQUID      2130.0 K -  2945.0 K 1

18  CU              <---   S1  FCC          270.0 K -  1356.0 K 1
19  CU2O                   S1  CUPRITE      298.0 K -  1509.0 K 1
20  CU*O                   S1  TENORITE     298.0 K -   300.0 K 2
21  CU                     S1  SOLID        298.0 K -  1357.0 K 1
22  CR8O21                 S1  SOLID        298.0 K -   700.0 K 1
23  CR5O12                 S1  SOLID        298.0 K -   700.0 K 1
24  CR2O3                  S1  SOLID        298.0 K -  1800.0 K 1
25  CR*O3                  S1  SOLID        298.0 K -   470.0 K 1
26  CR*O2                  S1  SOLID        298.0 K -   600.0 K 1
27  CR                     S1  SOLID        298.0 K -  2130.0 K 2
```

ENTER CODE NUMBERS, ENTER "LIST" TO DISPLAY, OR ENTER "HELP"
****(CHANGES TO EQUILIB REQUIRE THE USE OF SLASHES ABOUT GASEOUS SPECIES)***
>/2-11/15,17/19-27

PRESS "RETURN" WHEN READY FOR OUTPUT
ENTER AN OPTION NUMBER, OR ENTER "0" FOR OPTIONS MENU

Fig. 4.   All possible species in the Cu - Cr - O system.
Species automatically drawn from the user's
private file are marked with <---.   The species
selection includes all gas species from the main
database in an ideal solution, liquid Cu and Cr
in a solution phase and 9 solids.

```
OPTIONAL INPUT/OUTPUT CONTROL
*****************************

(A) SPECIES SELECTION SPECIFIC FEATURES
    ***********************************

    (1) SPECIFY ACTIVITIES OF SPECIES IN PRODUCT EQUILIBRIUM
    (2) SPECIFY ACTIVITY COEFFICIENTS OF SPECIFIED SPECIES
    (3) EXCLUDE SPECIFIED SPECIES FROM THE MASS BALANCE
    (4) SPECIFY VIRIAL COEFFICIENTS FOR GAS PHASE SPECIES

(B) SEMI-PERMANENT OUTPUT SPECIFICATIONS
    ************************************
    (5) INTERPRET MOLES (INPUT AND OUTPUT) AS MASSES IN GRAMS
    (6) SUPPRESS PRINTING BELOW A SPECIFIED CONCENTRATION
    (7) EXTRAPOLATE DATA FOR PURE CONDENSED SPECIES
    (8) ENERGY UNIT FOR EXTENSIVE PROPERTY CALCULATIONS
    (9) NUMBER OF LINES BETWEEN PAUSES IN OUTPUT

ALL OPTIONS IN GROUP "A" ARE AUTOMATICALLY CLEARED
WHEN A NEW SPECIES SELECTION (OR REACTION) IS ENTERED

ENTRY OF "B" WILL RESET ALL OPTIONS IN GROUP "B"
(I/O IN MOLES, PRINT ALL, NO EXTRAPOLATIONS, JOULES)

PRESS "RETURN" WHEN READY FOR OUTPUT
ENTER AN OPTION NUMBER, OR ENTER "O" FOR OPTIONS MENU
>2
ENTER SPECIES NUMBER AND "A" AND "B" IN LOG10(GAMMA) = A/T(K) + B
>17   7482   -4.199
ENTER SPECIES NUMBER AND "A" AND "B" IN LOG10(GAMMA) = A/T(K) + B
>
PRESS "RETURN" WHEN READY FOR OUTPUT
ENTER AN OPTION NUMBER, OR ENTER "O" FOR OPTIONS MENU
```

Fig. 5. The menu of output options now available to the
EQUILIB program. Option 2 is selected to introduce
an activity coefficient expression for Cr in the
molten Cu-Cr phase specified in Fig. 4.

```
>
      0.6  CU     +      0.4   CR     +     0.15 O2 =

          0.00000E+00          (   0.73165E-06        Cu                          T
                               +   0.36925E-07        Cr                          T
                               +   0.11524E-09        Cu2
                               +   0.43254E-11        CrO
                               +   0.10597E-15        O
                               +   0.29985E-19        O2)
                               (  1373.0,  1.00     ,G  ,0.769E-06)

      +   0.61864              (   0.96987            Cu
                               +   0.30128E-01        Cr                      #   T)
                               (  1373.0,  1.00     ,SOLN  2)

                               +   0.18136            Cr
                               (  1373.0,  1.00     ,S1,   1.0000    )

                               +   0.10000E+00        Cr2O3
                               (  1373.0,  1.00     ,S1,   1.0000    )

                               +   0.00000E+00        Cu                          T
                               (  1373.0,  1.00     ,S1,  0.95631    )

                               +   0.00000E+00        CrO2                        T
                               (  1373.0,  1.00     ,S1,  0.12324E-04)

                               +   0.00000E+00        Cu2O
                               (  1373.0,  1.00     ,S1,  0.79310E-07)

                               +   0.00000E+00        CuO                         T
                               (  1373.0,  1.00     ,S1,  0.22578E-08)
```

GASEOUS IONIC SPECIES ARE SUPPRESSED BELOW 3000 K

THE CUTOFF CONCENTRATION HAS BEEN SPECIFIED TO 0.100E-20

DATA ON 8 PRODUCT SPECIES IDENTIFIED WITH 'T' HAVE BEEN EXTRAPOLATED

DATA ON 1 PRODUCT SPECIES IDENTIFIED WITH '#' HAVE BEEN CALCULATED WITH A
USER-SUPPLIED ACTIVITY-COEFFICIENT EXPRESSION

Fig. 6. The EQUILIB output associated with Fig. 4 and 5.
No gas phase forms however the partial pressures
of the selected gas phase species are calculated.
The sum is 0.769E-06 atm. The molten alloy phase
contains 0.30128E-01 mole fraction Cr and coexists
with pure solid Cr (0.18136 moles) and solid Cr2O3
(0.1 moles). The activity of CrO2 which does not
form is 0.12324E-04 with respect to pure solid
(allotrope 1) at 1 atm.

```
****** ENTER REACTANTS ******
(OR PRESS "RETURN" FOR LAST ENTRY)
NUMBERS BEFORE SPECIES ARE MASSES IN GRAMS (OPTION 5)

>        <100-A> FE   +  <A>  C   +   1  C*O   =

ENTER SUBSCRIPTS, ENTER "HELP", OR PRESS "RETURN"
>
**********************************************************************************
   <A>   T PROD   P PROD   V GAS   DELTA H  DELTA G  DELTA S  DELTA U  DELTA A
         (K)      (ATM)    (L)    <-- MAY SPECIFY IF SUBSCRIPTS PROVIDED --->
**********************************************************************************

> 0.2     1790      1

THERE MAY BE A DELAY WHILE THE LOCATION OF
ALL DATA ON COMPOUNDS CONTAINING THE REACTANT ELEMENTS
IS DETERMINED.

GASEOUS  SPECIES    1 -  18
LIQUID   SPECIES   19 -  23
SOLID    SPECIES   24 -  38

Fe liquid.  Modified interaction parameters.
   39  FE
   40  C
   41  O2

ENTER CODE NUMBERS, ENTER "LIST" TO DISPLAY, OR ENTER "HELP"
****(CHANGES TO EQUILIB REQUIRE THE USE OF SLASHES ABOUT GASEOUS SPECIES)***
>LIST
```

Fig. 7. Input to examine the solubility of CO in molten
steel containing <A> percent carbon initially.
The carbon level is here set to 0.2 percent
at 1790 and 1 atm. total gas pressure. The
species appropriate to a non-ideal solution of
Fe-C-O are automatically introduced.

POSSIBLE PRODUCT COMPOUNDS FOUND ("<---" WILL IDENTIFY YOUR PRIVATE DATA)
UNNUMBERED SPECIES MEANS INCOMPLETE DATA SET

```
  1   FE(C*O)5                      G1   GAS          380.0 K -   1000.0 K 1
  2   FE                            G1   GAS         3135.0 K -   3600.0 K 1
  3   C3O2                          G1   GAS          298.0 K -   6000.0 K 2
  4   C*O2[-]                       G1   GAS          298.0 K -   6000.0 K 2
  5   C*O2                          G1   GAS          298.0 K -   2500.0 K 1
  6   C*O                           G1   GAS          298.0 K -   2500.0 K 1
  7   O3                            G1   GAS          298.0 K -   2000.0 K 1
  8   O2[-]                         G1   GAS          298.0 K -   6000.0 K 2
  9   O2                            G1   GAS          298.0 K -   3000.0 K 1
  .
  .
 21   FE*O                          L1   LIQUID      1650.0 K -   3687.0 K 1
 22   FE3C                          L1   LIQUID      1500.0 K -   2000.0 K 1
 23   FE                            L1   LIQUID      1809.0 K -   3135.0 K 1

 24   C                  <--- S3    DIS.DELTFE   1650.0 K -   1810.0 K 1
 25   FE*C*O3                       S1   SIDERITE     298.0 K -    800.0 K 1
 26   FE3O4                         S1   SOLID-A      298.0 K -    866.0 K 1
 27   FE3O4                         S2   SOLID-B      866.0 K -   1870.0 K 1
 28   FE2O3                         S1   SOLID-A      298.0 K -    953.0 K 1
 29   FE2O3                         S2   SOLID-B      953.0 K -   1053.0 K 1
 30   FE2O3                         S3   SOLID-C     1053.0 K -   1735.0 K 1
 31   FE*O                          S1   WUESTITE     298.0 K -   1650.0 K 1
 32   FE3C                          S1   SOLID-A      273.0 K -    463.0 K 1
 33   FE3C                          S2   SOLID-B      463.0 K -   1500.0 K 1
 34   FE                            S1   ALPHA(MAG)   298.0 K -   1184.0 K 5
 35   FE                            S2   GAMMA       1184.0 K -   1665.0 K 1
 36   FE                            S3   DELTA       1665.0 K -   1809.0 K 1
 37   C                             S1   GRAPHITE     298.0 K -   4073.0 K 2
 38   C                             S2   DIAMOND      298.0 K -   1200.0 K 1
```

Fe liquid. Modified interaction parameters.
```
 39   FE
 40   C
 41   O2
```

ENTER CODE NUMBERS, ENTER "LIST" TO DISPLAY, OR ENTER "HELP"
****(CHANGES TO EQUILIB REQUIRE THE USE OF SLASHES ABOUT GASEOUS SPECIES)***
>/2,5,6,9/39-41/36,24/21,37

PRESS "RETURN" WHEN READY FOR OUTPUT
ENTER AN OPTION NUMBER, OR ENTER "0" FOR OPTIONS MENU

Fig. 8. The listing of species for the Fe-C-O system.
Species 24 has been user-supplied to represent
the state of carbon in dilute solution with
delta iron.

```
    <100-A>  FE    +  <A>  C     +    1  C*O     =
     5.1137        litre (   98.949      vol% CO
                      +   1.0488      vol% CO2
                      +   0.26824E-02 vol% Fe                           T
                      +   0.39607E-09 vol% O2)
                        ( 1790.0, 1.00    ,G)


 +   82.274      gram  (   99.747      wt.% Fe
                      +   0.24336      wt.% C
                      +   0.10082E-01 wt.% O2)
                        ( 1790.0, 1.00    ,SOLN 2)


 +   17.745      gram  (   99.941      wt.% Fe
                      +   0.58549E-01 wt.% C                        <---)
                        ( 1790.0, 1.00    ,SOLN 3)


                      +   0.00000E+00 gram FeO
                        ( 1790.0, 1.00    ,L1, 0.90814E-01)


                      +   0.00000E+00 gram C
                        ( 1790.0, 1.00    ,S1, 0.73004E-02)


WHERE 'A' ON THE REACTANT SIDE IS 0.200
```

Fig. 9. Equilibrium for a 0.2 weight percent iron-carbon alloy at 1790 when in contact with carbon monoxide rich gas. No FeO or solid C exist in the phase assemblage but the activity of these species appears in the subscript. The numbers preceding reactants are masses in grams

```
****** ENTER REACTANTS ******
(OR PRESS "RETURN" FOR LAST ENTRY)
NUMBERS BEFORE SPECIES ARE MASSES IN GRAMS (OPTION 5)
>    94 FE   +   1.1 SI  +  0.8 MN  +  4.1   C   +

ENTER SUBSCRIPTS, ENTER "HELP", OR PRESS "RETURN"
>

CONTINUE REACTANT INPUT OR PRESS "RETURN" FOR LAST ENTRY
>     3.5 CA*O   +   3.5 MG*O      +   3 SI*O2   +   <A> O2 =

************************************************************************************
    <A>   T PROD   P PROD   V GAS   DELTA H  DELTA G  DELTA S  DELTA U  DELTA A
          (K)      (ATM)    (L)    <-- MAY SPECIFY IF SUBSCRIPTS PROVIDED --->
************************************************************************************
>   5      1900       1

THERE MAY BE A DELAY WHILE THE LOCATION OF
ALL DATA ON COMPOUNDS CONTAINING THE REACTANT ELEMENTS
IS DETERMINED.

GASEOUS   SPECIES   1 -  32
LIQUID    SPECIES  33 -  58
SOLID     SPECIES  59 - 147

Fe liquid.  Modified interaction parameters.
 148   FE
 149   C
 150   MN
 151   O2
 152   SI

Quasichemical model for molten oxides.
 153   SI*O2
 154   CA*O
 155   MG*O
 156   FE*O
 157   MN*O

ENTER CODE NUMBERS, ENTER "LIST" TO DISPLAY, OR ENTER "HELP"
****(CHANGES TO EQUILIB REQUIRE THE USE OF SLASHES ABOUT GASEOUS SPECIES)***
>/2,7,16,19,20,23,26/148-152/153-157/123,128,136,142,146
```

Fig. 10. EQUILIB input for a hot metal - slag - gas
          computation. The input has been set so that
          numbers preceding the reactants are taken to
          be masses in grams. The presentation of the
          appropriate non-ideal solution data is auto-
          matically performed.

```
94 FE   +   1.1 SI  +  0.8 MN  +  4.1   C   +

 3.5  CA*O   +   3.5 MG*O       +   3  SI*O2    +   <A> O2 =

     40.259        litre (    99.829       vol% CO
                          +  0.90492E-01  vol% CO2
                          +  0.61372E-01  vol% Mg
                          +  0.10751E-01  vol% Fe
                          +  0.81663E-02  vol% SiO
                          +  0.15032E-07  vol% O
                          +  0.25419E-10  vol% O2)
                          ( 1900.0,  1.00      ,G)

 +   96.017       gram  (    97.868       wt.% Fe
                          +   1.0422       wt.% C
                          +  0.68872       wt.% Mn
                          +  0.39743       wt.% Si
                          +  0.36930E-02  wt.% O2)
                          ( 1900.0,  1.00      ,SOLN 2)

 +   11.745       gram  (    38.617       wt.% SiO2
                          +   29.799       wt.% CaO
                          +   29.745       wt.% MgO
                          +   1.5250       wt.% MnO
                          +  0.31444       wt.% FeO)
                          ( 1900.0,  1.00      ,SOLN 3)

                   +  0.00000E+00 gram MgO
                      ( 1900.0,  1.00      ,S1,  0.71337      )

                   +  0.00000E+00 gram (MgO)2(SiO2)
                      ( 1900.0,  1.00      ,S1,  0.55722      )

                   +  0.00000E+00 gram C
                      ( 1900.0,  1.00      ,S1,  0.45984E-01)

                   +  0.00000E+00 gram CaO
                      ( 1900.0,  1.00      ,S1,  0.25611E-01)

                   +  0.00000E+00 gram SiO2
                      ( 1900.0,  1.00      ,S4,  0.23669E-01)
```

WHERE 'A' ON THE REACTANT SIDE IS 5.00

Fig. 11. Hot metal - slag - gas equilibrium allowing for
         the non-ideal behaviour of the condensed phases.
         The numbers preceding the reactants have been
         declared to be masses in grams.

```
59.4 H2   + 23.3 H2O   + 9.8 C*O   + 7.5 C*O2   + 0.3 H2S   +

0.5 LI2C*O3     +    0.5 K2C*O3   =

   85.693              (   0.41715        H2
                       +   0.37728        H2O
                       +   0.85225E-01    CH4
                       +   0.67365E-01    CO2
                       +   0.49471E-01    CO
                       +   0.34813E-02    H2S
                       +   0.19470E-04    COS
                       +   0.75727E-06    C2H6
                       +   0.37899E-06    K(OH)
                       +   0.29513E-06    CH2O
                       +   0.27378E-06    C4H8
                       +   0.13968E-06    C2H4
                       +   0.53979E-07    LiOH
                       +   0.34454E-07    CH3SH
                       +   0.26071E-07    CH3OH)
                       (   973.0,  10.0     ,G)

 +   1.0172            (   0.48309        Li2CO3
                       +   0.48308        K2CO3
                       +   0.16913E-01    LiOH
                       +   0.16912E-01    KOH
                       +   0.63395E-07    Li2SO4
                       +   0.63393E-07    K2SO4)
                       (   973.0,  10.0     ,SOLN 2)

                       +   0.00000E+00    C
                       (   973.0,  10.0     ,S1, 0.36639    )

                       +   0.00000E+00    K2CO3
                       (   973.0,  10.0     ,S1, 0.31160    )

                       +   0.00000E+00    C
                       (   973.0,  10.0     ,S2, 0.17661    )
```

Fig. 12. Fuel gas - molten carbonate electrolyte equilibrium.
The numbers preceding each of the reactants have been
declared to be moles. The computation has incorporated
non-ideal mixing terms in the treatment of the molten
molten carbonate phase.

```
****** ENTER REACTANTS ******
(OR PRESS "RETURN" FOR LAST ENTRY)
>       55.5   H2O      +     3  NA*CL     +     1  H*CL     +    C*02    =

ENTER SUBSCRIPTS, ENTER "HELP", OR PRESS "RETURN"
>

*****************************************************************************
   T PROD   P PROD   V GAS    DELTA H   DELTA G   DELTA S   DELTA U   DELTA A
     (K)     (ATM)    (L)     <---- MAY SPECIFY IF SUBSCRIPTS PROVIDED ----->
*****************************************************************************
>   298        *    500

THERE MAY BE A DELAY WHILE THE LOCATION OF
ALL DATA ON COMPOUNDS CONTAINING THE REACTANT ELEMENTS
IS DETERMINED.

GASEOUS   SPECIES   1 - 345
LIQUID    SPECIES 346 - 391
AQUEOUS   SPECIES 392 - 432
SOLID     SPECIES 433 - 456

Li,Na,K/OH,CO3,SO4 molten salt solution.
 457   NA*O*H
 458   NA2C*O3

Aqueous solution with Pitzer parameters.
< GO(298K) from F*A*C*T>
 459   H2O
 460   NA[+]
 461   H[+]
 462   CL[-]
 463   O*H[-]
 464   H*C*O3[-]
 465   C*O3[2-]
 466   C*O2

ENTER CODE NUMBERS, ENTER "LIST" TO DISPLAY, OR ENTER "HELP"
****(CHANGES TO EQUILIB REQUIRE THE USE OF SLASHES ABOUT GASEOUS SPECIES)***
>/1-345/459-466/433-456
```

Fig. 13. Input to compute the equilibrium of NaCl
         dissolved in 1 molar HCl when in contact with
         CO2. The product equilibrium is to computed
         at 298 K for a gas volume of 500 litres. All
         gases, aqueous species in the non-ideal phase
         and all solids are considered.

```
     55.5   H2O      +      3 NA*CL      +     1 H*CL      +    C*O2    =

     1.5287                (   0.65312          CO2
                           +   0.34683          H2O
                           +   0.47394E-04      HCl )
                             ( 298.0,0.748E-01,G)


  + 0.99029                (    55.509          H2O
                           +    4.0391          Cl[-]
                           +    3.0294          Na[+]
                           +    1.0097          H[+]
                           +   0.15797E-02      CO2)
                             ( 298.0,0.748E-01,AQ)
                             (Eh= 0.636 V, pH=-0.246)


                           +   0.00000E+00      H2O                          T
                             ( 298.0,0.748E-01,S1, 0.66118    )


                           +   0.00000E+00      NaCl
                             ( 298.0,0.748E-01,S1, 0.24936    )
```

Fig. 14. Output for a volume constrained equilibrium.
For the required gas volume of 500 litres
specified in Fig. 13, the pressure has been
computed to be 0.748E-01 atm. The treatment
of the aqueous phase has incorporated Pitzer-
Brewer coefficients. The redox potential
(SHE) and pH are provided immediately beneath
the aqueous phase composition. Only the more
concentrated of dissolved species actually
considered in the computation are shown here.

SESSION II

Pyrometallurgical and Process Applications

Co-Chairmen: F. AJERSCH
               Ecole Polytechnique
               Montreal, Quebec

               G. KAIURA
               Falconbridge Ltd.
               Sudbury, Ontario

EQUILIBRIUM THERMODYNAMICS OF THE VACUUM METALLOTHERMAL REDUCTION OF LITHIUM OXIDE AND SPODUMENE

A.A.J. Smeets, D.J. Fray and J.A. Charles

Department of Materials Science and Metallurgy
University of Cambridge, England

ABSTRACT

The extraction of lithium metal from spodumene ($Li_2O.Al_2O_3.4SiO_2$) is a complicated and expensive process. A possible novel extraction method is vacuum thermal reduction. Two promising reductants for such a process are aluminium and silicon. Thermodynamic equilibria calculated using the SOLGASMIX program show that with both reductants pure lithium vapour can be obtained at various combinations of temperature and pressure. Aluminium compares favourably with silicon both from the point of view of reduction temperature and of vapour purity. With silicon as a reductant, silicon monoxide is formed, which contaminates the lithium vapour. Its formation can however be suppressed through addition of CaO.

KEYWORDS

Lithium; pyrometallurgy; spodumene; thermal reduction; equilibrium thermodynamics.

INTRODUCTION

The element lithium is mainly used in the form of chemicals for a wide variety of applications. Metallic lithium which, so far, has only been used in minor quantities, is increasingly gaining importance, mainly in dry-cell batteries and aluminium-lithium alloys. Crozier (1986) reports that, although the rapid large-scale introduction of aluminium-lithium alloys that was once foreseen has not materialized, a steady growth in the demand for lithium metal is generally anticipated.

At present, lithium is extracted from two sources: brines and spodumene ($Li_2O.Al_2O_3.4SiO_2$). In both cases, the extraction process yields lithium carbonate (Li_2CO_3), which is both an important lithium chemical and the starting material for the production of

other lithium chemicals and lithium metal. Extraction of lithium from brines is through solar evaporation and treatment with various chemicals to precipitate impurities and finally lithium carbonate. Extraction from spodumene is more complicated: after concentration through froth flotation, the concentrate is heated to about 1373 K to convert the naturally occurring α-spodumene to the more reactive β-structure. Subsequent treatment with sulphuric acid at 523 K followed by water leaching yields an impure lithium sulphate solution, which is purified and finally treated with sodium carbonate to precipitate lithium carbonate.
To produce lithium metal, lithium carbonate is reacted with hydrochloric acid to give lithium chloride, which is dried and mixed with potassium chloride. The salt mixture is electrolysed at 733 K to give lithium metal.

Compared to other light metals like aluminium and magnesium, lithium is extremely expensive (US\$55/kg for lithium vs. \$1.85 for aluminium and \$3.30 for magnesium), and there appears to be room for a cheaper, less complicated extraction method. The vast majority of proposed alternative lithium extraction processes, mainly based on ion exchange, lead to the formation of lithium carbonate or lithium salts, and still require (expensive) electrolysis to produce lithium metal. One process that potentially gives lithium metal in a single step from the ore concentrate is thermal reduction.

Thermal reduction is used in the Pidgeon process for the production of magnesium, where magnesium oxide reacts under vacuum with (ferro) silicon in the presence of CaO, forming magnesium vapour which is condensed in a cooler part of the furnace. In a similar way, lithium could be produced by reduction of either the oxide (made from the carbonate by vacuum decomposition) or spodumene concentrate. Ideally, the reductant would cause the formation of a pure lithium vapour phase at a suitable temperature. Fig. 1 shows the free energy of formation of the oxides of the most likely reductants for lithium oxide. Calcium and magnesium are the thermodynamically most suitable elements, but apart from being expensive, both have boiling points close to that of lithium, and will produce a vapour mixture rather than pure lithium vapour. Reduction with carbon will produce lithium vapour containing CO and CO_2, causing the lithium to back-react upon cooling. Aluminium and silicon have neither of these disadvantages, although a vacuum is needed to lower the equilibrium reduction temperatures to more acceptable levels, particularly for silicon.

This paper describes a thermodynamic investigation of the vacuum reduction of lithium oxide and spodumene by both silicon and aluminium, and the effects of CaO additions.

COMPUTER CALCULATIONS

The program used for calculating the thermodynamic equilibria is an updated version of SOLGASMIX, a program originally written by Eriksson (1975). It calculates the equilibrium composition and various thermodynamic quantities of chemical systems consisting of up to 150 species built up of up to 20 elements and containing up to 20 separate (ideal) mixtures. The program contains an empty subroutine for user-written software describing non-ideality. No previous knowledge of the reactions involved is necessary; from the

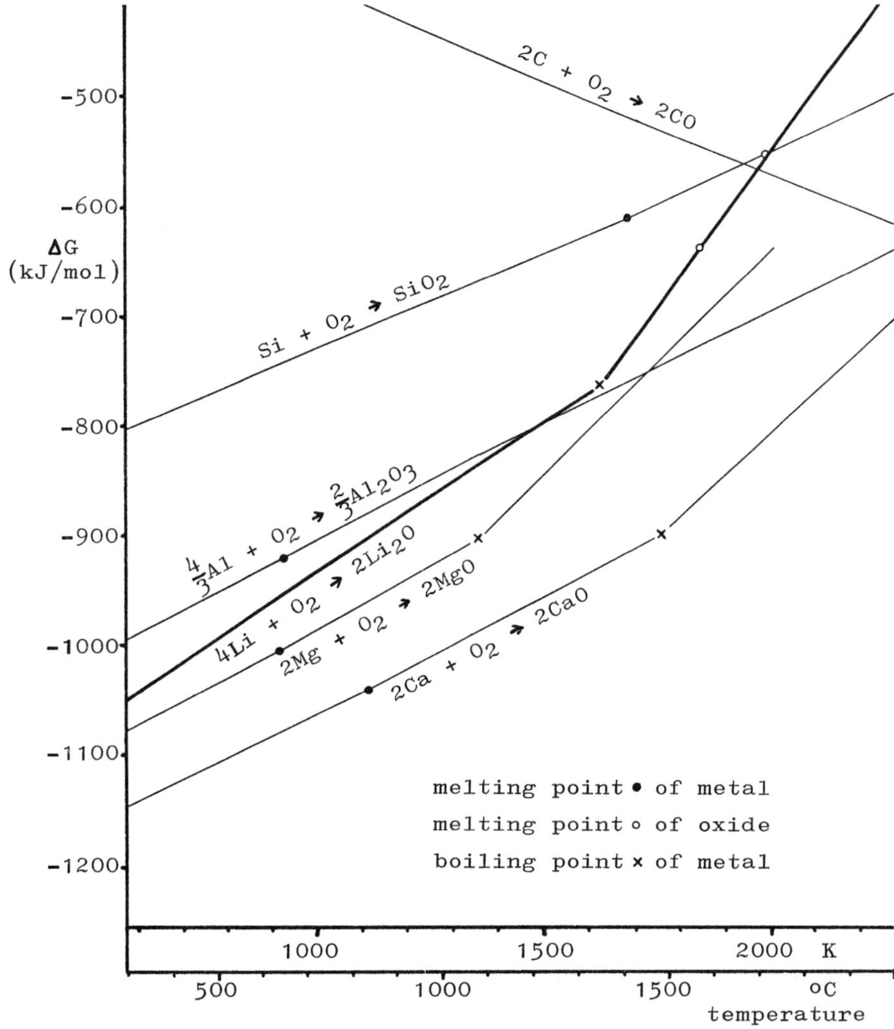

Fig. 1. Free energy of formation of the oxides of
       lithium and several reductants.

thermodynamic data of all possible species and the general initial
system characteristics the program finds the composition with the
lowest total free energy which satisfies the material balances.

For this study, all thermodynamic data was taken from Barin, Knacke
and Kubaschewski (1973, 1977). Where a choice had to be made, only
the data for the most stable modification of a species within a
phase was used. A list of all phases and species considered is given
in Table 1. Since little or no relevant activity data is available,
all activities were arbitrarily set as unity, i.e. the gas, liquid
metal and liquid oxide phases were all considered to be ideal
mixtures, and all solid species were considered separately, without

mixing or non-stoichiometry. In all cases, the pressure was 1 mPa, and the temperature varied to obtain composition versus temperature plots for the various systems.

TABLE 1   Phases and Species used in calculations

Gas	Liquid Metal	Liquid Oxide	Solids
Li	Li	Li_2O	Li
Li_2	Al	$Li_2O.Al_2O_3$	Li_2O
LiO	Si	$Li_2O.SiO_2$	Li_2O_2
Li_2O	Ca	$Li_2O.2SiO_2$	$Li_2O.Al_2O_3$
Li_2O_2		$2Li_2O.SiO_2$	$Li_2O.SiO_2$
Al		Al_2O_3	$Li_2O.2SiO_2$
AlO		SiO_2	$2Li_2O.SiO_2$
Al_2O		CaO	$Li_2O.Al_2O_3.4SiO_2$
Al_2O_2		$CaO.SiO_2$	$Li_2O.Al_2O_3.2SiO_2$
AlO_2		$CaO.Al_2O_3.2SiO_2$	Al
Si			Al_2O_3
SiO			$Al_2O_3.SiO_2$
Ca			$Al_2O_3.2SiO_2$
O			$3Al_2O_3.2SiO_2$
O_2			Si
			SiO_2
			Ca
			CaO
			$CaO.Al_2O_3$
			$CaO.2Al_2O_3$
			$12CaO.7Al_2O_3$
			$CaO.Al_2O_3.SiO_2$
			$CaO.Al_2O_3.2SiO_2$
			$2CaO.Al_2O_3.SiO_2$
			$3CaO.Al_2O_3.3SiO_2$
			$CaO.SiO_2$
			$2CaO.SiO_2$
			$3CaO.SiO_2$
			$3CaO.2SiO_2$
			$3CaO.Al_2O_3$

RESULTS AND DISCUSSION

A large number of systems of varying compositions were studied, all at a pressure of 1 mPa. Selected results in the form of composition versus temperature diagrams are given in Figs. 2-9. Obviously, since this is a thermodynamic model and not a kinetic one, not all of the reactions that follow from the diagrams can be expected to occur in practice, particularly at low temperatures.

The reduction of lithium oxide by aluminium, as shown in Fig. 2, is very straightforward; at low temperatures, lithium meta aluminate $(Li_2O.Al_2O_3)$ is the stable phase, allowing a maximum lithium extraction of 75%, whereas above 900 K Al_2O_3 is stable, freeing all the lithium. The vapour phase is pure up to the temperature where Al_2O_3 starts to decompose at this pressure. By contrast, reduction by silicon shows a much more complicated picture (Fig. 3); three lithium silicates are formed apart from SiO_2, forming a liquid oxide phase at intermediate temperatures. The most important observation however is the occurrence of gaseous SiO, which in a practical

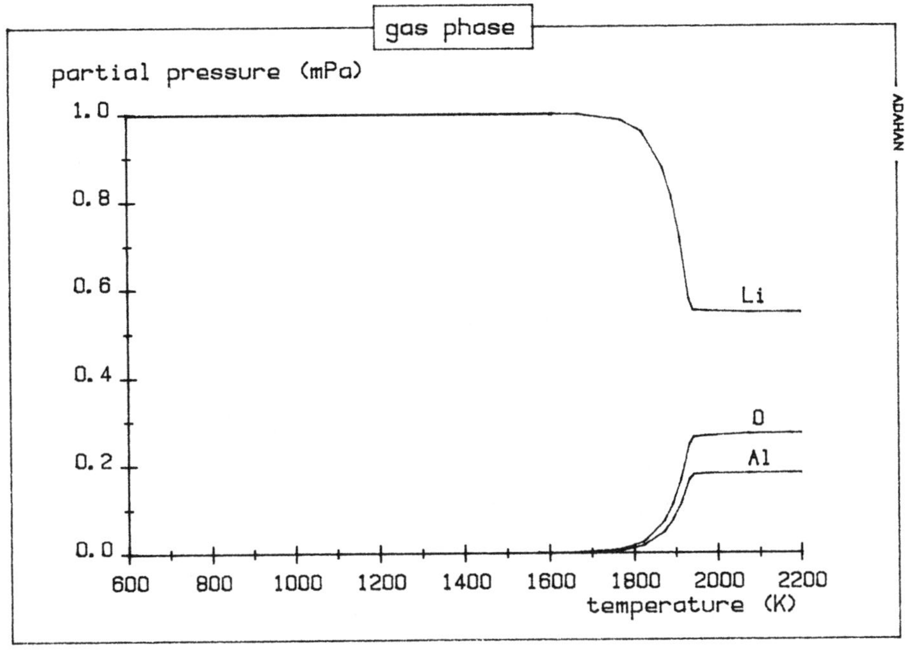

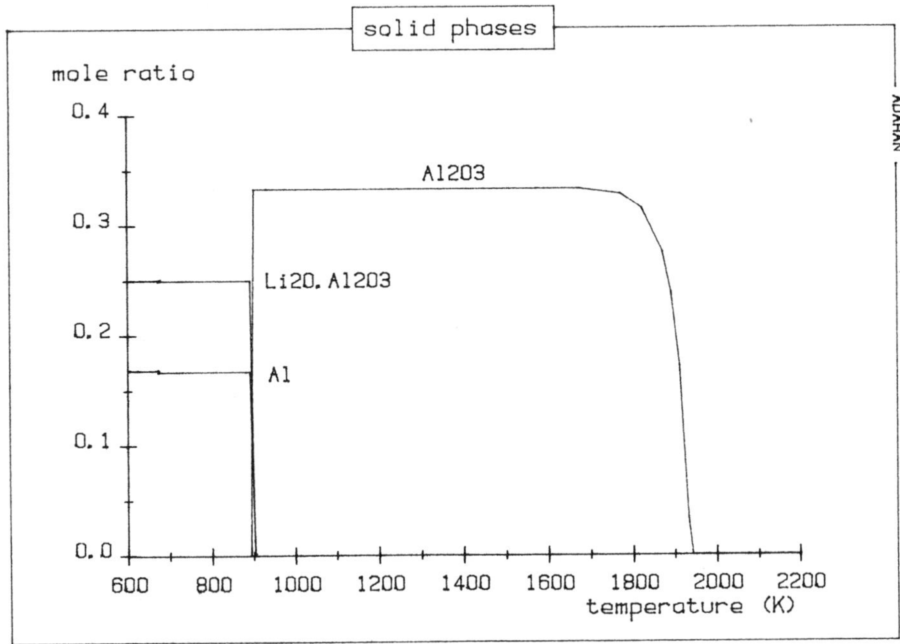

Fig. 2.   Equilibrium of the system Li_2O + 2/3 Al at 1 mPa

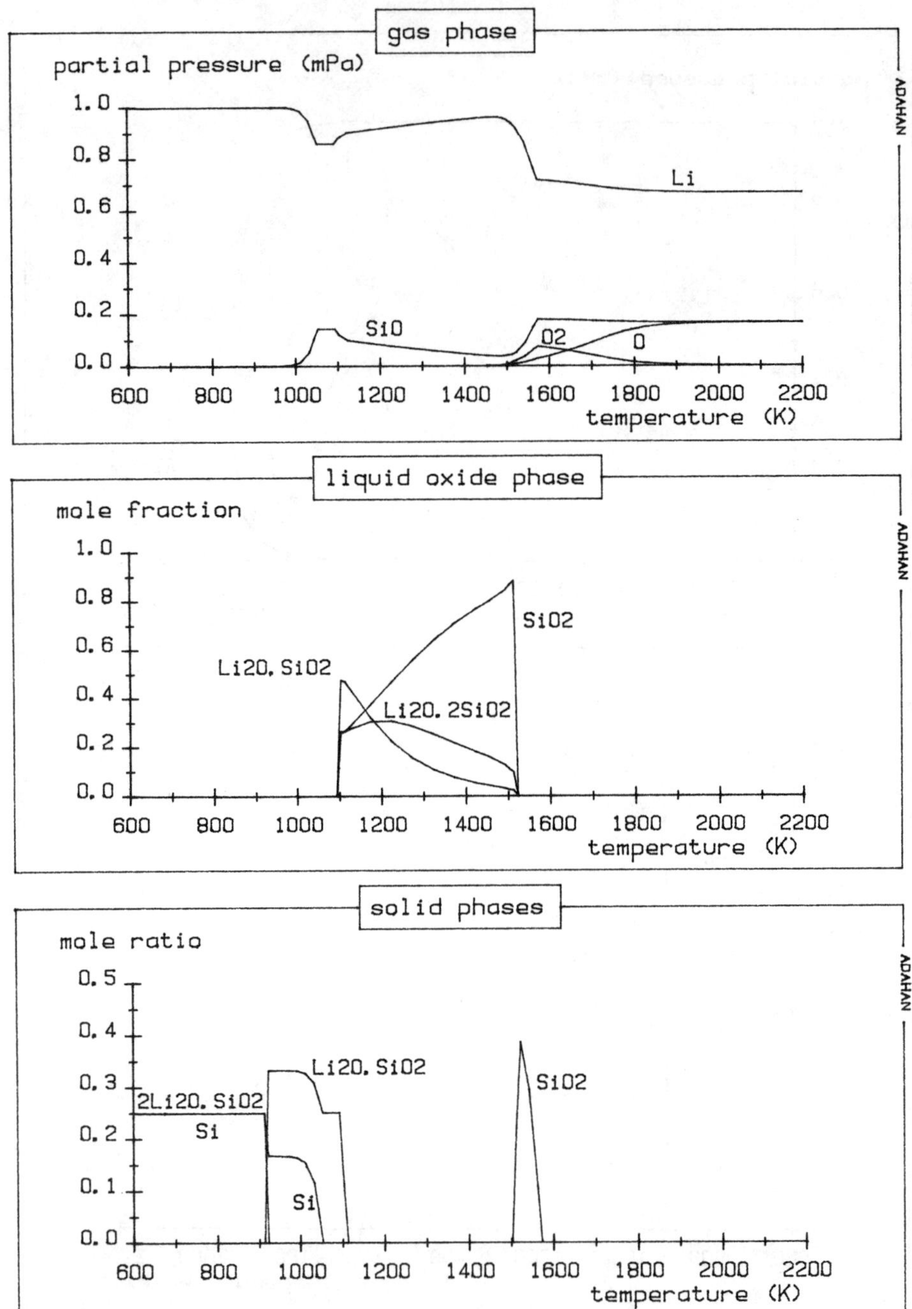

Fig. 3.   Equilibrium of the system $Li_2O + \frac{1}{2}Si$ at 1 mPa

situation can be expected to back-react with the lithium vapour upon cooling, re-forming Li_2O. The SiO formation can however in principle be completely suppressed through addition of sufficient CaO, as is shown in Figure 4. This leads to the formation of the thermodynamically more stable calcium orthosilicate ($2CaO.SiO_2$), and also lowers the minimum temperature for complete reduction to around 800 K. In the same way, CaO added to the Li_2O/Al system lowers the minimum temperature for complete reduction by around 100 K through the formation of a more stable final product (Fig. 5).

The reduction of spodumene shows characteristics similar to those of the reduction of Li_2O. One important difference is that with aluminium as a reductant, the SiO_2 component of the spodumene is also reduced. Not only does this sharply increase the amount of aluminium necessary for complete reduction, but it also causes the formation of metallic silicon which reacts with part of the Al_2O_3 to form gaseous SiO at higher temperatures. Thus, in Fig. 6 the temperature interval over which a pure lithium vapour phase is formed extends only up to 1100 K. Addition of CaO in this case has little effect, since it is reduced by the silicon to give not only SiO, but calcium vapour as well (Fig. 7). Again, however, the minimum temperature for complete reduction is lowered by around 100 K. Figures 8 and 9 show that, as in the case of Li_2O, reduction of spodumene with silicon alone leads to a gas phase containing SiO, and CaO is necessary to give pure lithium vapour, which in this case is formed over a relatively large temperature range of 880-1600 K.

If these processes are to be used to extract lithium in a practical situation, the right balance needs to be found between temperature and pressure to produce lithium vapour of sufficient purity at an acceptable rate. From the point of view of thermodynamics, these results show that aluminium as a reductant allows lower temperatures and/or higher pressures than silicon. More speculatively, with silicon and CaO a pure vapour phase is in practice probably more difficult to achieve than these results suggest; unless there is perfect mixing of the reagents some SiO formation is bound to occur.

CONCLUSION

Spodumene can be reduced by both aluminium and silicon in the presence of CaO to give a pure lithium vapour phase. At 1 mPa, with aluminium as a reductant pure lithium vapour is obtained between 800 and 1100 K, and with silicon between 880 and 1600 K.

ACKNOWLEDGEMENT

The authors wish to thank BP Minerals, who provided support for the work, for permission to publish this paper. The opinions expressed are those of the authors and not necessarily those of BP Minerals.

REFERENCES

Barin, I., O. Knacke and O. Kubaschewski (1973 and 1977). Thermochemical Properties of Inorganic Substances. Springer, Berlin.
Crozier, R.D. (1986). Mining Magazine, February 1986, 148-152
Eriksson, G. (1975). Chemica Scripta, 8, 100-103

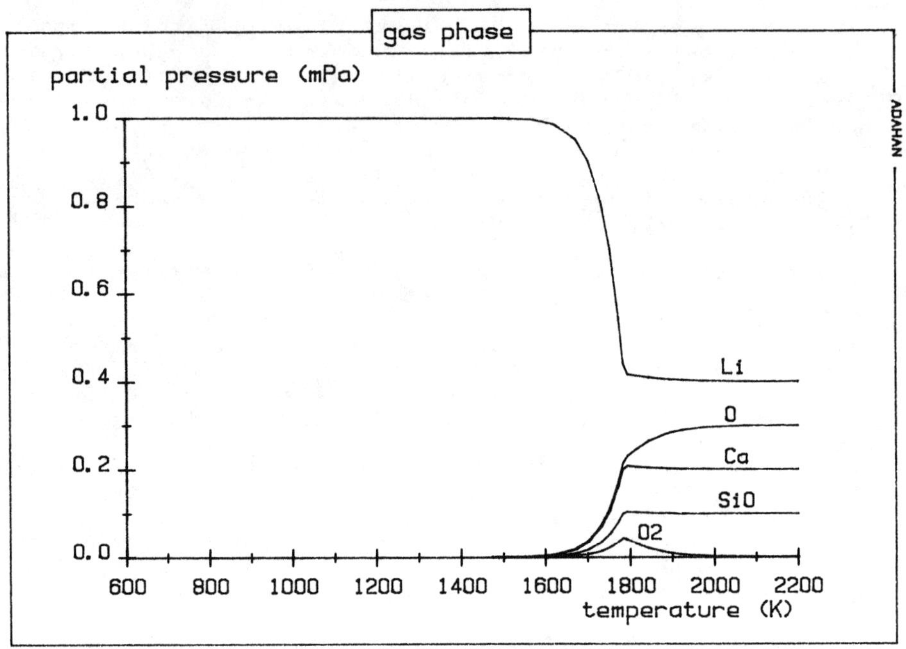

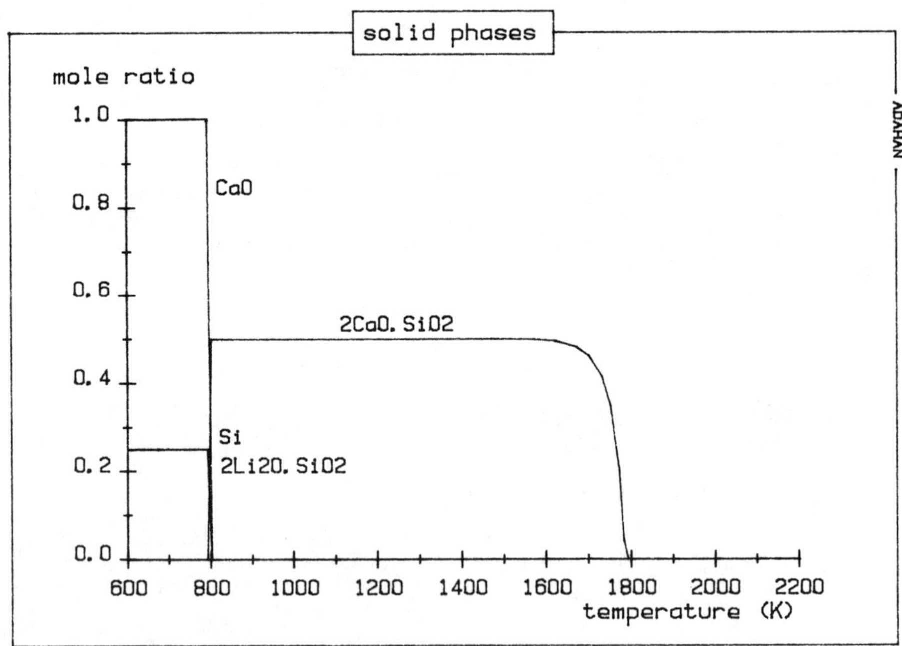

Fig.4. Equilibrium of the system $Li_2O + \tfrac{1}{2}Si + CaO$ at 1 mPa

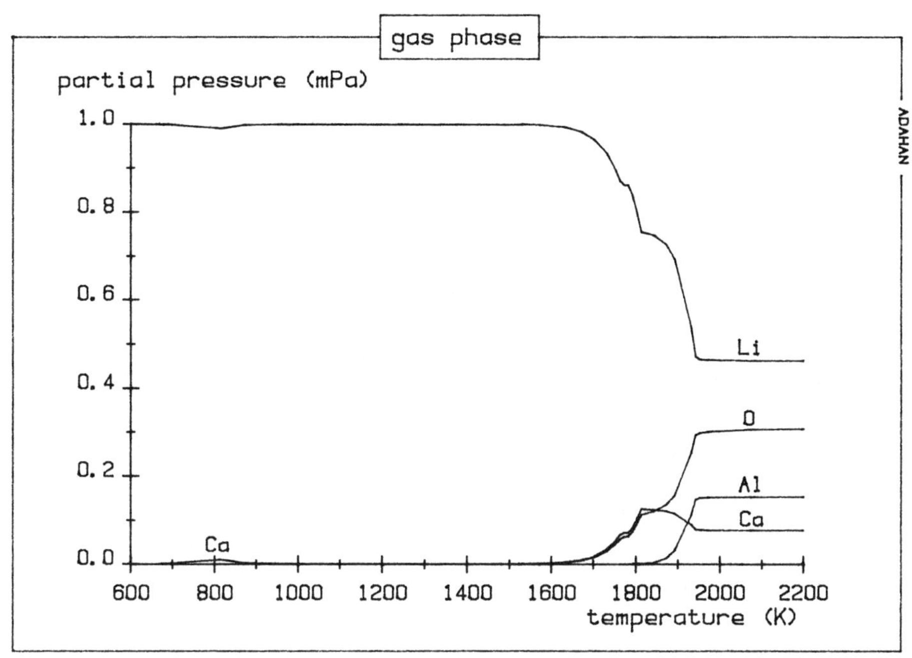

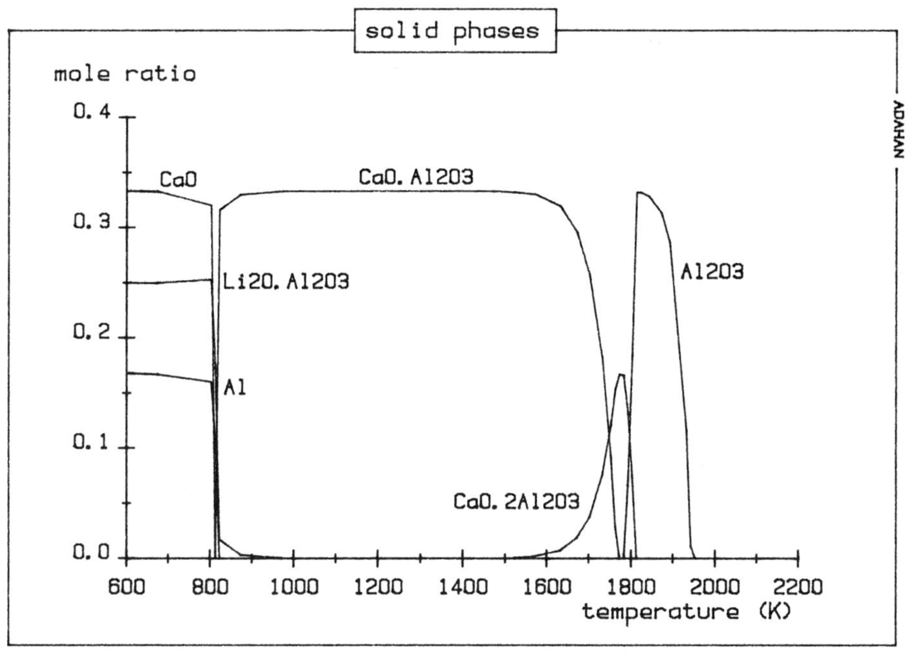

Fig. 5. Equilibrium of the system Li_2O + 2/3 Al + 1/3 CaO
at 1 mPa

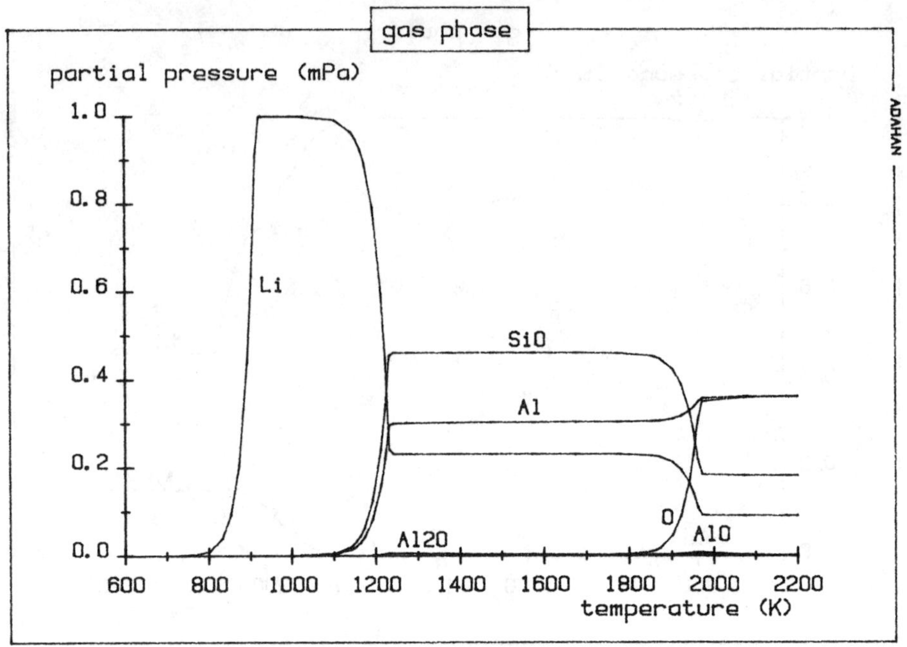

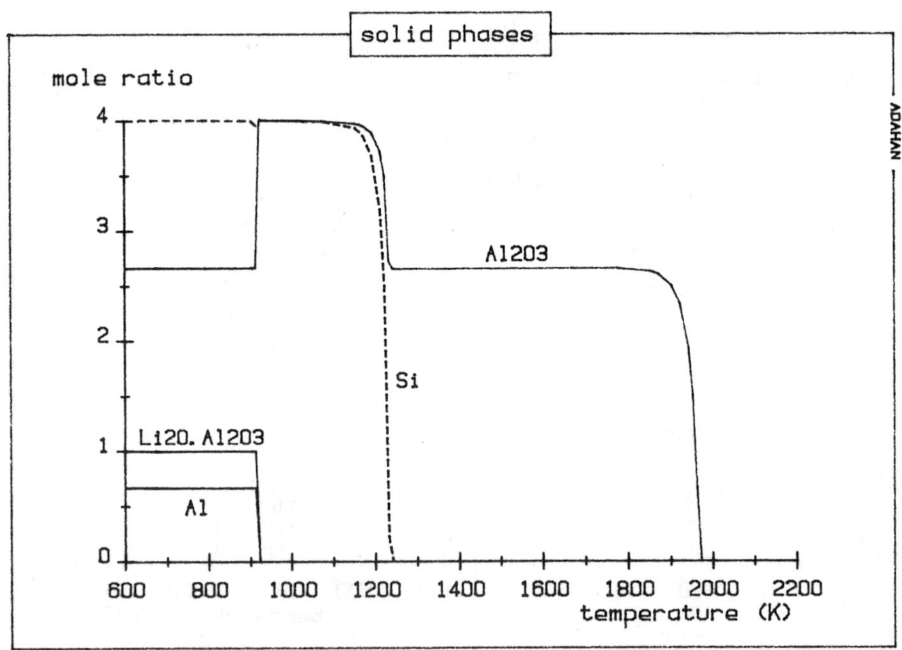

Fig. 6.   Equilibrium of the system $Li_2O.Al_2O_3.4SiO_2 + 6\ Al$ at 1 mPa

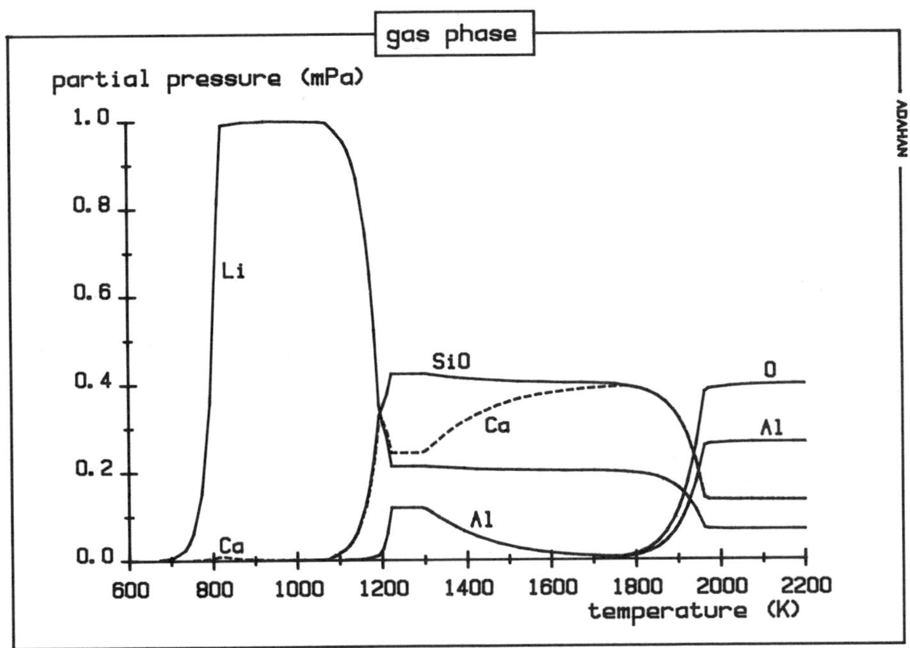

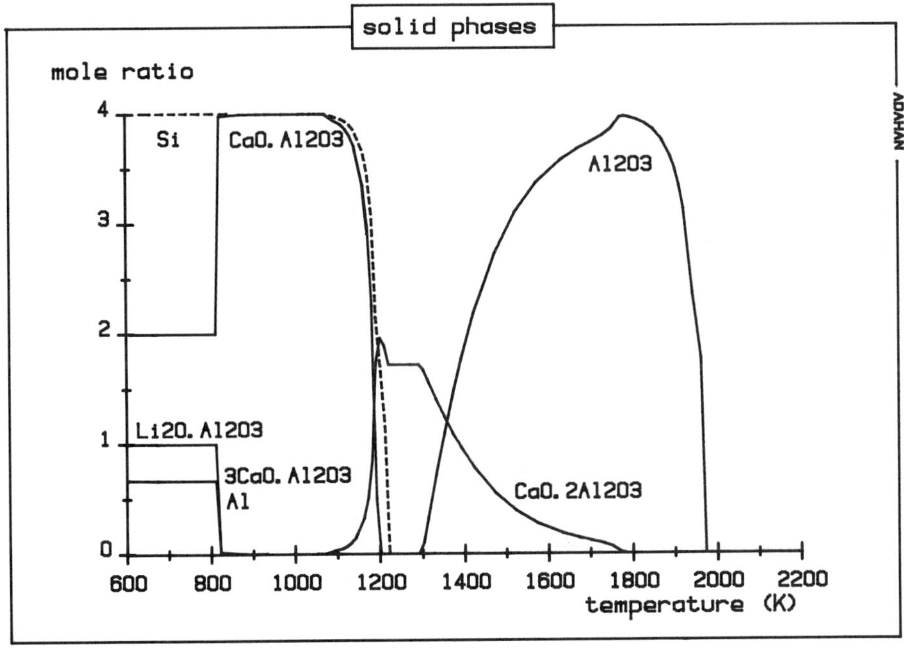

Fig. 7. Equilibrium of the system
$Li_2O.Al_2O_3.4SiO_2$ + 6 Al + 4 CaO at 1 mPa

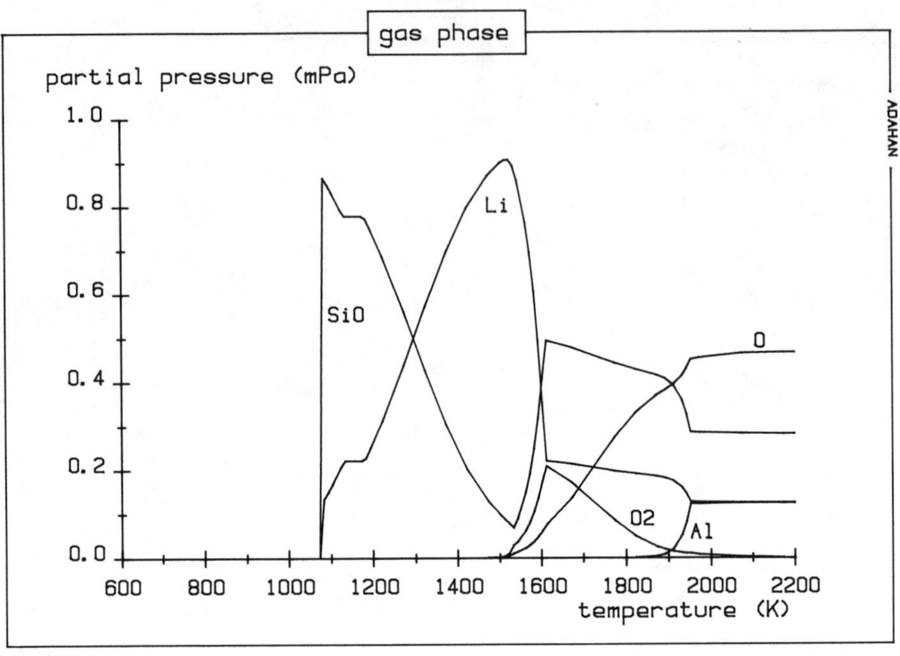

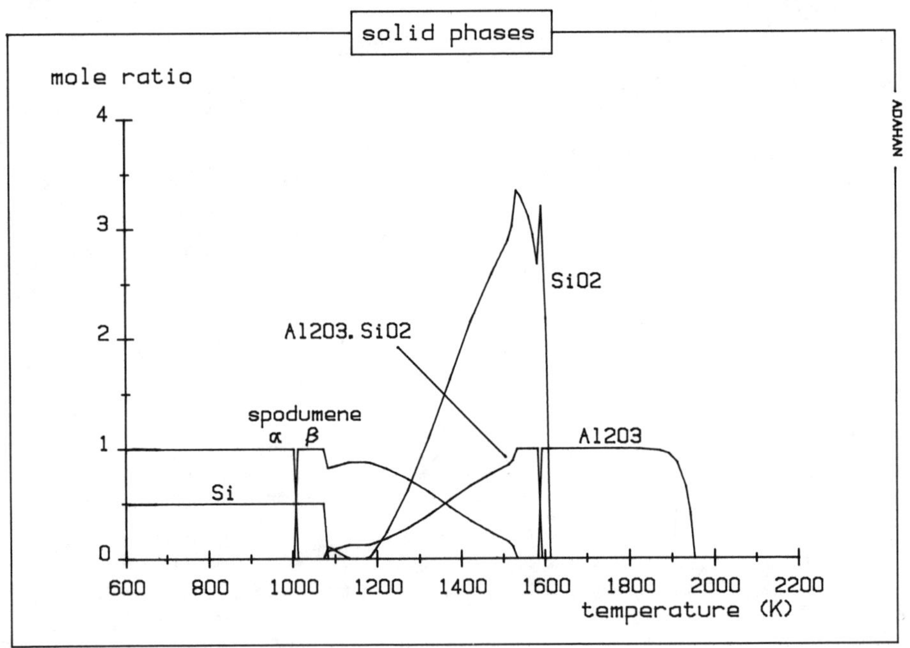

Fig. 8.   Equilibrium of the system $Li_2O.Al_2O_3.4SiO_2 + \frac{1}{2} Si$ at 1 mPa

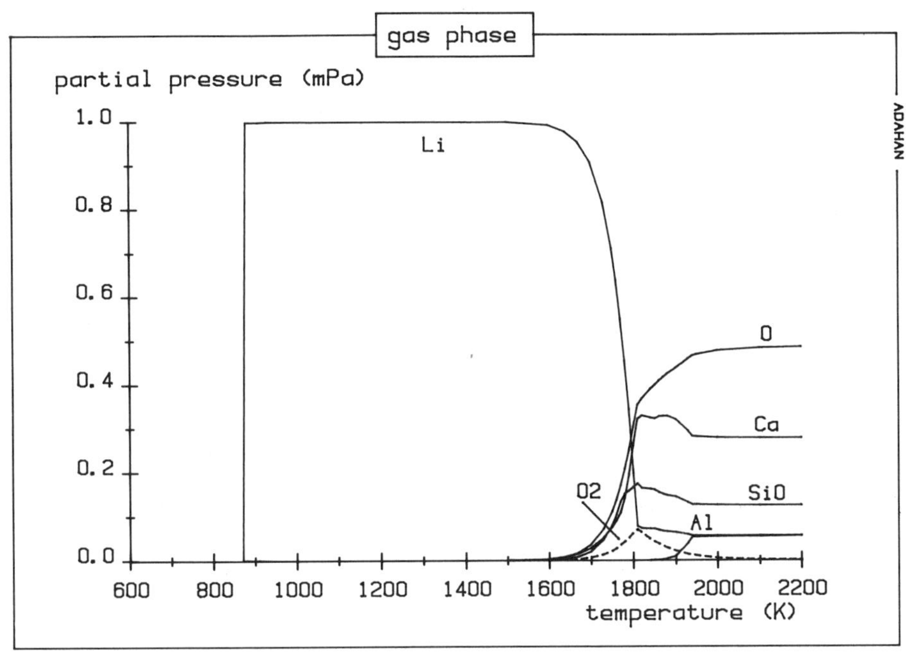

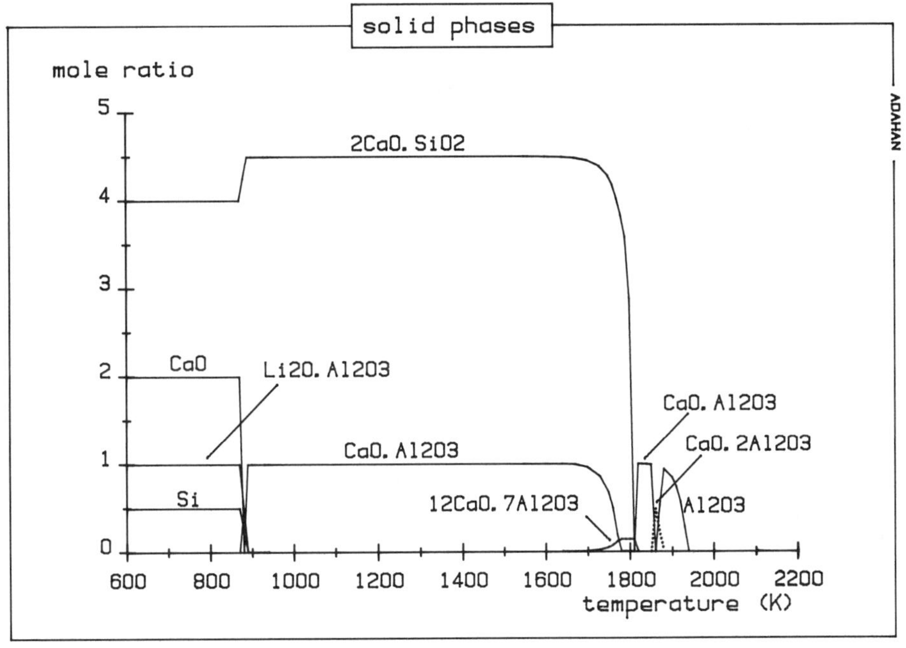

Fig. 9. Equilibrium of the system
$Li_2O.Al_2O_3.4SiO_2 + \frac{1}{2} Si + 10\ CaO$ at 1 mPa

ENERGY REQUIREMENTS FOR THE HYDROGEN GAS PRODUCTION
FROM DECOMPOSITION OF HEATED ASBESTOS TAILINGS

F. AJERSCH
Dép. de génie métallurgique
S. ADCOCK
Centre de recherche en calcul thermochimique
ÉCOLE POLYTECHNIQUE
C.P. 6079, succ. A, Montréal, Qc., H3C 3A7

M. MORENCY, R. MINEAU
Dép. des Sciences de la Terre
UNIVERSITÉ DU QUÉBEC À MONTRÉAL
C.P. 8888, succ. A, Montréal, Qc., H3C 3P8

ABSTRACT

Ultramafic rocks such as asbestos tailings consisting mainly of brucite, $(Mg_{.95} - Fe_{.05})(OH)_2$, and serpentine, $(Fe, Mg)_3 (Si_2O_5)(OH)_4$, can be treated in the presence of iron particles of equivalent granulometry at temperatures between 250-700C to liberate gaseous products rich in H_2 and CO. This process can be used as an alternative method of producing hydrogen (*).

The F*A*C*T system of thermochemical computation was used to optimize the gaseous reaction products and the energy requirements for the process by varying the input conditions of the solid mixture.

The coumputed reaction products are in agreement with experiments carried out in a laboratory batch reactor using 100 g samples and which have also shown favorable reaction kinetics.

(*) Domestic and foreign patents favoring T.H.E. Corp. are pending on a mineral processing technology (of which this process is a part) principally developed by M. Morency.

INTRODUCTION

This study reports the detailed thermochemical computation of the variability of reaction products and energy requirements for the treatment of asbestos tailings based on the Morency process (1). The tailings consist of hydrated serpentine and brucite minerals, in which a certain amount of stoichiometric water in the form of (OH) has been retained in the lattice structure of the mineral. On heating, hydrogen is liberated from the decomposition of these products. Any other constituents in the mineral, such as carbonates, also decompose, resulting in a complex gas mixture consisting primarily of H_2, CO, H_2O, CO_2 as well as other hydrocarbon products, principally methane.

Laboratory experiments leading to the development of the process have shown that an addition of iron powder to the mineral in amounts up to 55% iron results in increased production of hydrogen at 700°C, with a maximum value at about 40% Fe, generating 16.7 litres at standard temperature and pressure per 100 grams of mineral. It was also observed that the hydrogen production increased with decreasing particle size of the mineral, a phenomenon which is characteristic of the kinetics of solid-solid particle reactions.

An optimization of the process was undertaken in order to evaluate the energy balance of the process from thermodynamic considerations, and also to investigate the possibilities of reducing the cost of hydrogen production using this method. The principal parameters considered in this study are therefore temperature and the ratio of the constituents of the solid reaction mixture.

To optimize these parameters it is necessary to calculate the hydrogen yield and solid reaction products at different temperatures and with different initial solid mixture ratios in order to determine the optimal iron-mineral ratio, temperature and other possible materials that can be added to yield optimum values.

Since magnesium, silica as well as other metals can be recovered from the solid reaction product by a leaching process, the production of hydrogen should be considered as only one of the products, thus adding to the potential benefits of the process.

The method employed in this optimization calculation uses the F*A*C*T thermochemical computation technique developed at Ecole Polytechnique (2). This system consists of a data bank and different software packages which calculate the equilibrium reaction products for different input conditions based on free energy minimisation.

The flexibility of these programs allows one to vary pressure, temperature and reactant compositions. The computed result presents not only the thermodynamic limitation of the reaction system but can also give an indication of the related kinetic difficulties that need to be overcome, such as gas and solids volumes, in order to achieve the desired yield.

METHODOLOGY

Thermochemical computation

The equilibrium product calculations are carried out using the "EQUILIB" program where the gaseous, liquid and solid species are calculated on the basis of a thermodynamic data bank. All the various reactants are introduced as predetermined quantities and all the species that could be formed are subsequently retrieved from the F*A*C*T database, or from any additional private database that can be easily incorporated.

The gaseous products are considered to be an ideal mixture and the liquid and solid products and are considered to be pure products with unit activity.

A selection can be made of the most important species by suppressing the presence of the trace constituents for the calculations under study; thus reducing the computation time.

In combination with the "EQUILIB" program, an additional program, "REACTION" was used to calculate the thermal characteristics of the product. This calculation gives the temperature, pressure, enthalpy of reaction (ΔH), free energy of formation (ΔG), change in volume (ΔV), the entropy (ΔS), the change in internal energy (ΔU) and the Helmholz energy (ΔA) of the reaction.

EXPERIMENTAL

The laboratory experiments for optimization of the process and for verification of the thermochemical computation were carried out in the glass apparatus shown schematically in Figure (1). The charge consisting of 100 g of mineral of a granulometry 50% less than 100 mesh, together with the various additions of iron and carbon of about 50% less than 325 mesh, were heated progressively from room temperature to 700°C in a quartz reaction flask.

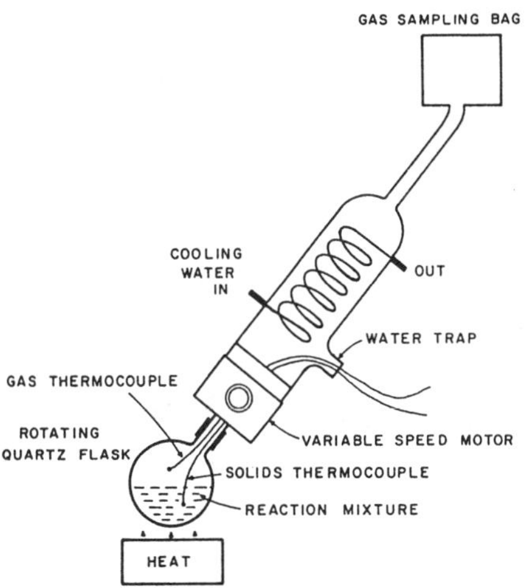

Figure 1: Schematic diagram of experimental apparatus

The quartz flask was constantly rotated by means of a Roto-Vap system which ensures mixing of the reactants throughout the reaction period. A water cooled condenser mounted to the flask served to trap the water vapor produced and the non-condensable gases were collected in a gas sampling bag. The total volume of collected gases was measured and subsequently analysed by gas chromatography in order to determine the composition of the gases produced.

Thermocouples were inserted into the gas region above the reaction mixture and into the mixed solid in order to monitor the reaction temperature.

After complete reaction, the solid samples were cooled and removed for analysis and further treatment.

RESULTS

The calculations are based on an average starting composition of typical asbestos tailings shown in Table I which however does not identify the mineralogical species that are present.

TABLE I: AVERAGE ASBESTOS TAILINGS ANALYSIS

	Wt % Dry
MgO	42.07
SiO_2	39.9
H_2O comb	16.68
Fe	3.80
Ni	0.20
Co	0.02
MnO	0.08
Cr	0.10
Others	0.07
H_2O	--
	100.00

A first series of calculations was made on the basis of the above chemical analysis assuming a wet mineral. The possible species that can be formed based on the above analysis are given in the Appendix I. Using X-ray fluorescence analysis of the products, the "EQUILIB" program predicts all constituents that are stable at 298°K and 1 atm total pressure. For 100 g of mineral the major constituents were calculated to be 75 g $Mg_3Si_2O_5(OH)_4$ (serpentine), 11 g Fe_2O_3 (hematite), 10 g $Mg_3Si_4O_{10}(OH)_2$ (talc), and 3 g H_2O for a total of 99 g of mineral. This assemblage was used for the "REACTION" calculations.

A second series of calculations was made on the assumption that the mineral is dry, containing only structural water. For this case the mineral was assumed to be 90% serpentine, $(FeMg)_3 Si_2O_5(OH)_4$, by weight and 10% Brucite, $Mg_{.95}Fe_{.05}(OH)_2$. The iron in the mineral is in a ferrous state and the total iron content of the tailing material is approximately 10% Fe_2O_3. The results of this calculation are indicated with an asterisk (*) in all the tables that are presented. As can be seen from the calculated results, the initial water content does not change the results to any great extent for the reaction product gases. The solid products however show some difference in composition.

The results for different reaction temperatures and different starting mixtures were then computed. A typical calculation for a mixture of 15 g iron and 70 g mineral and 15 g carbon is shown in Appendix II.

Table II summarizes the result of the effect of increased iron addition to the mineral for reactions carried out at $427^\circ C$. The hydrogen content of the gas increases from 35 to 94.3% H_2 with increasing iron content, the other gas being H_2O. In addition, one can see that it is not necessary to use more than 30% Fe in the initial mixture, since any amount above this value does not react, and is found in the solid products. As would be expected, the enthalpy requirements for 100 g of mixture go down with decreasing amount of mineral because of the endothermic decomposition reaction, and the increase of the bulk thermal conductivity of the mixture.

TABLE II: EFFECT OF IRON-MINERAL RATIO

Mixture grams		Temp.	Enthalpy change	Gaseous Products					Solid Products (grams)				
				Total volume litres at $427^\circ C$	% H_2	% H_2O	Litres H_2**	Litres H_2O**	FeO	$MgSiO_3$	Mg_2SiO_4	Fe	Fe_3O_4
Fe	Min	$^\circ C$	Joules per 100 g mixture										
15	85	427	88666.9	38.5	35.0	65.0	6.18	11.47	22.4	34.3	29.2	---	5.7
15	85*	427	84301.3	36.9	36.7	63.3	6.25	10.76	26.4	10.3	55.5	---	---
20	80	427	77787.2	36.3	48,0	52.0	8.49	9.19	33.6	32.3	27.5	---	---
30	70	427	65503.2	31.7	88.5	11.5	15.63	2.03	45.5	28.3	24.1	---	---
30	70*	427	66063.8	30.4	94.3	5.7	15.97	0.96	43.7	8.5	45.7	0.6	---
40	60	427	58552.4	27.2	94.3	5.7	16.67	1.01	41.0	24.2	20.6	12.7	---
40	60*	427	59681.1	26.1	94.3	5.7	15.99	0.94	37.4	7.3	39.1	14.8	---
50	50	427	52357.8	22.7	94.3	6.7	16.73	1.19	34.2	20.2	17.2	27.3	---

* Dry Mineral
** Litres at $0^\circ C$, 1 atm per 100 g of mineral

The calculated volume of hydrogen at standard temperature and pressure increases from 6.18 litres to 15.63 litres for an increase of 15 to 30% iron in the mixture and only increases slightly to 16.73 litres when the mixture consists of 50% iron. These values are calculated on a basis of 100 grams of mineral in the mixture in order to facilitate the comparison of the reactivity, rather than on a basis of 100 grams of mixture. The figures clearly show that the quantity of unreacted iron only becomes a thermal load.

A mixture of 30% iron with 70% mineral was selected to show the effect of temperature on the calculated conversion rate. As expected, the enthalpy requirements go up with increasing temperature from about 38000 joules at $227^\circ C$, to 97500 joules 97600 joules at $700^\circ C$ while the hydrogen content decreases from 98.5% to 85.1% for the same interval of temperature. The volume of hydrogen is almost constant at 11 litres per 100 grams mixture, or almost 16 litres per 100 grams of mineral, with no significant variation with temperature. The results are shown in Table III.

TABLE III EFFECT OF TEMPERATURE

Mixture g		Temp.	Enthalpy change	Gaseous Products (Vol %)				Solid Products (grams)				
Fe	Min	°C	Joules per 100 g mixture	Total volume litres	H_2	H_2O	Litres H_2**	FeO	$MgSiO_3$	Mg_2SiO_4	Fe	Fe_3O_4
30	70	227	37919.4	20.4	98.5	1.5	15.67	45.5	---	32.0	---	21.4
30	70	327	54479.3	27.2	88.5	11.5	15.64	45.5	28.3	24.1	---	---
30	70	427	65503.7	31.7	88.5	11.5	15.62	45.5	28.3	24.1	---	---
30	70	527	76817.6	36.3	88.5	11.5	15.65	45.5	28.3	24.1	---	---
30	70	627	88498.8	40.8	87.6	12.4	15.48	45.2	28.3	24.1	0.27	---
30	70	700	97574.1	44.7	85.1	14.9	15.24	44.2	28.3	24.1	1.0	---

** Litres at 0°C, 1 atm per 100 g of mineral

Since the calculations do not give any information on the kinetics of the reaction this could be misleading in the interpretation of the results. As has been shown in the experimental work by Morency (1), the laboratory results indicate that brucite releases its structural water at a temperature lower than 500°C producing hydrogen with relativeley fast reaction kinetics. However, the amount of brucite is small (10%) so that only a small proportion of the total hydrogen is generated from this decomposition.

The bench scale laboratory tests indicated that a good overall reaction rate is obtained at 700°C, but above this temperature other undesirable reactions can take place, which will again reduce the hydrogen evolution.

In order to evaluate the effect of carbon addition to the iron-mineral mixture, amounts of 15 and 30% carbon were added to mixtures containing 5 to 15 % iron at reaction temperatures of 427°C and 627°C. The hydrogen content of the reaction gas was found to be only 15-20% at 427°C whereas it increases to more than 50% at 627°C as shown in Table IV.

The calculations have clearly shown that an increase in temperature increases the hydrogen yield from an average of about 2.5 litres to about 10 litres. What is more significant however, is that with carbon, additional combustible reaction gases are produced, principally methane (CH_4) and carbon monoxide (CO). Generally, at 427°C, 16 to 36 volume % CH_4 is found in the product gas but only 5 to 10% CH_4 was calculated at 627°C. At 627°C however, about 12-18 vol. % CO are also formed, whereas little CO is produced at 427°C. As would be expected from thermodynamic considerations, the addition of carbon has a considerable effect on the equilibrium gas mixture of the products.

For example, a 15% iron-70% mineral-15% carbon mixture at 427°C yields only 2.61 liters of hydrogen but also gives 4.71 litres of CH_4 and traces of CO, whereas at 627°C the same mixture gives 9.82 litres of H_2, 1.7 litres of CH_4 and 2.25 litres of CO.

The heat requirements at 627°C are almost twice the amount required at 427°C due to the larger gas volume production as a result of the higher conversion rate and the higher sensible heat content of the products.

TABLE IV: EFFECT OF CARBON ADDITION

Mixture grams			Temp °C	Enthalpy change joules per 100 g mixture	Gaseous Products (vol %)							Solid Products grams						
Fe	Min	C			Total volume litres	H_2	H_2O	CH_4	CO_2	CO	Litres H_2**	FeO	$MgSiO_3$	Mg_2SiO_4	Fe	Fe_3O_4	C	$MgCO_3$
5	80	15	427	86063.8	36.2	14.8	46.0	19.8	18.8	0.7	2.61	--	32.2	27.5	--	15.4	12.0	---
5	80*	15	427	85577.8	34.0	15.1	45.7	20.7	17.8	0.7	2.50	--	9.7	52.2	--	14.1	12.2	---
10	60	30	427	63382.4	22.9	18.1	41.5	29.6	10.2	0.5	2.69	--	24.2	20.6	--	20.2	28.1	---
0	70	30	427	90520.5	34.6	13.3	46.3	16.0	23.6	0.8	2.56	--	28.3	24.1	--	7.4	27.1	---
15	70	15	427	59860.6	24.7	19.9	37.1	35.8	6.7	0.4	2.74	1.8	28.3	24.1	--	26.3	12.8	---
15	70*	15	427	62396.6	23.6	19.9	37.1	35.8	6.7	0.4	2.61	8.3	8.5	45.7	--	18.1	12.9	---
5	80	15	627	143912.0	64.4	42.1	18.7	5.8	16.3	17.0	10.28	14.4	32.3	27.5	--	---	10.9	---
5	80*	15	627	141488.3	60.6	43.0	18.7	6.0	15.6	16.7	9.88	13.1	9.7	52.2	--	---	11.2	---
10	60	30	627	115428.8	43.4	47.6	18.2	7.4	12.1	14.7	10.44	18.8	24.2	20.6	--	---	27.6	---
10	60*	30	627	113616.1	40.5	49.0	18.0	7.8	11.1	14.0	10.03	17.9	7.3	39.2	--	---	27.8	---
-	70	30	627	144925.4	59.8	39.4	18.7	5.1	18.6	18.2	10.20	6.9	28.3	24.1	--	---	25.9	---
-	70*	30	627	142802.6	54.6	40.0	18.7	5.2	18.1	17.9	9.46	5.9	8.5	45.7	--	---	26.2	---
15	70	15	627	114437.6	48.0	50.6	17.6	8.4	10.0	13.4	10.52	26.2	28.3	24.1	--	---	12.5	---
15	70*	15	627	108741.8	42.7	53.1	17.0	9.2	8.4	12.2	9.82	24.3	10.2	43.3	--	---	12.9	---

** Litres at $0^{\circ}C$, 1 atm per 100 g of mineral

Since the specific reaction products for the cases of carbon addition were not previously known it was not possible to predict its effect in replacing iron as a reducing agent. Calculations with no iron addition and 30% carbon showed little conversion (2.56 litres H_2 and 3.07 litres CH_4) at 427°C. This same mixture yields 10.20 litres H_2 and 1.22 litres CH_4 at 627°C, which is still not as good as the cases of higher iron/ mineral ratios. The calculations however clearly showed that only about 3% of the carbon is actually used during the reaction and the remaining amount is found unreacted in the solid product.

It can therefore be deduced that the iron mineral mixture should be at least 15% iron and that a quantity of carbon in the vicinity of only 3% needs to be added in order to optimize the hydrogen and combustible gas yield. The experimental results of the H_2 generation rates indicated that the reaction temperature should be raised to 700°C in order to obtain both high rates and high yieds of hydrogen.

A summary of the calculations for the optimum range of 15, 20 and 25% Fe and 0, 3 and 5% carbon additions at 700°C is shown in Table V.

TABLE V: OPTIMIZED REACTION MIXTURES AT 700°C

Mixture grams			Temp °C	Enthalpy change joules per 100 g mixture	Gaseous Products											Solid Products grams				
Fe	Min	C			Total volume litres at 700°C	% H_2	% H_2O	% CH_4	% CO_2	% CO	Litres H_2**	Litres H_2O**	Litres CH_4**	Litres CO_2**	Litres CO**	FeO	MgSiO_3	Mg_2SiO_4	Fe	C
15	85*	0	700	117781.7	51.3	36.7	63.3	--	--	-	6.21	10.76	--	--	--	26.4	10.3	55.5	--	--
15	82*	3	700	143111.1	65.7	57.3	12.3	3.8	6.7	20.8	12.88	2.76	0.62	1.51	4.63	26.2	9.9	53.5	--	--
15	80*	5	700	150040.8	68.2	53.7	9.4	3.8	6.7	26.2	12.84	2.24	0.90	1.65	6.24	22.1	9.7	52.2	3.1	1.2
20	80*	0	700	111045.8	48.3	53.1	45.9	--	--	--	9.16	7.77	--	--	--	32.4	9.7	52.2	--	--
20	77*	3	700	137241.0	61.7	57.6	10.1	3.8	6.0	22.6	12.68	2.22	0.85	1.32	5.08	25.7	9.3	50.2	5.0	--
20	75*	5	700	143400.4	64.0	53.7	9.4	3.8	6.9	26.2	12.85	2.24	0.91	1.65	6.26	20.7	9.1	48.9	8.8	1.4
25	75*	0	700	104309.9	45.3	73.8	26.2	--	--	--	12.50	4.34	--	--	--	38.4	9.1	48.9	--	--
25	72*	3	700	133418.2	58.9	56.3	9.9	3.8	6.3	23.8	12.92	2.27	0.88	1.45	5.45	22.7	8.7	47.1	12.1	--
25	70*	5	700	136760.0	59.7	53.7	9.4	3.8	6.9	26.3	12.84	2.24	0.90	1.64	6.28	19.3	8.5	45.7	14.6	1.7

** Litres at 0°C, 1 atm per 100 g of mineral

Table V clearly shows that in all cases only 3% carbon needs to be added to the reaction mixture since the volume and composition of gases do not change significantly above this value. When comparing these results with the no carbon addition case, an addition of 25% iron is required to yield volumes equal to the volumes for the case of 15% iron with 3% carbon. In fact, it can be observed that the hydrogen production is almost constant at about 12.8 litres per 100 g of mineral for each of the 15, 20 and 25% iron addition cases.

The experimental results for a reaction temperature of 750°C are shown in Figure (2) and compare quite well with the values of hydrogen volume predicted by computation at 700°C. Between 40 and 50% iron the results can be considered to be almost identical, taking into account the experimental variability, particularly since the measurement of the experimental temperature was only approximate. The gas temperature was found to be about 100°C colder than the solids temperatures in the batch reactor. With 15% iron addition, 12.2 litres of hydrogen were produced compared to only 6.21 litres for the calculated value. At higher values of iron content the experimental results are much closer to the predicted values.

The calculated results of the hydrogen yield for the 3% carbon addition are also shown in Figure (2) indicating the marked increase of H_2 generation over the case of no carbon addition. An experimental result for the case of 3% carbon addition to a 15% iron mixture is also shown on this Figure and gives a value of about 11 litres as compared to the calculated values of 12.88 litres. More experiments need to be carried out in order to establish the yields that are attained for mixtures containing 3% carbon addition.

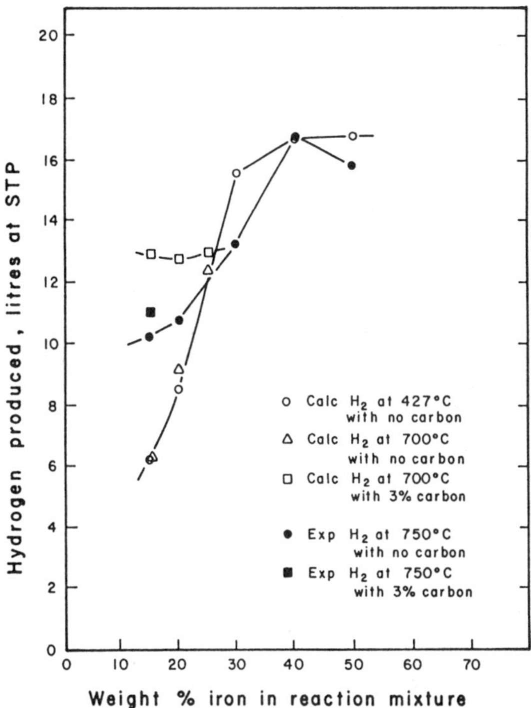

Figure 2:   Comparison of experimental and calculated results

The enthalpy requirements for the iron-mineral-carbon mixtures are considerably higher (133,000 to 150,000 joules per 100 g of mixture) than for the no carbon addition cases (104,00 to 118,000 joules), as would be expected from thermodynamic considerations, but the total yield of combustible gases is much improved.

A feed mixture of 15 wt.% Fe, 82% mineral and 3%C results in a calculation of 12.88 litres of H_2, 0.62 litres of CH_4 and 4.63 litres of CO per 100 g of mineral. If 5% carbon is added to the mixture containing 15% iron, 12.84 litres of H_2, 0.90 litres of CH_4 and 6.24 litres of CO are produced per 100 g of mineral. In this case however, 3.1 grams of iron and 1.2 grams of carbon also remain unreacted in the final solid products for a 100 grams starting mixture.

In order to evaluate the energy demand of the process, a simple calculation of the heat content of the products was carried out. If the gases produced (H_2, CH_4 and CO) are separated and used as a combustion fuel, a maximum value of about 200,000 joules can be recovered (3) (combustion gas products at 60°F and 30 cm Hg) from the 15% Fe-82% Min - 3% C mixture, and 210,870 joules from the 15% Fe-80% Min-5% C mixture respectively.

The calculated amount of solids recovered for the two cases above are 89.6 and 88.3 grams respectively. Assuming an average specific heat of 0.20 cal g^{-1} $^{\circ}$C, the additional heat recoverable is about 44,000 joules if the solids are cooled from 700°C to 100°C.

This totals to about 244,000 and 255,000 joules theoretically recoverable from the separated products as compared to 143,000 and 150,000 joules calculated to produce the final reaction product mixture.

The yield of product gases can be used for hydrogen production as well as for supplying part of the heat requirements of a viable commercial process. Iron from the solid products can also be recovered and recycled after reduction by the separated reducing gas products.

A complete energy balance connot be carried out without reference to a specific process design. From practical considerations, the laboratory batch reaction results are very promising in that the values closely approach thermodynamic equilibrium conditions. Larger scale equipment should give even better results because operating conditions can be much more closely controlled.

CONCLUSIONS

The method of thermochemical computation of the reaction products of various iron-mineral carbon mixtures has been shown to be very effective in optimizing the operating conditions of an industrial process for the production of hydrogen from asbestos tailings.

The results of the calculations present the limiting thermodynamic values that can be achieved. Experiments on a bench scale using a batch reactor with a 100 gram reaction mixture produced results which approached the values that were computed, particularly for temperatures at 700°C. With larger scale equipment these results should improve due to a better control of experimental conditions.

The heat generated from the combustion of the reducing gases produced together with the sensible heat of the solid products are factors that must be incorporated into the design of a viable process.

The kinetics of the process are strongly influenced by reaction temperature, and particle size distribution of the reactants. More experiments need to be carried out in order to establish the process reaction kinetics on a pilot plant scale.

REFERENCE

(1)   M. Morency, U.S. Patent Appl., filed September 12, 1985, "Production of Elements and Compounds by Deserpentinisation of Ultramafic Rock".

(2)   A.D. Pelton, C.W. Bale, and W. Thompson, "Facility for the Analysis of Chemical Thermodynamics (F*A*C*T) User's Guide", 1st Edition, École Polytechnique–McGill University, Montreal (1979).

(3)   Chemical Engineers'Handbook, R.H. Perry and C.H. Chilton Ed., 5th Editon, McGraw-Hill (1973).

APPENDIX I

INITIAL MINERAL ASSEMBLAGE AT 298°K


```
fact

*In progress


F*A*C*T SYSTEM, MONTREAL, QUEBEC, CANADA
COPYRIGHT 1986 THERMFACT LTD/LTEE
A.D.PELTON, C.W.BALE, W.T.THOMPSON

****** ENTER A PROGRAM NAME OR PRESS "RETURN" TO EXAMINE F*A*C*T LIBRARY ******
?
equi

EQUILIBRIUM PRODUCT CALCULATION
*******************************

****** ENTER REACTANTS ******(OR PRESS "RETURN" FOR LAST ENTRY)(07OCT87)
NUMBERS BEFORE SPECIES ARE MASSES IN GRAMS (OPTION 5)
?
39 Si*O2 + 11 Fe2O3 + 36 Mg*O + 13 H2O

ENTER SUBSCRIPTS, ENTER "HELP", OR PRESS "RETURN"
?
(298) (298) (298) (298)

*********************************************************************************
  T PROD    P PROD    DELTA H    DELTA G    DELTA V    DELTA S    DELTA U    DELTA A
   (K)       (ATM)      (J)        (J)        (L)       (J/K)       (J)        (J)
*********************************************************************************
---S1 S1 S  L  -----------------------------------------------------------------
?
298 1

THERE MAY BE A DELAY WHILE THE LOCATION OF
ALL DATA ON COMPOUNDS CONTAINING THE REACTANT ELEMENTS
IS DETERMINED.
GASEOUS   SPECIES    1 -  28
LIQUID    SPECIES   29 -  42
AQUEOUS   SPECIES   43 -  58
SOLID     SPECIES   59 - 101
LIQ. FE WAGNER SOLUTION FROM JAPANESE COMPILATION.  THESE SHOULD BE
CORRECT VALUES.
  102   FE
  103   O2
  104   SI
QUASICHEM SLAG MODEL, BINARY PARAMS ONLY, NO AL2O3 YET, REF PHASES ARE
HIGHEST SOLID PHASE IN MAINBASE PLUS ADDED DELTA G FUSION
  105   SI*O2
  106   MG*O
  107   FE*O
PITZER PARAMETERS FROM HARVIE ET AL.
   GO(298) FROM F*A*C*T
  108   H2O
  109   H[+]
  110   O*H[-]

ENTER CODE NUMBERS, ENTER "LIST" TO DISPLAY, OR ENTER "HELP"
****(CHANGES TO EQUILIB REQUIRE THE USE OF SLASHES ABOUT GASEOUS SPECIES)****
?
```

```
PRESS "RETURN" WHEN READY FOR OUTPUT
ENTER AN OPTION NUMBER, OR ENTER "O" FOR OPTIONS MENU
?

39 SI*O2 + 11 FE2O3 + 36 MG*O + 13 H2O
(298) (298) (298) (298)

          0.00000E+00          (  0.30934E-01      H2O
                               +  0.23758E-27      H2
                               +  0.11879E-27      O2
                               +  0.12251E-33      OH
                               +  0.12049E-36      H2O2
                               +  0.22917E-47      HO2
                               +  0.35343E-49      H
                               +  0.25732E-54      O
                               +  0.32436E-70      O3

                                 (  298.0, 1.00    ,G ,0.309E-01)

                               +   75.077    gram (MgO)3(SiO2)2(H2O)2
                                 (  298.0, 1.00    ,S1, 1.0000    )

                               +   11.000    gram Fe2O3
                                 (  298.0, 1.00    ,S1, 1.0000    )

                               +   10.167    gram (MgO)3(SiO2)4(H2O)
                                 (  298.0, 1.00    ,S1, 1.0000    )

                               +   2.7555    gram H2O
                                 (  298.0, 1.00    ,L1, 1.0000    )


GASEOUS IONIC SPECIES ARE SUPPRESSED BELOW 3000 K

DATA ON  9 PRODUCT SPECIES IDENTIFIED WITH 'X' HAVE NOT BEEN EXTRAPOLATED IN
COMPUTING THE PHASE ASSEMBLAGE

DATA ON  3 PRODUCT SPECIES IDENTIFIED WITH 'T' HAVE BEEN EXTRAPOLATED

**********************************************************************************
  DELTA H     DELTA G     DELTA V     DELTA S     DELTA U     DELTA A    REACT V
    (J)         (J)         (L)        (J/K)         (J)         (J)        (L)
**********************************************************************************
---S1 S1 S  L  -----------------------------------------------------------------
  -50116.2    -43091.6   -0.350E-01   -23.573     -50112.6    -43088.0  0.399E-01
```


APPENDIX II

"EQUILIB" PRODUCT CALCULATION FOR 15% Fe- 70% MINERAL-15% CARBON


```
ENTER CODE NUMBERS, ENTER "LIST" TO DISPLAY, OR ENTER "HELP"
****(CHANGES TO EQUILIB REQUIRE THE USE OF SLASHES ABOUT GASEOUS SPECIES)****
?
/1-86/123,161-203,208-217

PRESS "RETURN" WHEN READY FOR OUTPUT
ENTER AN OPTION NUMBER, OR ENTER "O" FOR OPTIONS MENU
?

52.5 MG3SI2O5(O*H)4 + 7.7 FE2O3 + 7 MG3SI4O10(O*H)2 + 2.8 H2O + 15 FE + 15 C
(298) (298) (298) (298) (298) (298)
```

```
24 670      litre (    37.071       vol% H2O
                  +    35.842       vol% CH4
                  +    19.936       vol% H2
                  +     6.7444      vol% CO2
                  +     0.40607     vol% CO
                  +     0.19487E-03 vol% C2H6            T
                  +     0.67885E-04 vol% C4H8
                  +     0.33988E-06 vol% C2H4
                  +     0.20957E-06 vol% CH2O
                  +     0.76030E-07 vol% CH3OH
                  +     0.41124E-07 vol% HCOOH
                  +     0.50635E-08 vol% C3H8            T
                  +     0.20022E-08 vol% CH3COOH         T
                  +     0.35314E-09 vol% C2H4O           T
                  +     0.26789E-09 vol% CH2CO           T
                  +     0.20827E-09 vol% CH3
                  +     0.64381E-10 vol% HCO
                  +     0.54061E-10 vol% CH3CH2OH
                  +     0.11193E-11 vol% H
                  +     0.54063E-12 vol% C4H10
                  +     0.25863E-12 vol% C2H2
                  +     0.87567E-13 vol% C3H6
                  +     0.75710E-15 vol% CH3CH2OH        T
                  +     0.68741E-15 vol% C6H6
                  +     0.14671E-15 vol% OH
                  +     0.48199E-16 vol% C3H4
                  +     0.66872E-17 vol% C5H12           T
                  +     0.14848E-17 vol% C7H8
                  +     0.64548E-18 vol% C2H4O
                  +     0.17519E-18 vol% C3O2
                  +     0.10118E-18 vol% C4H6
                  +     0.17188E-19 vol% C6H12
                  +     0.97875E-21 vol% C6H14
                  +     0.20080E-21 vol% C8H10
                  +     0.97385E-22 vol% C8H10
                  +     0.60184E-22 vol% Fe              T
                  +     0.21870E-23 vol% C5H8
                  +     0.11805E-23 vol% C7H14
                  +     0.12217E-24 vol% H2O2
                  +     0.31954E-25 vol% C7H16

                  +    28.283       gram (MgO)(SiO2)
                       ( 700.0, 1.00     ,S1,   1.0000      )

                  +    26.278       gram Fe3O4
                       ( 700.0, 1.00     ,S1,   1.0000      )

                  +    24.058       gram (MgO)2(SiO2)
                       ( 700.0, 1.00     ,S1,   1.0000      )

                  +    12.782       gram C
                       ( 700.0, 1.00     ,S1,   1.0000      )

                  +     1.7636      gram FeO
                       ( 700.0, 1.00     ,S1,   1.0000      )
```

GASEOUS IONIC SPECIES ARE SUPPRESSED BELOW 3000 K

DATA ON 10 PRODUCT SPECIES IDENTIFIED WITH 'X' HAVE NOT BEEN EXTRAPOLATED IN
COMPUTING THE PHASE ASSEMBLAGE

DATA ON 26 PRODUCT SPECIES IDENTIFIED WITH 'T' HAVE BEEN EXTRAPOLATED

```
*********************************************************************************
  DELTA H     DELTA G     DELTA V      DELTA S      DELTA U      DELTA A    REACT V
    (J)         (J)         (L)         (J/K)         (J)          (J)        (L)
*********************************************************************************
---S  S1 S  L  S1 S1 ------------------------------------------------------------
  59860.6    -78754.9   0.247E+02    154.123      57359.4     -81256.1  0.128E-01
```

SIMULATING GRAVITY CIRCUITS: THE COMPROMISE BETWEEN
ACCURACY AND SIMPLICITY

A.R Laplante and Y. Shu
Department of Mining and Metallurgical Engineering
McGill University
3450 University St.
Montréal, Québec, Canada

ABSTRACT

Simple --two or three phase-- models of gravity circuits are presented.
Their use and limitations are discussed using two case studies. It is
argued that for systems where the two major phases are liberated,
simple models can accurately predict circuit performance and be used to
assess the need for more intensive models.

KEY WORDS

Gravity separation; two or three-phase models; grade-recovery curve;
recovery-recovery curve.

INTRODUCTION

Simulation of industrial mineral processing circuits began in the early
sixties, first with comminution (crushing and grinding) and
classification (hydraulic and screens) (Lynch, 1977), then flotation
(Lynch and others, 1981), and now other less common processes. Gravity
separation, which treats more tonnes than any other separation process,
including flotation, has received much attention only recently, partly
because most commodities processed are of little value, partly because
gravity separation is still considered 'low-tech.' Gravity separation
is now enjoying a surge of interest, for a number of reasons:

- the advent of high-capacity units such as the Reichert cone

- the development of lower-cost, higher-performance units such
  as spirals

- the advent of fine particle recovery units such as the GEC
  Duplex or the Knelson centrifugal concentrators

- a better understanding of gravity recovery mechanisms,
  especially for fine particles

- the emphasis on environmentally safe technology

- the incentive for preconcentration, with the steadily
  decreasing head grades and increasing grinding and
  separation costs

- an increased precious metal production, with gravity being
  used as the sole recovery process (placers), or in
  conjunction with flotation or cyanidation.

Modelling gravity units is steadily on the rise. Thin and flowing film
concentrators lend themselves well to fundamental simulation, because
of their simple geometry (Subasinghe and Kelly, 1984). Jigs have been
modelled with mostly phenomenological models, because of the complexity
of particle-particle interactions (Burt and Wills, 1984). Spiral
separators exhibit complex geometries, and their modelling is mostly
empirical (Tucker and others, 1985). Heavy-medium units should be
approached phenomenologically, very much like hydrocyclones. The
modelling activity of the eighties in gravity is comparable to that of
the late sixties/early seventies for comminution and flotation. This
activity preceeded the industrial application phase, which, arguably,
has not reached its promised potential, and has yet to gain complete
acceptance by plant personel.

Broadly speaking, the most attractive application of modelling today to
industrial mineral processing is scale-up. Gravity does not escape
this trend, although the main thrust here is the use of empirical
models to predict plant performance of complex circuits, typically for
cones and spirals (Holland Batt, 1979). The rationale is that
laboratory or pilot plant testwork on a single separation stage yields
information of adequate accuracy, such that error propagation through
multistage simulation does not threaten the validity of the simulator.
The alternative to simulation is often bleak: Either a costly
pilot-plant (if the scope of the project warrants it) or scale-up based
on less reliable information, such as operating experience with similar
projects or rules of thumb

Static simulators can assume two basic structures, shown in
Fig. 1. First, and most common, is the 'feedforward' simulator, in
which feed and circuit characteristics are inputted; each unit is
sequentially simulated to generate final products and recycle streams;
and successive circuit iterations integrating recycle streams are
performed until steady-state is achieved. Analytical solutions are
seldom possible, except for very simple models. Second, the 'feedback'
simulator assumes a certain response, typically a grade-recovery or a
recovery-recovery curve for separation processes, or a general size
distribution for comminution processes (McIvor, 1984). A material
balance is then established based on these curves, and operating
conditions and/or equipment size inferred from the mass balance. The
'feedback' model is simpler, but obviously less robust.

Simplifications in the feed description can sometimes be used.
Normally, a matrix is used, with classification according to size and
mineral composition. For comminution circuits, typical simulations
include a single species and discrete size distributions following
Tyler's progression. For complex sulphides, there is evidence that at
least two species, lights and heavies, should be used to describe both

classification and breakage phenomena (Laplante et Finch, 1984).
Simplifications are often possible, especially if simulation is to be
confined to the existing circuit flowsheet. For flowsheet
modifications, there is a growing body of evidence that even the
'unsimplified' models are often inadequate (Laplante, Finch, and del
Villar, 1987).

In this work, two case studies will demonstrate how gravity circuits
can be simulated with simple models. Uses and limitations of the
simple models will be correlated to mineralogical and metallurgical
considerations.

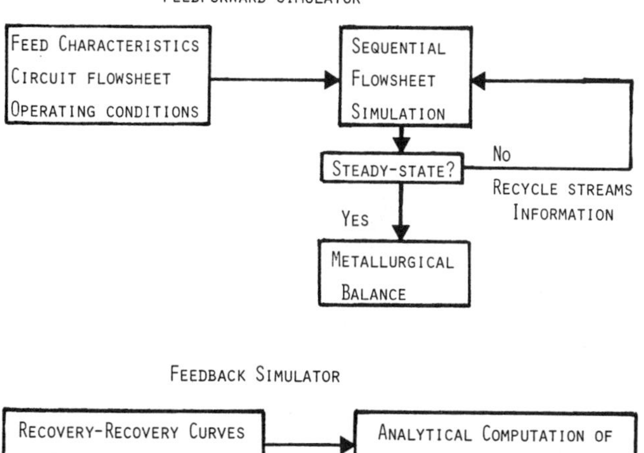

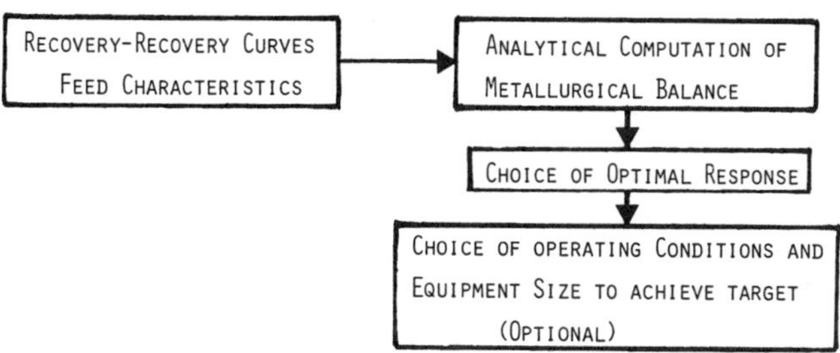

Fig. 1. Feedforward and feedback simulators

FIRST CASE STUDY: A SPIRAL CIRCUIT

The circuit is a typical rougher/cleaner/scavenger circuit which
upgrades a feed of 70-75% oxides to 87-89% oxides; the main gangue
component is anorthite. To gather data for a simple model, the circuit
was sampled four times for a two-hour period. Each sample was assayed
for oxide and silicate content. Samples were deslimed to 45 μm and the
+45 μm separated at a s.g. of 2.96 by dense liquid.

Two phase Model

A first simple model is generated by considering the recovery of
silicates as a function of that of oxides, as shown in Fig. 2.

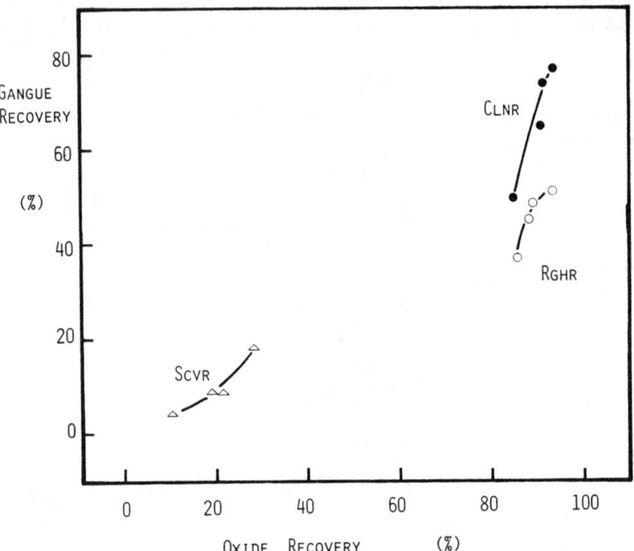

Fig. 2. Gangue recovery vs. oxide recovery
for all four plant tests

The rougher and cleaner performance is comparable, as operating
conditions and feeds are similar. The behaviour of the scavenger
differs markedly, with poor oxide recovery. This stems from the very
low feed density, 15-20% solids, and from the fineness of the feed
oxides, mostly below 100 μm.

The simulator selects five points on each of the three
recovery-recovery curves, and calculates overall recovery, Rt, equal
to:

$$Rt = 1 - \frac{(1 - R1) * (1 - R3)}{[1 - R1 * (1 - R2) - R3 * (1 - R1)]} \tag{1}$$

where R1, R2, and R3 are the fractional recoveries of the roughers,
cleaners, and scavengers, respectively. The simulator examines all 125
combinations of stage recoveries and selects, for a given concentrate

grade increment, the combination yielding the highest overall recovery.
Figure 3 shows typical results for three different head grades. Each
curve shows that from a maximum recovery of 95% (when all stage
recoveries are at a maximum), recovery first decreases slowly with
increasing concentrate grade, as cleaner recovery slowly decreases at
the maximum rougher recovery. At the lowest cleaner recovery, further
increases in concentrate grade can only be achieved by lowering rougher
recovery, but overall recovery then drops rapidly. In the last
segment, overall recovery slowly decreases with decreasing scavenger
recovery.

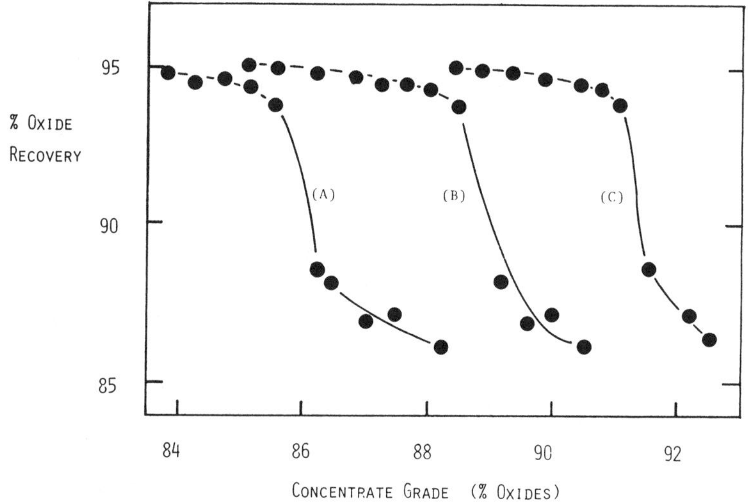

Fig. 3. Simulation results of the two-phase model for feed
grades of 70% (A), 75% (B), and 80% (C) oxides.

This first simple model reflects important aspects of circuit
behaviour, but fails to take into account important factors:

1. The effect of head grade on concentrate grade was confirmed from
plant data. However, a similar effect of head grade on recovery is not
shown by the model.

2. The simulator suggests that selectivity should be achieved in the
cleaner, rather than rougher stage. This has long been recognised, for
example in flotation circuits (Canas, 1980). It is particularly
obvious here because of poor scavenger performance.

3. As recovery is mostly controlled by feed density, the simulator
correctly predicts that the roughers should have a high density, 40-45%
solids, for high recovery, whereas the cleaners should be operated at
35-40% solids for increased selectivity.

The Three Phase Model

The simulator can be modified to take into account liberation while
retaining its simplicity. Dense liquid data were used to estimate the
performance of the float, which assays 5% oxides, and was considered
'free gangue'. The sink was considered to be made of 'free oxides',
assaying 95% oxides, and middlings, assaying 50% oxides. Their
relative proportion in the sink was estimated from its oxide assay.
The performance of the free gangue and middling was then represented as
a function of that of free oxides, as shown in Fig. 4.

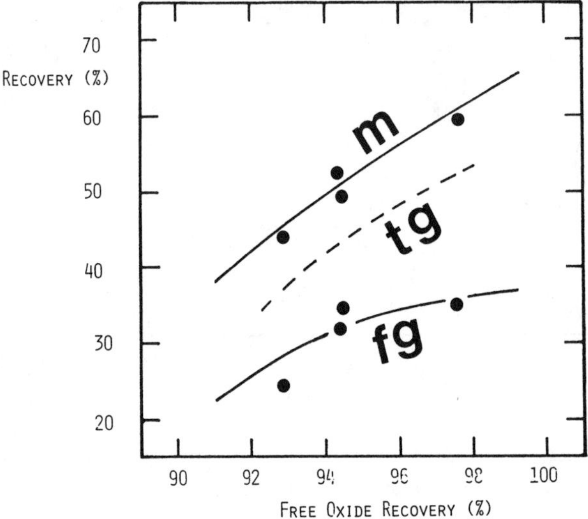

Fig. 4. Recovery of free gangue and middlings vs. recovery of
free oxides for the rougher stage of all tests

The recovery of free oxides is higher than that of oxides for the
corresponding test, that of free gangue lower than that of total
gangue, and that of middlings intermediate. Data of Fig. 4 are more
erratic than those of Fig. 2, as their estimation cumulates the errors
of more manipulation.

Figure 5 shows the fraction of free oxides, free gangue, and middlings
as a function of head grade. A decrease in head grade corresponds to
an increase in middlings content, rather than free gangue. This
compounds the problem of lower feed grades, as an increased middling
content lowers both concentrate grade and recovery.

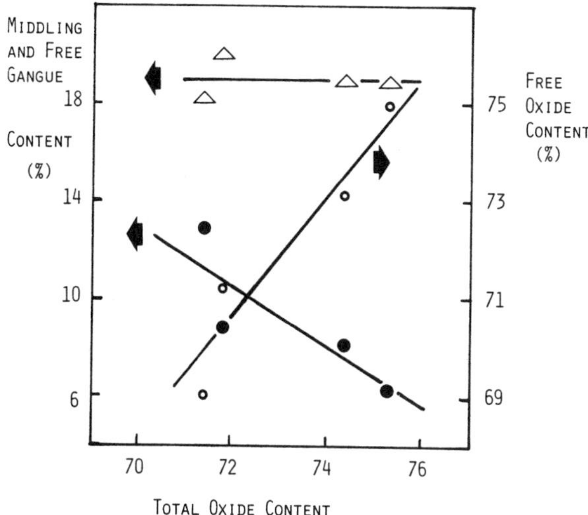

Fig. 5. Free oxide, free gangue, and middlings content vs.
total oxide assay for all four tests (fresh feed)

The overall circuit was simulated as with the first model. Figure 6
shows that although results are similar to those of Fig. 3, a greater
measure of realism is achieved. First, oxide recovery decreases with
decreasing head grade, although the effect is slight. Second, the
grade-recovery curves are no longer comprised of three sections, but
rather two, as most grade-recovery curves are, with the apex
corresponding to the minimum cleaner recovery and maximum rougher
recovery.

The three-phase model, while being more realistic than two phases,
still has serious limitations which are inherent not only to its
simplicity, but also to the data base it is derived from. Thus, the
effect of important operating variables such as feedrate, wash water
rate, and concentrate splitter position are not taken into account.
These could possibly be integrated to the model from previous work, but
this type of data is scarce. Additional data could be generated,
preferably from bench scale testing rather than plant surveys. Despite
these shortcomings, both simple models make good use of the available
data, yield valuable information, and are extremely cost-effective.

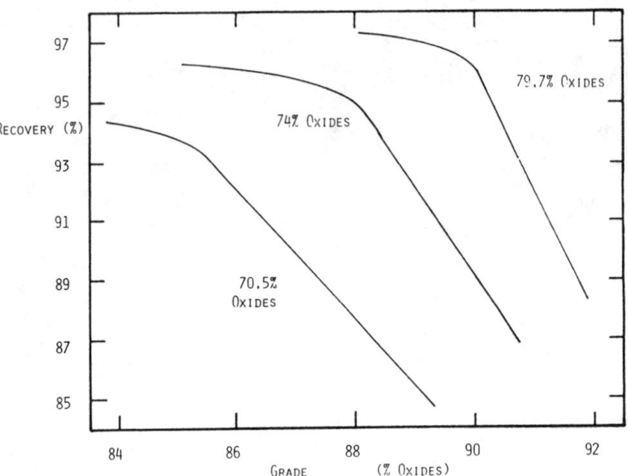

Fig. 6. Simulation results of the three-phase model

SECOND CASE STUDY: REICHERT TRAYS FOR
FINE HEMATITE SCAVENGING

This second case study considers the recovery of fine hematite lost in
the tailings of the Mount Wright mill of Mines Québec Cartier. The
losses are largely in the -100 μm because of the inherent inefficiency
of the GEC spirals to recover fines (Chong, 1978). The proposed
recovery method was upgrading with three stages of Reichert cones
followed by cleaning with spirals designed to recover fines. As a cone
will handle 60 to 100 tonnes per hour, it is hardly suitable for
pilot-plant work. The Reichert tray, which is a succession of pinched
sluices, each one twentieth of a full cone, offers the advantage of
reduced feedrate and multistage separation.

Details of the testwork and subsequent simulation are in Laplante, Nudo
and del Villar (1987). Testwork was labour-intensive and time
consuming, partly because three stages of 2DS upgrading were performed.
Thus, about 2 tonnes of concentrate had to be produced in the first
stage to provide feed material for the second and third stage.

The separation of each tray was modelled using the following empirical
equation:

$$R = b0 + b1 * D + b2 * D^2 + b3 * Q + b4 * Q^2 + b5 * (Q/S) \qquad (2)$$

where R: recovery (%)
    d: percent solids of the feed (v/v)
    Q: volumetric flowrate (L/min)
    S: slot setting (9 to 5)
b0,..,b5: model parameters

The simulator uses Eq. 2 in each of the separation stages. In a
typical 3-stage circuit of 4DS cones, there is 24 such stages. For
each stage, the recovery of hematite, silica, and water is computed as
a function of the solids and water rate, and concentrate slot setting.
The simulator was found to predict both upgrading and recovery
reliably. The question is: could similar results have been achieved
with a simpler approach? This would consist of a single upgrading
stage, a simpler sampling procedure, simpler data analysis, and a
simpler simulator. Table 1 compares the simpler approach to that which
was used by Laplante, Nudo and del Villar (1987).

TABLE 1 Comparing the Original and Simplified Approaches
          to Simulation for the Reichert Tray

ORIGINAL	SIMPLIFIED
EXPERIMENTAL	
Three stages are tested	Only the first stage is tested
All streams (8) sampled	Only the feed, total concentrate and second double tail sampled
DATA PROCESSING	
Recovery of each of the four stages calculated	Only overall recovery calculated
Effect of the three input variables included	Recovery-recovery curve used
SIMULATOR	
Complete freedom to choose circuit and slot settings	Each cone a 4DS, 2DS for concentrate, 2DS for middlings
Recovery of hematite, silica and water computed for each stage	recovery-recovery curve used

First, the overall performance of the 2DS separator for the first

separation stage was computed, as shown in Fig. 7. The correlation
between silica and hematite recovery is good, and can be fitted to:

$$Rs = 0.1994 - 0.6551 * Rh + 1.3911 * Rh^2 \qquad (3)$$

where   Rh: fractional hematite recovery
        Rs: fractional silica recovery

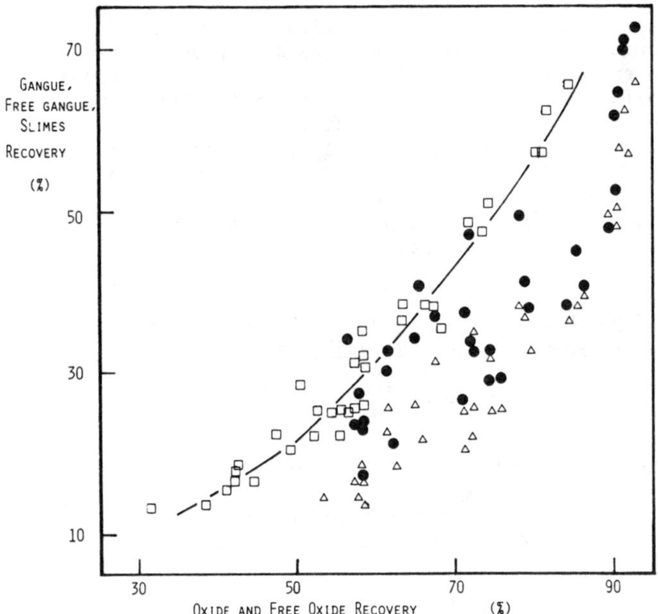

Fig. 7. Gangue recovery vs. oxide recovery (squares);
         float (triangles) and slimes (circles) recovery
         vs. sink recovery

Figure 7 also shows float and slimes recovery as a function of sink
recovery (slimes: -37 µm; float: -3.3 s.g. in the +37 µm; sink: +3.3
s.g. in the +37 µm). Data are more scattered, although they represent
a more phenomenological relation; thus, the hematite-silica relation
was retained for the simple model. The simulator considers each cone
to be a 4DS, the first 2DS unit producing a concentrate for the next
stage and the second 2DS a middling recycled to the head of the same
cone. Figure 8 summarizes how the whole circuit is simulated.
Recovery for each stage is function of recovery for its two 2DS units.
                        For example,

$$R1 = \frac{r1}{1 - r2 * (1 - r1)} \tag{4}$$

where R1: fractional recovery in the first
         4DS cone (rougher)
      r1: fractional recovery in the first 2DS
         of the rougher cone
      r2: fractional recovery in the second 2DS
         of the rougher cone

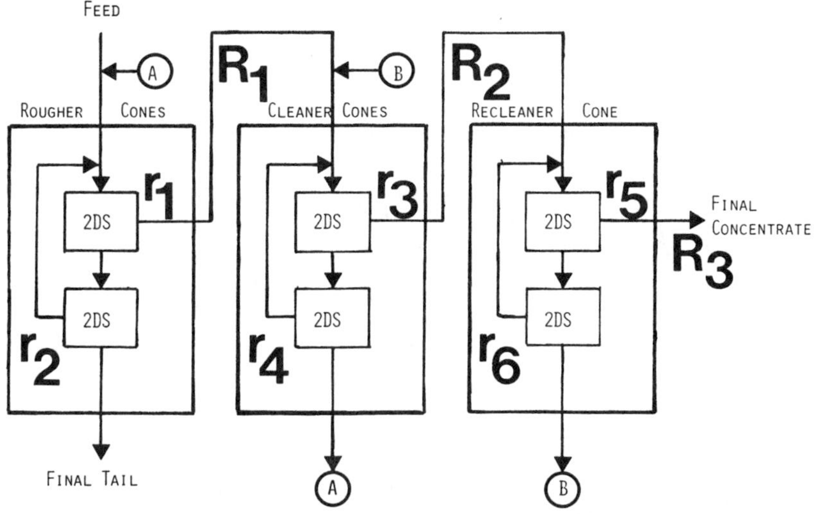

Fig. 8.   Schematic of the circuit model

Overall recovery, Rt, is equal to:

$$Rt = \frac{R1 * R2 * R3}{[1 + R2 * (R1 + R3) - R1 - R2]} \tag{5}$$

where R1, R2, and R3 are the recoveries of the rougher, cleaner, and
recleaner cones, respectively. Figure 9 compares simulation results
for the original (Eq. 2) and simplified (Eqs. 3-5) models. For the
original model, results are presented as a function of feedrate and
slot setting, using 5 rougher trays, 2 cleaner trays, and 1 recleaner
tray. Details of the slot settings are shown in Table 2.

TABLE 2   Slot Settings for the Original Model
(1: maximum opening; 9: minimum opening
First number: double cone;
Second number: single cone
A: first 2DS; b: second 2DS)

	First Cone		Second Cone		Third Cone	
	A	B	A	B	A	B
Curve 1	8-9	5-7	8-9	6-8	8-9	7-9
Curve 2	7-9	5-7	7-9	6-8	7-9	7-9
Curve 3	6-8	5-7	6-8	6-8	6-8	6-8
Curve 4	5-3	5-3	5-3	5-3	5-3	5-3

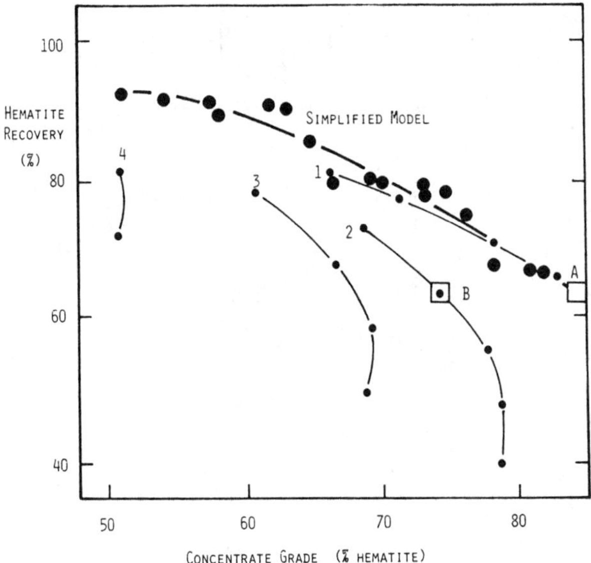

Fig. 9. Comparing the original (curves 1 to 4, Table 2) and
simplified (large circles) models

For the simplified model, results are based on an evaluation of all
possible combinations of r1 to r6 at the following hematite recoveries:
40, 55, 70, and 85%. The simplified model yields an unconstrained
solution to the overall recovery problem -i.e. it assumes that a
combination of slot setting, pulp density, and throughput can always be
found to achieve the optimum vector of 2DS recoveries selected. Thus,
it gives an ideal 'best result', which may or may not be achievable in
practice. In this case, the result can be achieved with a relatively

selective operation, as shown in Table 3, which gives the 2DS hematite recoveries for each point on the grade-recovery curve. It shows that recovery should be minimum in the first cleaner and recleaner 2DS, and maximum in the second rougher 2DS. Moving along the optimum grade-recovery curve is achieved by changing recovery mostly in the first rougher 2DS, and to a lesser extent in the second cleaner 2DS. Figure 9 shows that the optimum line has a lot of scatter, which could be reduced by defining more points on the recovery-recovery curve.

TABLE 3   Details of the Optimum Grade-Recovery Curve

Grade (%)	Recov. (%)	STAGE RECOVERIES (%)					
		1	2	3	4	5	6
51.2	86.9	85	70	40	55	40	55
54.2	85.9	55	85	40	85	55	40
57.4	85.7	55	85	40	85	40	70
58.1	84.7	85	55	40	55	40	40
61.1	85.5	55	85	40	85	40	55
63.0	85.1	55	85	40	85	40	40
64.8	82.7	55	85	55	55	40	40
66.8	80.8	55	85	55	40	40	40
69.2	80.3	55	85	40	70	40	55
70.1	79.7	55	85	40	55	40	70
73.4	79.3	55	85	40	70	40	70
74.9	78.1	55	85	40	55	40	40
76.3	75.0	55	85	40	40	40	40
78.2	68.1	40	85	40	55	40	70
81.0	67.6	40	85	40	55	40	55
82.1	66.1	40	85	40	55	40	55

DISCUSSION

Use of Simplified Models

For many applications, simplified models seek not to replace but to supplement more phenomenological models. In the second case study, trying to optimize separation with the original model proved awkward. In retrospect, it would have been easier with the information generated by the simplified model. Knowledge of the unconstrained grade-recovery curve also provides a standard to determine if, for given concentrate grade, the chosen circuit and operating conditions yield a recovery close to what may be expected in the best of circumstances. For example, point A in Fig. 9 was achieved with a circuit of 5 roughers, 2 cleaners and 1 recleaner trays. Point b was achieved with a circuit of 2 roughers, 2 cleaners and 1 recleaner. Both circuits were operated at different loadings and slot positions to optimize recovery. Clearly, the first circuit is near optimal, whereas the second is not.

Agar and Kipkie (1978) propose a simple simulator to evaluate the steady-state metallurgy of a locked cycle test from the results of an open cycle test. They use split factors rather than recovery-recovery

curves, which means they choose a single point of the recovery curve of
each stage. The predictor is not used to replace locked cycle tests,
but rather to ascertain whether these costly tests can be justified.
The simplified models could be used for similar exploratory work. They
may also help choosing the type of more evolved testwork or model to
use. In the first case study, test results suggest that cleaner
selectivity could be improved to further increase grade along the first
segment of the grade-recovery curve, where significant increases in
concentrate grade can be achieved at only a small drop in recovery.

Simplified models can also be used to emphasize specific effects.
Finch and Matwijenko (1977) retained a simplified model of breakage
phenomena to focus on the differential breakage rate and classification
behaviour of various minerals. The same exercise with the classical
breakage model would have been tedious.

When technology transfer is the objective, simplified models can prove
invaluable. Plant personel is far more receptive to simple models,
because they require less effort in data generation and are easier to
master.

Limitations of Simplified Models

Simplified models work because they are based on data derived from the
circuit they seek to model. Looking at Fig. 2, it becomes obvious that
if performance for a rougher-cleaner circuit had been used to predict
that of the scavenger stage, and hence that of the existing circuit,
predictions would have been in serious error. Simple models do not
explicitly incorporate important variables such as size distribution or
mineral liberation, but their effect is implicit in the performance
curves used for simulation. Whenever these are substantially changed,
predictive errors can occur. Thus, the use of simplified models should
be limited to operating conditions and circuits from which they are
derived.

In the first case study, scavenger recovery could be improved
substantially by increasing slurry density. The effect of such an
increase may be difficult to model with the simplified models, which do
not incorporate the effect of particle size. Figure 10 shows overall
circuit recovery of oxides as a function of particle size. Recovery
below 100 μm decreases steadily with size; this is an indication that
fine losses in the roughers are important. Improving scavenger
recovery would contribute to an increased amount of fines to the
roughers, which would eventually create a circulating load of fine
liberated oxides. This load would alter the size distribution of the
oxides in the rougher and cleaner feeds, and modify the
recovery-recovery curve.

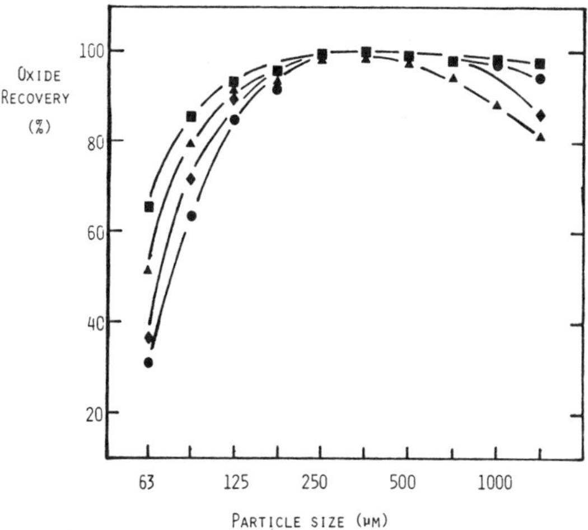

Fig. 10. Oxide recovery vs. size for the first case study

In the second case study, both the effect of particle size and
liberation are neglected. The feed material used is well liberated, as
is shown in Table 4, which presents heavy liquid test results on the
concentrate of the third stage of tray tests.

TABLE 4   Heavy Liquid Separation of the Tray Concentrate

Product	%Mass	%Fe	%Rec. Fe
Float s.g. 2.96	52.5	2.5	4.3
Float s.g. 3.12	3.0	20.6	2.1
Float s.g. 3.59	3.5	57.6	6.7
Float s.g. 4.12	1.6	38.3	2.6
Sink s.g. 4.12	39.1	64.8	84.4

The effect of particle size is small because the study focussed on lost
hematite in the range of 37-105 µm, for which Reichert cone and tray
recovery is approximately constant. Below 37 µm, recovery decreases
strongly with decreasing particle size, an effect which would have to
be incorporated in the simulator.

CONCLUSION

The two case studies illustrate that simplified models can be used
either when the data base is inadequate for more comprehensive models,
or as an aid in their development and use. Their reliability relies
heavily on use close to the data base from which they are derived.
This restriction is particularly tight when liberation or particle size
effects are important.

ACKNOWLEDGEMENTS

The authors wish to thank les Mines Québec Cartier and QIT - Fer et
Titane for technical and financial support.

REFERENCES

Agar, G.E. and W.B. Kipkie (1978). Can. Inst. Min. Metall. Bull.,
   Nov.. pp. 119-125
Burt, R. and C. Mills (1984). Gravity Concentration Technology.
   Adv. Ed., D.W. Fuerstenau, Developments in Mineral Processing, 5,
   Elsevier Scientific, Amsterdam
Canas, F. (1980). Chem. Engng. April, pp. 95-96
Chong, S.P. (1978). Mining Engrs.. Dec., pp. 1639-1643
Finch, J.A., and O. Matwijenko (1977). Can. Inst. Min. Metall. Bull.,
   Nov., 9p.
Holland Batt, A.B. (1979). Design of Gravity Concentration Circuits by
   use of Empirical Mathematical Models. Proc. XI Commonwealth Min.
   Metall. Cong., Hong Kong, 1978, London, The Instn. Min. Metall.,
   pp. 133-143
Laplante, A.R. and J.A. Finch (1984). Intern. J. Min. Proc.. 13,
   pp. 1-11
Laplante, A.R., V. Nudo, and R. del Villar (1987). Proc. 19th Ann.
   Meet. Can. Min. Proc., Paper 28, pp. 752-786.
Laplante, A.R., J.A. Finch, and R. del Villar (1987). Trans. Instn.
   Min. Metall. (Sect. C: Mineral Process. Extr. Metall.), 96.
   C108-112
Lynch, A.J. (1977). Mineral Crushing and Grinding Circuits. Their
   Simulation, Optimization, Design and Control. Vol. 1, Developments
   in Mineral Processing, D.W. Fuerstenau, adv. ed., Elsevier
   Scientific, Amsterdam.
Lynch, A.J., N.W. Johnson, E.V. Manlapig, and C.G. Thorne (1981).
   Mineral and Coal Flotation Circuits. Their Simulation and
   Control. Vol. 3, Developments in Mineral Processing, D.W.
   Fuerstenau, adv. ed., Elsevier Scientific, Amsterdam.
McIvor, R.E. (1984).CIM Bull., 77(872), pp. 50-53
Subasinghe, G.K.N.S., and E.G. Kelly. Control '84. Ed. J.A. Herbst,
   1984.
Tucker, P., K.A. Lewis, W.J. Hobba, and D. Wells (1985). A
   mathematical model of spiral concentration, as part of a generalised
   gravity-process simulation model, and its application at two Cornish
   tin operations. Proceedings 15th Intern. Min. Proc. Cong., 3-15.

SOLUTIONS TO THE SYSTEM OF BATCH BALL MILL GRINDING EQUATIONS

E. M. Sanchez* and W. Amauta**

*Salt Lake City, Utah

**Post-graduate Student, University of Oruro, Bolivia

ABSTRACT

Batch ball mill grinding has been described, for most engineering applications, by
a mathematical model formulated in a particle size-discrete time-continuous domain
as a system of ordinary differential equations. Two analytical methods to solve
this system are reviewed. Also two numerical methods are applied for the solution, and a method based on the approximation of a well mixed batch grinding
process by a N-mixer-in-series model. The solutions are implemented using a
computer, tested, and compared with actual experimental results. Practical
engineering implications are discussed.

KEYWORDS

Ball mill; batch grinding; grinding model; breakage and selection function
parameters; vector; matrix; system of initial value ordinary differential
equations; closed form of analytical solution; numerical solution; parameter
estimation; Newton-Raphson optimization program.

INTRODUCTION

The grinding process occurring in a ball mill is complex not only due to the wide
range of physical and physicochemical properties of the multiphase multiparticu-
late material undergoing size reduction, but also due to its dependence on the
type of fracturing forces that are generated by the ball load during the operation
of the mill. For the past several decades ball mill grinding was described by
simple particle size-energy models (Bond, 1952, 1962; Charles, 1957; Agar and
Brown, 1962), which roughly reflected the size reduction process occurring in a
mill. With the advent of the digital computer the trend has been toward the
introduction of more modelling details such as (i) the use of several mass size
fraction intervals to characterize the ore particulate material in the mill;
(ii) formulation of size reduction subprocess for particles in each size interval;
and (iii) formulation of particulate material transport through the mill in case of
continuous grinding operation. Now that the digital computer is a common
engineering tool, further refinement of grinding models is still in progress and
is limited only by the complexity of mathematical algorithms that can still
meaningfully reflect the physical features of the process, and by computational

time, which can be critical for modelling applications such as to process control. Much of the knowledge of ball mill grinding systems has been achieved through models formulated on population balance principles. The physical significance of these models has become a solid framework for understanding the size reduction process occurring in a mill.

For the case of batch grinding, a population balance model, formulated as a system of initial value ordinary differential equations in the particle size-discrete time-continuous domain has been found useful for practical applications, e.g. (Reid, 1965; Herbst and Fuerstenau, 1968; Austin, 1971-72). This is probably due to the fact that ore particulates are usually characterized on a size discrete basis. In this paper, two analytical forms of solution to the above system of initial value ordinary differential equations are reviewed. Alternative solutions based on numerical methods are also presented, and a solution based on the analogy of a well mixed batch grinding process with a plug flow process, which is described by a N-mixer-in-series model. Finally, the potential and implications of the size-discrete time-continuous batch grinding model for the understanding of industrial grinding systems is discussed.

PARTICLE SIZE-DISCRETE TIME-CONTINUOUS BATCH GRINDING MODEL

Following Reid's formulation (1965), the mass balance in the particle size-discrete time-continuous domain for the 1st, 2nd, 3rd, and ith size intervals can be written as follows:

$$dm(1)/dt = -s(1)m(1)$$

$$dm(2)/dt = -s(2)m(2) + b(2,1)s(1)m(1)$$

$$dm(3)/dt = -s(3)m(3) + b(3,1)s(1)m(1) + b(3,2)s(2)m(2)$$

$$\cdots \qquad \cdots \qquad \cdots \qquad \cdots \tag{1}$$

$$dm(i)/dt = -s(i)m(i) + \sum_{j=1}^{i-1} b(i,j)s(j)m(j)$$

subject to

$$\text{I.C.} \quad t = 0; \quad m(i;t) = m(i;0), \quad i = 1,2,3, \ldots n$$

where $m(i)$ denotes the mass fraction in the ith size interval at time t; n the number of size intervals; $s(i)$ the ith size discretized selection function, or the rate at which particles are broken out of the ith size interval; and $b(i,j)$ the size discretized breakage function, or the fraction of particles from the jth size interval reporting to the ith size interval as a result of primary breakage events.

SOLUTIONS TO THE SYSTEM OF BATCH GRINDING EQUATIONS

With the assumption that the breakage and selection function parameters are constants, the first equation of the system of batch grinding equations (1) can be readily solved for m(1), which can then be replaced in the second equation to solve for m(2), and so on. Applying this recursive procedure the following solution to the system of batch grinding equations was found (Reid, 1965)

$$m(i) = \sum_{j=1}^{i} a(i,j)\exp(-s(j)t) \tag{2}$$

where

$$a(i,j) = \begin{cases} 0; & i < j \\[2ex] m(i;0) - \sum_{k=1}^{i-1} a(i,k); & i = j \\[2ex] \left(\sum_{k=j}^{i-1} b(i,k)s(k)a(k,j)\right)/(s(i)-s(j)); & i > j \end{cases}$$

The system of equations (1) in a more compact notation can be written as follows:

$$d(\underline{m})/dt = (I - B)\, S\, \underline{m} \tag{3}$$

subject to

$$\text{I.C. } t = 0; \quad \underline{m}(t) = \underline{m}(0)$$

where $\underline{m}$ denotes the vector of n mass fractions at time t; and I, B, and S denote the identity matrix, breakage function matrix, and selection function matrix of order n, respectively. Note that (I - B)S is a matrix of order n that through similarity transformation can be transformed to a diagonal form. Then the ordinary differential equations denoted by Equation (3) are unlocked leading to the following analytical solution (Herbst and Fuerstenau, 1968)

$$\underline{m} = T\, J(t)\, T^{-1}\, \underline{m}(0) \tag{4}$$

where

$$J(t) = J \text{ is a diagonal matrix with elements:}$$

$$J(i,i) = \exp(-s(i)\,t), \quad i = 1,2,\ldots,n$$

$$T(i,j) = \begin{cases} 0; & i < j \\[4pt] 1; & i = j \\[4pt] \left(\sum_{k=1}^{i-1} b(i,k)\,s(k)\,T(k,j)\right)/(s(i)-s(j)); & i > j \end{cases}$$

Numerical methods can also be applied to solve the above system of batch grinding equations. For instance, Heun's method and the 4th order Runge-Kutta method were used in this paper to solve the initial value problem given by Equation (1). These methods are treated in detail in Numerical Analysis textbooks, e.g. (Burden, Faires and Reynolds, 1978), and are not described here.

The N mixer-in-series solution of the system of batch grinding equations (1) is based on the analogy of a well mixed batch process with a plug flow process, which can then be approximated by a N mixer-in-series model with a sufficiently large number of perfect mixers N. The mass fraction for the ith size interval in the product from the kth mixer is given by:

$$m(i;k) = \left(mf(i;k) + \sum_{j=1}^{i-1} b(i,j)m(j;k)t(k)\right)/(1+s(i)t(k)) \tag{5}$$

$$i = 1,2,\ldots,n; \quad k = 1,2,\ldots,N;$$

where $mf(i;k)$ denotes the mass fraction of the ith size interval in the feed to the kth mixer; and $t(k)$ the mean residence time in the kth mixer. The implication of an analytical approach of this solution has been discussed in detail (Morozov and Shumailov, 1983), and is not given here.

MODEL PARAMETERS

The number of size discretized breakage and selection function parameters, which characterize the fracture properties of the material undergoing size reduction, have been found, for most mineral particulates of uniform fracture properties, to be generated by functional expressions dependent on fewer parameters. Most investigators, e.g. (Herbst and Fuerstenau, 1973; Herbst, Rajamani and Kinneberg, 1977), found the following correlations sufficiently accurate:

$$B(i,j) = \alpha_1 (x(i)/x(j+1))^{\alpha_2} + (1-\alpha_1)(x(i)/x(j+1))^{\alpha_3} \tag{6}$$

$$b(i,j) = B(i,j) - B(i+1,j) \tag{7}$$

$$B(i,j) = \sum_{k=1}^{i} b(k,j) \tag{8}$$

where α_1, α_2, and α_3 are the breakage function correlation parameters.

$$s(i) = s(1)\exp\left(\sum_{k=1}^{K} z(k)(k \ln(SQRT(x(i)x(i+1)))/(SQRT(x(1)x(2)))\right)$$

$$k \leq 2; \quad i = 1,2,....,n-1 \tag{9}$$

where $s(1)$, $z(1)$, and $z(2)$ are the selection function correlation parameters, and
$K = 1$ or 2 depending on the ratio of the largest particle size to maximum ball
size.

The above breakage and selection function correlation parameters are usually
determined fitting the grinding model to a set of batch grinding results, using a
nonlinear optimization algorithm.

BATCH BRINDING SIMULATION

Computer programs for batch grinding simulation were designed based on the
solutions to the system of batch grinding equations already described. The programs were coded in Fortran language and executed in a personal computer to
simulate actual batch grinding experimental results. For this purpose, two sets
of limestone batch grinding experimental results were selected (Siddique, 1977):
namely, (i), a set of 10x14-mesh limestone grinding results with 0.5, 2, and 4
minute grind times, and (ii), a set of minus 10-mesh limestone grinding results
with 0.5, 2, and 4 minute grind times. As described by Siddique (1977) a
25.40x29.21 cm ball mill was used in the batch experiments at the following
operating conditions: 60% solids, 60% of critical speed, 50% of mill volume
filling with a seasoned minus 3.81 cm ball size distribution. The number of size
discretized intervals used in characterizing the particulate limestone was 12,
i.e., from 10 mesh to 400 mesh, following the square root of two sieve series.

The set of 10x14-mesh grinding results was used to simultaneously estimate the
limestone breakage and selection function parameters, using a numerical Newton-
Raphson nonlinear optimization program developed for this purpose (Sanchez and
Amauta, 1988). After estimating the breakage function parameters, the set of
minus 10-mesh grinding results was used to estimate the optimum limestone
selection function parameters, using the previously mentioned numerical Newton-
Raphson parameter optimization program in its selection function parameter estimation mode. The optimum breakage and selection function parameters were used in
conjunction with batch grinding simulation programs, which were designed based on
the solutions already described, to simulate the minus 10-mesh limestone batch
grinding results already mentioned. It should be pointed out that the numerical
Newton-Raphson parameter optimization program was designed to work, among other
algorithms, in conjunction with Reid's analytical solution. Since the most
accurate batch grinding simulation should be that based on Reid's solution,
grinding simulation results based on this solution were taken as reference. The
adequacy of prediction was measured in terms of the mean square of residuals deviation (RMS), and the residual sum of squares (RSS) given as follows:

$$RSS = \sum_{k=1}^{nx} \sum_{i=1}^{n} (m(i,k), model - m(i,k),exp.)^2 \tag{10}$$

$$RMS = SQRT(RSS/(n\ nx)) \tag{11}$$

where nx denotes the number of batch grinding experiments in the set. A summary of the comparison of experimental and model predicted product mass size fractions, based on the solutions to the system of batch grinding equations described in this paper, is given in Table 1.

TABLE 1   Comparison of Model Simulated Product Mass Size Fractions with Those Determined from Experiments Carried Out in a 25.40x29.21 cm Ball Mill, Using a Minus 10-Mesh Limestone Feed and 0.5, 2.0, and 4.0 Minute Grind Times

Solution	RSS	RMS	Remarks
Reid	.9831E-3	.522584E-2	Data fitting
Sim. Transf.	.9831E-3	.522585E-2	
Heun	.9965E-3	.526130E-2	20 Intervals
	.9679 "	.518416 "	50 "
	.9694 "	.518919 "	70 "
	.9722 "	.519668 "	100 "
	.9769 "	.520927 "	200 "
	.9805 "	.521875 "	500 "
	.9818	.522222 "	1000 "
Runge-Kutta, fourth order	.9821E-3	.522298E-2	5 Intervals
	.9831 "	.522570 "	10 "
	.9831 "	.522584 "	20 "
	.9831 "	.522585 "	30 "
	.9831 "	.522585 "	50 "
N-mixer	.5652E-2	.125304E-1	1 mixer
	.1396 "	.622628E-2	5 "
	.1122 "	.558288 "	10 "
	.9878E-3	.526472 "	50 "
	.9896 "	.524291 "	100 "
	.9861 "	.522887 "	200 "
	.9843 "	.522887 "	500 "
	.9839 "	.522799 "	700 "
	.9837 "	.522734 "	1000 "

DISCUSSION

From work carried out in this study, and from consideration of the computational results shown in Table 1, the following is inferred: (i) both the batch grinding simulations carried out in a computer with algorithms based on the closed form of analytical solution of Reid, and the similarity transformation method have the same predictive accuracy; however, the latter is less efficient than the former due to additional computation required with the similarity transformation matrix and its inverse, and additional matrix operations; (ii) batch grinding simulations carried out with algorithms based on Heun and fourth order Runge-Kutta numerical methods have unstable solutions below a certain critical time interval, which is 12 seconds for the fourth order Runge-Kutta method. Heun's solution never overcomes this instability, and still converges extremely slowly to the analytical solution, even for time intervals as short as 0.08 second. Therefore, numerical methods with equal or more accuracy than the fourth order Runge-Kutta method should be applied in batch grinding simulation; (iii) computer simulations carried out with the algorithm based on the N-mixer-in-series solution converge slowly to the analytical solution for increasing values of N. For N = 1000, the solution is still converging to the analytical solution. Batch grinding simulations based on the fourth order Runge-Kutta solution are more efficient and more accurate than that based on the N-mixer-in-series solution.

It should be pointed out that the analytical solutions of Reid and similarity transformation fail when there are two or more equal size discretized selection function parameters, i.e., when $s(i) = s(j)$, $i \neq j$. Herbst and Kim (1973) showed that the solution is still unstable for nearly equal selection functions in the range of ± 0.0001. In this case, any sufficiently accurate numerical method can be used to solve the problem. On the other hand, the probability of having two identical size-discretized selection functions is extremely low, and the analytical solution is to be preferred due to its computational efficiency.

It is important to point out that the solutions reviewed assume the invariability of the breakage and selection function parameters with the size distribution of the material in the mill, i.e., linear breakage kinetics. Although dry grinding, and short grind time wet ball mill grinding, of a predominantly homogeneous particulate material at a normal range of operating conditions, can be approximately described by a linear model, in general, ball mill grinding is a nonlinear process. This is not encouraging, since most of the available programs for simulation of grinding circuits are based on linear models. This means that in practical circuit design applications serious errors can be introduced in the analysis, or design of grinding circuits. If, for instance, a linear ball mill grinding model is used in conjunction with a hydrocyclone model to simulate several circuit configurations of ball mills and hydrocyclones, the simulation of a nonlinear ball mill grinding process by a linear model can lead to appreciable simulation errors; therefore, making the choice of an appropriate circuit configuration more difficult. For the case of grinding systems with a not so pronounced kinetic nonlinearity the linear model has been used within a concept of similarity of fineness approach to successfully analyze industrial grinding systems (Siddique, 1977; Herbst and co-workers, 1982, 1985). A drawback of this approach is that it involves intensive laboratory work. For the case of strictly nonlinear kinetics the system of batch grinding equations (1) may well take an inappropriate form for an analytical solution. In this case, a numerical method can always be used to find an approximate solution, provided the nonlinear function of the kinetic parameters is known.

In summary, computer simulations with algorithms based on the solutions to the system of batch grinding equations presented in this paper suggest the following: (i) grinding simulations based on the closed forms of analytical solution, and the fourth order Runge-Kutta solution are sufficiently accurate and computationally efficient; (ii) batch grinding simulations based on Heun's method and the N-mixer in series solution, using both a larger number of time intervals and a large number of mixers, N, respectively, converge extremely slowly to the analytical solution. This makes the solutions based on these methods inefficient.

To conclude, for computer simulations with the particle size-discrete time-continuous batch grinding model, based on any of the solutions presented in this paper, linear size reduction kinetics for the material should be experimentally verified, so that the model can be meaningfully used in realistic grinding simulations.

REFERENCES

Agar, G. E. and J. H. Brown (1964). Energy requirements in size reduction. Can. Min. and Met. Bull., 57, No. 622, 147-151.

Austin, L. G. (1971-72). A review introduction to the description grinding as a rate process. Powder Tech., 5, 1-7.

Bond, F. C. (1952). The third theory of comminution. Trans. AIME, 193, 484-494.

Bond. F. C. (1962). Crushing and grinding calculations. Allis Chalmers Publication, Wis., Milwaukee.

Burden, L. R., J. D. Faires and A. C. Reynolds (1978). Numerical Analysis, Prindle, Weber & Schmidt, Boston.

Charles, R. J. (1957). Energy-size reduction relationships in comminution. Trans. AIME, 208, 80-88.

Herbst, J. A. and D. W. Fuerstenau (1968). The zero order production of fine sizes in comminution and its implications in simulations. Trans. AIME, 241, 538-548.

Herbst, J. A. and H. I. Kim (1973). Nearly equal selection functions in linear size-discretized grinding models, Trans. IMM Sec.C, 82, C169-C173.

Herbst, J. A. and D. W. Fuerstenau (1973). Mathematical simulation of dry ball milling using specific power information, Trans. AIME, 254, 343-348.

Herbst, J. A., K. Rajamani and D. J. Kinneberg (1977). Estimill: a program for grinding simulation and parameter estimation with linear models. Program description and user manual., University of Utah, Salt Lake City, Utah.

Herbst, J. A., M. Siddique, K. Rajamani and E. Sanchez (1982). Population balance approach to ball mill scale-up: bench and pilot scale investigations, Trans. AIME, 272, 1945-1954.

Herbst, J. A., Y. C. Lo and K. Rajamani (1985). Population balance model predictions of the performance of large diameter mills, Min. & Met. Processing, May, 114-120.

Morozov, E. F. and V. K. Shumailov (1983). Modified solution of the batch grinding equation, Sov. Min. Science, 19, 43-47.

Reid, K. J. (1965). Solutions to the batch grinding equation, Chem. Eng. Sci., 20 , 953-963.

Sanchez, E. M. and W. Amauta (1988). Numerical Newton-Raphson model parameter estimation program, in preparation.

Siddique, M. (1977). Ball mill grinding kinetics., M.S. thesis, University of Utah, Salt Lake City, Utah.

MICROSEGREGATION IN CAST ALLOYS

Thomas P. Battle and Robert D. Pehlke

Department of Materials Science and Engineering
The University of Michigan
Ann Arbor, MI 48109-2136

ABSTRACT

The importance of microsegregation in solidifying metals is discussed, with particular emphasis on the mathematical models which have been developed to help in the determination of solute concentration profiles and second phase contents after solidification. The predictions of a recently-developed numerical model are compared with other models and experimental data in the Al-Cu system. Results indicate that only the two numerical models were able to reproduce the amount of non-equilibrium second phase accurately. A plot is presented by which the amount of second phase can be estimated for hypoeutectic alloys as a function of cooling rate.

KEYWORDS

Microsegregation; solidification; solute partitioning; coarsening; castings

I. INTRODUCTION

Microsegregation is a phenomenon well-known in the metals industry. Most castings show some evidence of this fine-scale solute segregation, whether or not it alters the physical properties of the product. The compositional inhomogeneities of microsegregation can be manifested as concentration gradients of solute on a micron scale, or it can result in the formation of a second phase or as porosity. Microsegregation is an important problem in many metallic systems, including steels (Hammer and Grunbaum, 1974; Rickinson and Kirkwood, 1979), nickel-based superalloys (Loria, 1988), aluminum alloys (Hunsicker, 1967), and copper anodes (Forsen, Hettula, and Lilius, 1985).

The basic causes of microsegregation are well known. Referring to the schematic phase diagram of the Al-Cu system (Fig. 1), a liquid alloy with a given bulk composition will cool until the liquidus temperature is reached. At this time, solid will begin to precipitate. The concentration of solute in the first-formed solid will depend on the thermodynamics of the system - essentially, the affinity of the solute species (copper) for the solvent (aluminum). In most cases the solute will prefer to be in the liquid, which has a more open structure than the solid and can more easily accommodate atoms of a different size. In systems with like atoms, such as Fe-Ni, little partitioning occurs; the equilibrium partition coefficient, which is defined as the ratio of solute concentration in the solid to that in the liquid at equilibrium during solidification, is near unity. On the other hand, copper is a much larger atom than aluminum and is much more stable in the liquid phase - thus, the partition coefficient of copper in aluminum is closer to 0.2 than to 1.

Regardless of the degree of partitioning during solidification, the presence of microsegregation after the completion of cooling depends mainly on two other factors - the solute diffusivity in the solid state and the local cooling rate. Equilibrium cooling (as demonstrated by the phase diagram) assumes very rapid solid-state diffusion or infinitely slow cooling. The result is that all concentration gradients developed during solidification disappear instantaneously, and the alloy below its solidus temperature is homogeneous. Mass transfer during equilibrium freezing is mathematically described by the lever rule, equation (1),

$$C_L^i = \frac{C_o}{f_s(k-1)+1} \tag{1}$$

where C_o is the bulk solute concentration, f_s the fraction solid, k the partition coefficient, and C_L^i the liquid concentration at the solid-liquid interface. Under most casting conditions, equilibrium cooling does not take place, except perhaps where the solute is interstitial and diffuses rapidly through the solid (such as carbon in ferritic iron). For the last forty or fifty years segregation models more realistic than equilibrium freezing have been developed by several investigators. In the next section, several of these models will be discussed.

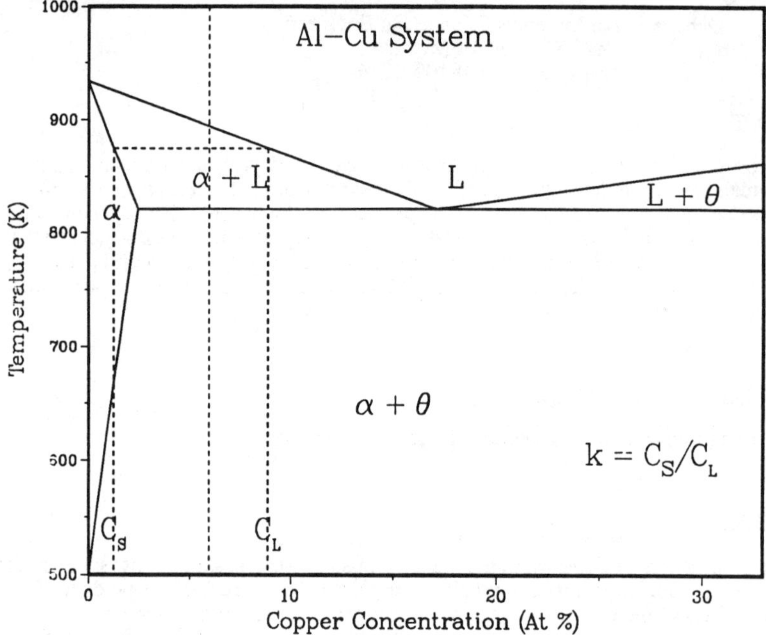

Fig. 1. Aluminum-rich end of Al-Cu phase diagram

II. REVIEW OF MASS-BALANCE MICROSEGREGATION MODELS

Equilibrium freezing represents one limiting case for microsegregation in the absence of liquid-phase concentration gradients. The other limiting case is described by the Scheil equation, which assumes no solid-state diffusion whatsoever (equation (2)).

$$C_L^i = C_o(1-f_s)^{k-1} \tag{2}$$

In the past twenty years there have been several attempts to create a model which can accurately calculate concentration profiles for those systems that are not at either limit. The physical situation being modelled is depicted in Fig. 2, which shows a metal solidifying dendritically. Primary dendrites are growing opposite the direction of heat flow, with secondary arms at right angles to the primaries. The models have been essentially one-dimensional, considering a region extending from the center of a dendrite arm to a point halfway to the next arm (this can be either the secondary arm - as pictured here - or the primary). This one-dimensional region is shown in Fig. 3. No mass transfer in or out of the system boundaries is allowed.

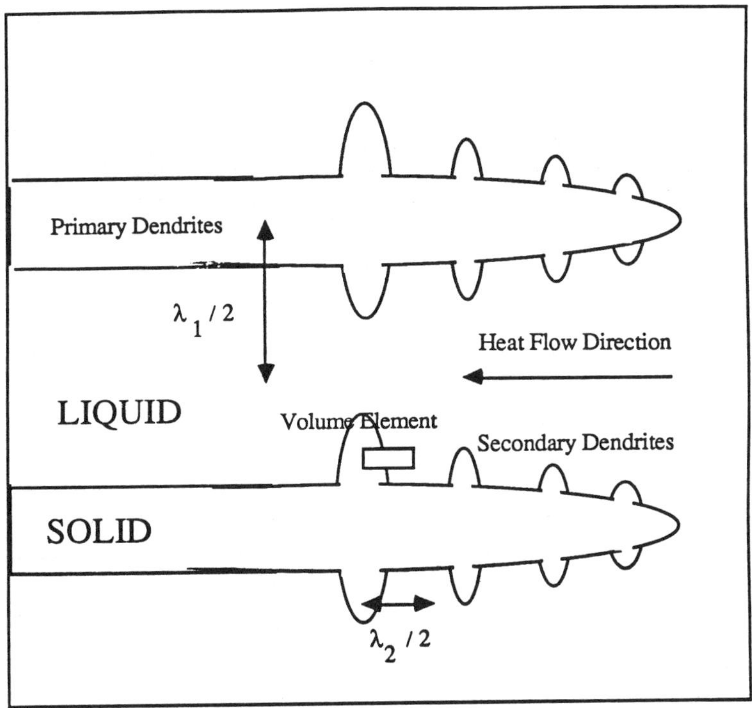

Fig. 2 Schematic diagram of solidifying region of a casting

Many of the mass-transfer models begin with a mass balance equation of the form of equation (3) (Ogilvy and Kirkwood, 1987),

$$C_L (1 - k) \frac{dh}{dt} = D_S (\frac{\partial C_S}{\partial x})_h + \frac{dC_L}{dt} (\frac{\lambda_2}{2} - h) + \frac{1}{2} (C_L - C_0) \frac{d\lambda_2}{dt} \qquad (3)$$

where h is the position of the interface, D_S the solute diffusion coefficient in the solid state, and λ_2 the secondary dendrite arm spacing. Physically, this describes mass transfer during solidification by assuming that the mass rejected by the growing solid partitions in one of three ways - back-diffusion into the already-existing solid, diffusion into a liquid of uniform concentration, and diffusion into a segment of the system newly created as a result of dendrite-arm coarsening. This equation, with the removal of the coarsening term, is identical to the model developed by Brody and Flemings (1966) twenty years ago. They made an analytical approximation for the solid-phase concentration gradient at the interface, and assumed a parabolic freezing rate. The result of this analysis was equation (4), where α is defined by equation (5),

$$C_L^i \; = \; C_o \left(1 - (1 - 2\alpha k) \, f_s \right)^{(k-1)/(1-2\alpha k)} \tag{4}$$

$$\alpha \; = \; 4 D_s \, t_f / \lambda^2 \tag{5}$$

with t_f the solidification time and λ the diffusion length. This model has been criticized because it does not actually conserve solute under conditions of appreciable solid-state diffusion, and has been supplanted by a relation developed by Clyne and Kurz (1981). Their model utilizes equation (4) with Ω replacing α (equation (6)). This substitution is essentially a spline fit which forces the Brody/Flemings equation to be correct at either limit solid-state diffusion coefficient. It has been criticized because the fit is only mathematical, not physical (Kirkwood, 1984).

$$\Omega \; = \; \alpha \left(1 - \exp\left(-\frac{1}{\alpha} \right) \right) - \frac{1}{2} \exp\left(-\frac{1}{2\alpha} \right) \tag{6}$$

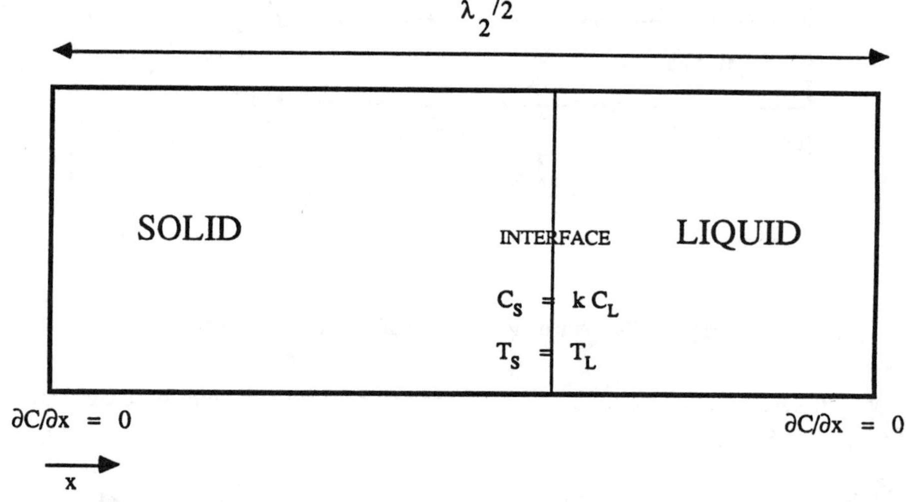

Fig. 3 One-dimensional region used for microsegregation calculations

Equations (3) through (6) represent only a few of many microsegregation models; others were reviewed by Kirkwood (1984). These models all share common features - they are based on a mass balance at the interface (some form of equation (3)), and all assume equilibrium between solid and liquid at the interface, no undercooling before solidification, and a uniform liquid concentration (Flemings, 1981). These are by no means the only types of models which have been developed to predict microsegregation in solidified metals. In fact, a bewildering variety of techniques have been published in recent years (see, for example, the proceedings of the *International Conference on Solidification Processing*, held in Sheffield, UK, 1987). There is not space to review these models here, but one general one-dimensional technique will be considered.

III. GENERAL ONE-DIMENSIONAL SOLUTION

A model has been developed by the present authors that solves the combined heat and mass transfer problem during and after solidification. Seven equations must be considered. The first two relations (equations (7) and (8)) are the equations for mass transfer and conductive heat transfer, respectively, in the bulk phases. Equations (9) through (13) are all heat or mass balances at the solid/liquid interface. Temperature is assumed equivalent on either side of the interface, but concentration is not, due to solute partitioning. The difference between the heat fluxes in the solid and liquid is a result of the latent heat of fusion, while the difference in mass fluxes is due to the solute partitioning (equations (11) and (12), respectively). Finally, equation (13) relates the temperature at the interface to the local liquid-phase composition.

$$\rho C_p \frac{\partial T}{\partial t} = \frac{\partial}{\partial x}\left(\kappa \frac{\partial T}{\partial x}\right) \tag{7}$$

$$\frac{\partial C}{\partial t} = \frac{\partial}{\partial x}\left(D \frac{\partial C}{\partial x}\right) \tag{8}$$

$$T_S = T_L = T^* \tag{9}$$

$$C_S = k\,C_L \tag{10}$$

$$\Delta H_f\, \rho_L \frac{dh}{dt} = \left(\kappa_S \frac{\partial T_S}{\partial x}\right) - \left(\kappa_L \frac{\partial T_L}{\partial x}\right) \tag{11}$$

$$\left(C_L - C_S\right)\frac{dh}{dt} = D_S \frac{\partial C_S}{\partial x} - D_L \frac{\partial C_L}{\partial x} \tag{12}$$

$$T^* = T_O - m\,C_L \tag{13}$$

In equations (7) through (13) ρ is the mass density, C_p the heat capacity, κ the thermal conductivity, T^* the temperature at the interface, ΔH_f the latent heat of fusion, m the slope of the liquidus equation, and T_O the melting point of the pure solvent.

The numerical algorithm developed to solve these equations involves the use of the Method of Lines and Invariant Imbedding (MOL/II) (Meyer, 1973, 1981a, 1981b) to convert the system of partial differential equations to ordinary differential equations, which can be solved by normal numerical techniques. The steps involved in the derivation of this algorithm have been described (Battle, 1988; Battle and Pehlke, 1987a and 1987b). In the following section predictions made by this model will be compared to experimental data and the mass-balance models discussed previously.

IV. COMPARISON OF MODEL PREDICTIONS

A considerable number of microsegregation measurements have been made on alloys in the aluminum-copper system, which was depicted in Fig. 1. This system is characterized by a long melting range (particularly in the presence of microsegregation), a low equilibrium partition coefficient (resulting in strong segregation during solidification), and the presence of a eutectic reaction at 821K. The physical properties of the aluminum-rich end of the Al-Cu system are summarized in Table 1. It should be noted that the complicated equations for physical properties given in this table were often not required. The only parameters that affected the calculated concentration profiles enough to require very accurate values were the solid-state diffusivity of the solute and the equilibrium partition coefficient.

Figure 4 shows a comparison of the lever rule, the Scheil equation, the Brody/Flemings (BF) relation, the Clyne/Kurz (CK) equation, and the Ogilvy/Kirkwood (OK) model in an Al-2.14 mole% Cu alloy. The difference between equilibrium freezing and the Scheil equation is apparent. In this alloy the other models all predict similar profiles, which lie close to the Scheil curve. All predicted varying amounts of second phase; if this alloy had undergone equilibrium freezing, no eutectic at all would have been found (if the sample were quenched from just below the eutectic temperature).

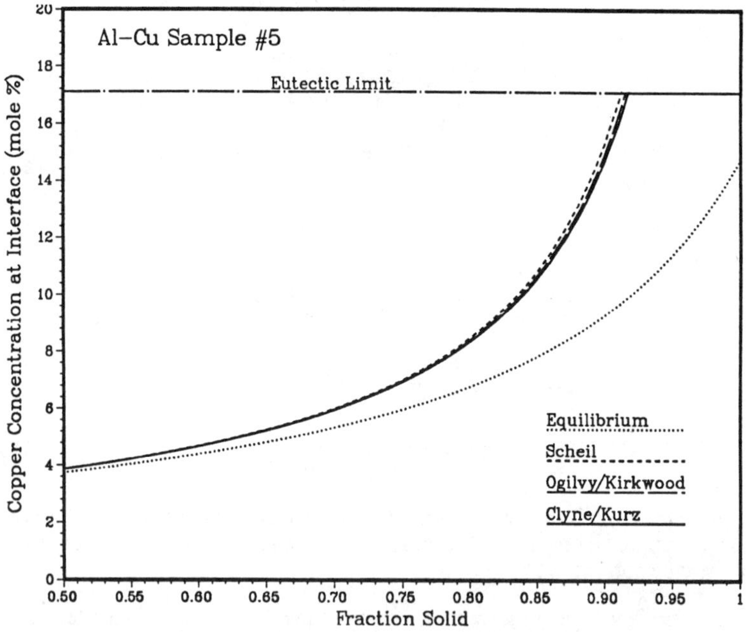

Fig. 4 Liquid phase composition at the interface as a function of fraction solid as predicted by several models

TABLE 1 Physical Properties of Al-Cu Alloys

Property	Phase[a]	Value[b]	Units	Ref
Chemical Diffusivity	L	$1.05*10^{-7}*exp(-2856/T)$	m^2/sec	Ejima and others, 1980
	α	$0.29*10^{-4}*exp(-15610/T)$	m^2/sec	Murphy, 1961
Mass Density	L	$2358.5 + 21.685*W_{Cu} + 7.2914*10^{-2}*(W_{Cu}^2) - 7.2351*10^{-4}*(W_{Cu}^3)$	kg/m^3	Ganesan and Poirier, 1987
	α	$2558.1 + 2.1743*W_{Cu,l}/k + 6.0443*10^{-2}*(W_{Cu,l}^2)/k^2$	kg/m^3	Ganesan and Poirier, 1987
	θ	4241	kg/m^3	Seshadri and Downie, 1979
Liquidus Temperature	-	$933.54 - 6.57*X_{Cu}$	K	Murray, 1985
Eutectic Temperature	-	821.2	K	Murray, 1985
Eutectic Composition	-	17.1	mole%	Murray, 1985
Partition Coefficient	α/L	0.145	mole%/mole%	Murray, 1985
Latent Heat of Fusion	-	$4000*X_{Al} + 2050*X_{Cu}$	J/kg	Hultgren and others, 1973
Heat Capacity	α	$766.1 + 0.459*T -4.099*W_{Cu} - 0.00359*T*W_{Cu}$	J/kgK	Kubaschewski and Alcock, 1979 (K&A)
	L	$1179 - 6.85*W_{Cu}$	J/kgK	K&A, 1979
Thermal Conductivity	L	$0.5*\kappa_\alpha$ (avg)	W/mK	Ho and others, 1978
	α	$(-5.327*10^{-4} - 6.794*10^{-3}*X_{Cu})*T + 153.2851 + 22.391*X_{Cu} - 1.186*X_{Cu}^2$	W/mK	Ho and others, 1974

(a) L = liquid α = primary solid phase θ = eutectic phase

(b) T is temperature in K, X_i is mole percent of species i, $W_{i,j}$ is weight percent of species i in phase j, k is partition coefficient

A qualitative test of the MOL/II model is to examine its predictions in the limits of solid-state diffusion. Figure 5 again shows equilibrium cooling and Scheil equation curves, along with the MOL/II model. The close correspondence between the current model and the Scheil curve would be even greater if D_S were decreased. If the diffusivity is increased by a factor of 100 in the MOL/II model, less segregation is predicted. If D_S is increased by a factor of 100,000, the results very closely approach the equilibrium cooling predictions.

Experimental measurements from two different investigators were examined and compared with model predictions (see Table 2). The first seven specimens come from a study by Sarreal and Abbaschian (1986). The samples were solidified at widely different cooling rates, and the authors measured the amount of the non-equilibrium second phase (θ-$CuAl_2$) present in the quenched samples. In contrast, Bennett (1978) measured solute concentration profiles in his samples, the first two of which were directionally solidified, as were the samples of Sarreal and Abbaschian. Samples 10 and 11 were obtained from a chill casting prepared by Bennett. Bulk compositions in all cases were low enough that no theta phase would precipitate during equilibrium solidification. In this presentation only the Sarreal and Abbaschian results will be discussed. See Battle (1988) for more details on other model comparisons.

TABLE 2 Experimental Measurements of Microsegregation in Al-Cu Alloys

Sample	Bulk Copper Concentration (mole %)	Temperature Gradient (K/mm)	Cooling Rate (K/s)	Volume Percent Second Phase	Secondary Arm Spacing (μm)
1	2.14	10	0.1	4.18	91
2	2.14	10.5	1.05	4.90	46
3	2.14	7.5	11.25	5.14	23
4	2.14	13	65	5.37	14
5	2.14	18.7	187	5.62	10
6	2.14	17	1700[a]	4.66	5.4
7	1.21	1200	37000[a]	1.31	2.2
8	1.81	2	0.5	-	42.7
9	1.72	1.7	1.7	-	29
10	2.16	2.2	1.9	-	37
11	2.16	1.31	3.5	-	24

(a) Estimated values, using the measured dendrite arm spacings and the cooling rate-arm spacing relationship derived from the first five samples.

Results of tests against the Sarreal/Abbaschian data can be seen in Fig. 6. Here the experimental measurements are represented by a curve connecting the data points from Table 2. The upper curve is the predicted amount of second phase if the Scheil equation held in this system. Three sets of predicted points are given at each cooling rate. The first is the predicted eutectic after cooling to the eutectic temperature, the second that predicted upon further cooling, to 600K. The difference between these two sets of points indicates the effect of re-dissolution of theta phase into the alpha-aluminum matrix. This effect is greatest at low cooling rates, where there is more time during cooling for reaction to occur. These calculations are not completely accurate, however, since theta phase may precipitate below the eutectic due to the dcreasing solubility of copper in aluminum as temperature decreases.

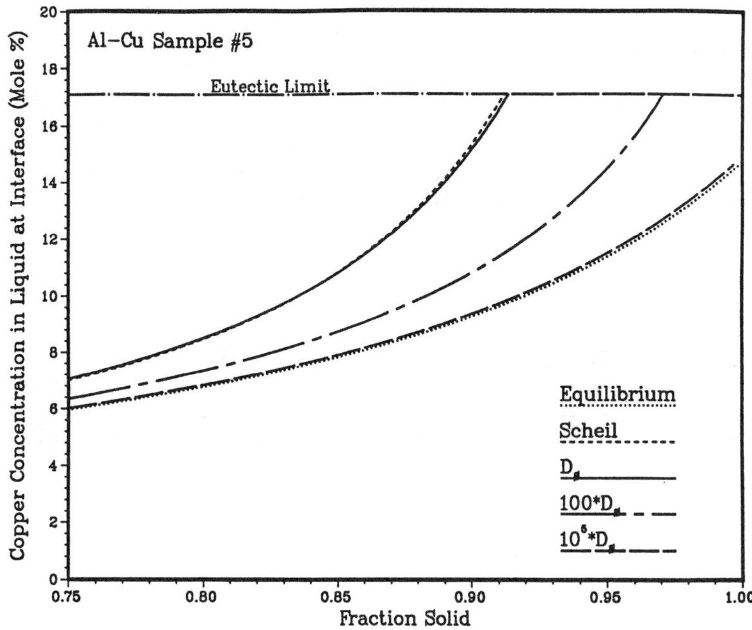

Fig. 5 Predicted liquid-phase concentrations as a function of fraction solid and solid-state solute diffusivity, MOL/II model

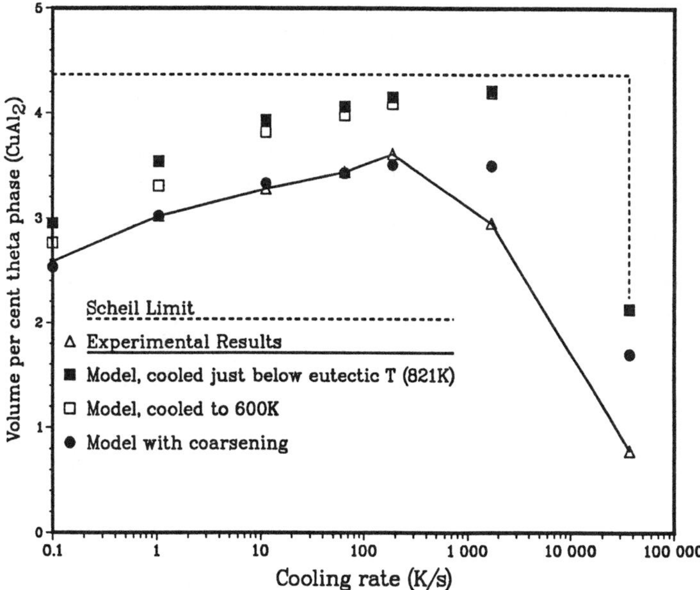

Fig. 6 Comparison of MOL/II model to experimental data of Sarreal and Abbaschian

The final set of points in Fig. 6 shows model predictions in the presence of dendrite-arm coarsening. This coarsening results from surface-area reduction during solidification, and leads to a significant change in the secondary dendrite arm spacing with time during solidification. In Fig. 7, the predicted coarsening rate for sample 5 is shown. The analysis is assumed to begin with a system size of 0.83 microns, which quickly increases with time. Final solidification is achieved when the spacing reaches its final value of 5 microns (this is 1/2 of the secondary arm spacing for sample 5 from Table 2). This coarsening rate is calculated from equation (14), which is an empirical relationship derived from experimental measurements of final arm spacing as a function of solidification time (for the samples of Sarreal and Abbaschian, $\alpha = 6.14$ microns, $n = 0.29$). More complex and physically realistic equations for coarsening could have been used (Kirkwood, 1984; Rappaz, Kurz, and Glicksman, 1985; Roosz, Halder, and Exner, 1986).

$$\lambda_2 = \alpha\, t^n \tag{14}$$

The mass-balance models of section II are compared against the Sarreal and Abbaschian data in Figs. 8 and 9. It is apparent from Fig. 8 that the analytical models are unable to accurately predict the amount of second phase present in the specimens. Fig. 9 shows the comparison with the numerical model of Ogilvy and Kirkwood. It is clear that with the inclusion of dendrite-arm coarsening the model matches experimental measurements quite well.

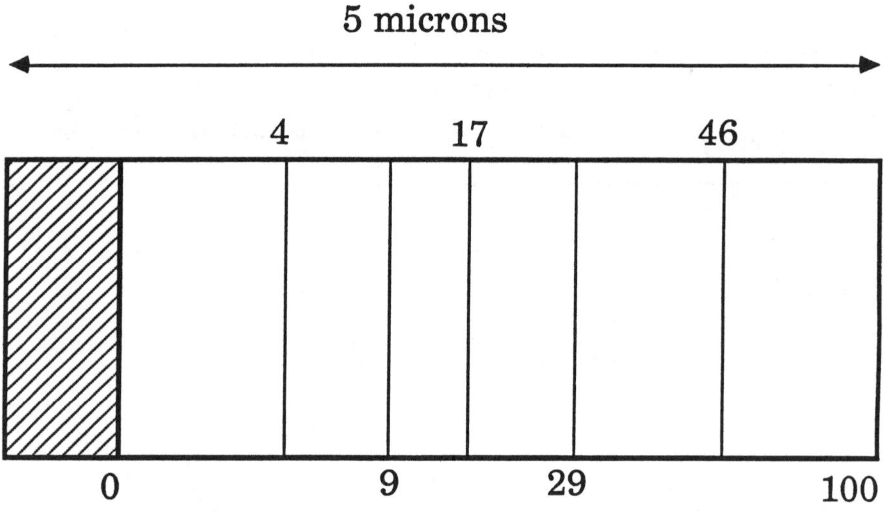

Fig. 7 Schematic diagram of dendrite arm coarsening, Al-Cu sample 5, showing region size as a function of elapsed time (time as a percentage of solidification time)

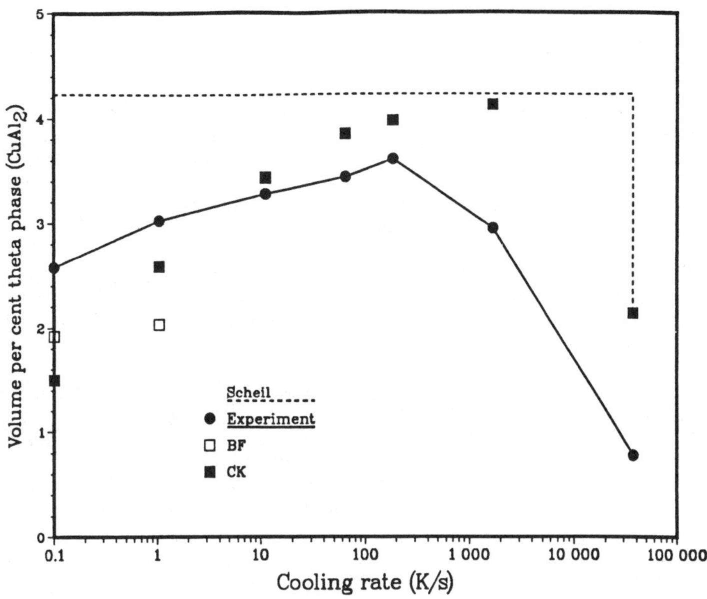

Fig 8. Comparison of analytical microsegregation equations to data of Sarreal and Abbaschian

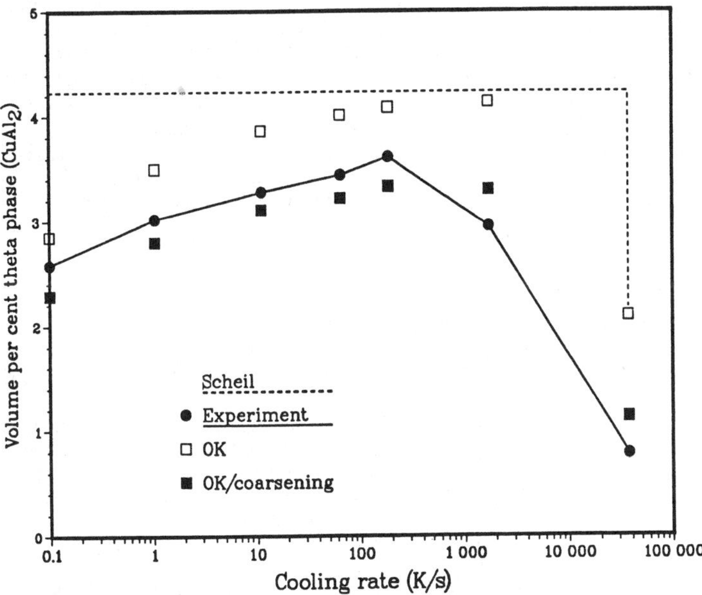

Fig. 9 Comparison of Ogilvy/Kirkwood numerical model to data of Sarreal and Abbaschian

V. DISCUSSION OF RESULTS

It is clear from Figs. 6, 8, and 9 that only the numerical models are able to reproduce the experimental measurements. This is due to the fact that the analytical models are unable to include the effects of time-dependent parameters in the basic equations. For the results of Sarreal and Abbaschian, these critical parameters are the solid-state diffusivity of copper in aluminum, and the inclusion of coarsening. In systems where the equilibrium partition coefficient is temperature-dependent, this parameter will also stronlgy affect the calculations.

The MOL/II and Ogilvy/Kirkwood models predict the percentage of non-equilibrium second phase quite accurately, and demonstrate the utility of a numerical model. The importance of coarsening is apparent from these results, where its inclusion decreased the amount of theta phase by up to 0.7% (approximately 20% less than the non-coarsened value).

It should be noted that neither numerical model made accurate predictions at the two highest cooling rates measured. At high cooling rates, the fundamental assumptions of nearly all microsegregation models begin to fall apart. The presumption of local equilibrium at the solid/liquid interface is no longer valid; neither is the condition that no mass enters or leaves the system. There is no general microsegregation model valid for rapid solidification conditions, although the analytical model of Solari and Biloni (1980) includes the effect of dendrite-tip undercooling, the major cause of mass flow through the volume element of interest.

Of the two numerical models discussed above, the MOL/II is more general, since the Ogilvy/Kirkwood assumes unlimited diffusion in the liquid and infinitely rapid heat transfer. However, due to these simplifications, the Ogilvy/Kirkwood model runs much faster and at this stage may be a more useful tool for the calculation of microsegregation. It has recently been extended to consider the peritectic reaction and multiple solutes (Howe, 1987).

Figure 10, following the discussion by Sarreal and Abbaschian (1986), plots the percentage of second phase as a function of cooling rate for the data sets of Table 2. In order to reduce the differences in percentage due to changes in bulk concentration, the results are plotted against the fraction of second phase relative to the Scheil equation prediction for that composition. The results match quite well, and show the peak in second-phase content with cooling rate discussed by Sarreal and Abbaschian. This is said to be a result of rapid solidification counteracting the tendency for segregation to approach the Scheil limit at higher cooling rates. It is possible to use Fig. 10 to predict the amount of second phase that would be present for an alloy of hypoeutectic composition, for solidification at a particular cooling rate. This must be verified with other experimental data, and is restricted to lower cooling rates, until a microsegregation model capable of making accurate predictions under rapid solidification conditions is developed.

VI. SUMMARY AND CONCLUSIONS

In this paper the importance of microsegregation in a variety of solidifying metal systems has been considered. The most commonly used analytical equations are presented, along with two numerical models. These models are compared to each other and to experimental measurements in the Al-Cu system.

It is evident that numerical models can more accurately predict microsegregation than the analytical solutions. However, the analytical approach is still quite popular due to its simplicity. These equations have been utilized recently in combined macroscopic/microscopic models to make more accurate predictions of solidification in complicated systems and processes. An example is the work of Clyne (1982a, 1982b) where the Clyne/Kurz equation was used to predict the relationship between fraction solid and temperature during solidification. This is more accurate than the traditional technique in macroscopic solidification models of assuming a linear change in fraction solid with temperature. However, as shown above, the analytical solutions generally cannot accurately predict second phase contents, particularly where coarsening is important.

The next stage in this type of analysis is to use a macroscopic model to predict local temperature gradients and cooling rates in a solidifying metal. These numbers can be input into a numerical microsegregation model which will then predict microsegregation at any point of interest. Thus it becomes possible to determine maximum/minimum solute concentrations, porosity and segregation as a function of physical properties and process variables. This is the ultimate goal of an engineering model - to be capable of making quantitative calculations and obviate the need for an extensive series of experimental measurements for a given process or alloy.

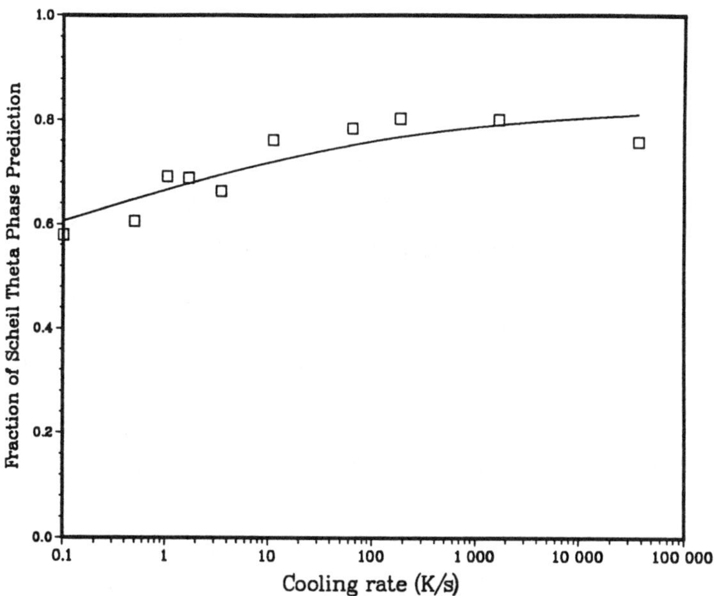

Fig. 10 Predicted second-phase percentage as a fraction of Scheil equation value, MOL/II model, data of Sarreal and Abbaschian and Bennett

REFERENCES

Battle, T.P. and R.D. Pehlke (1987a). Computer Simulation of Microsegregation During the Solidification of Steel. *Proceedings 70th Steelmaking Conference*, Pittsburgh, 121-128.

Battle, T.P. and R.D. Pehlke (1987b). Microsegregation in Cast Steels. Presented at International Conference on Solidification Processing, University of Sheffield, Sheffield, UK, September, 1987.

Battle, T.P. (1988). *Mathematical Modelling of Microsegregation in Binary Metallic Alloys*, PhD Thesis, University of Michigan.

Bennett, D.A. (1978). *Equilibrium and Solidification Studies in the Al-Cu and Al-Cu-Fe Systems*, PhD Thesis, University of Sheffield, UK.

Brody, H.D. and M.C. Flemings (1966). Solute Redistribution in Dendritic Solidification. *Trans. Met. Soc. AIME*, 236, 615-624.

Clyne, T.W. (1982a). Numerical Modelling of directional solidification of metaliic alloys. *Met. Sci.*, 16, 441-450.

Clyne, T.W. (1982b). The Use of Heat Flow Modeling to Explore Solidification Phenomena. *Metall. Trans. B*, 13B, 471-478

Clyne,T.W. and W. Kurz (1981). Solute Redistribution During Solidification with Rapid Solid State Diffusion. *Metall. Trans. A*, 12A, 965-971.

Ejima, T., T. Yamamura, N. Uchida, Y. Matsuzaki, and M. Nikaido (1980). Impurity Diffusion of Fourth Period Solutes (Fe, Co, Ni, Cu, and Ga) and Homovalent Solutes (In and Tl) into Molten Aluminum. *J. Jpn. Inst. Met.*, 44, 316-323.

Flemings, M.C. (1981). Segregation and Structure in Rapidly Solidified Cast Metals. In J.K. Tien and J.F. Elliott (Ed.), *Metallurgical Treatises*, TMS-AIME, New York.

Forsen, G., E. Hettula, and K. Lilius (1985). The Behaviour of Arsenic, Antimony, and Bismuth in the Solidification and Electrolysis of Nickel-Oxygen Bearing Copper Anodes. In V. Kudryk and Y.K. Rao (Ed.), *Physical Chemistry of Extractive Metallurgy*, TMS-AIME, New York.

Ganesan, S. and D.R. Poirier (1987). Densities of Aluminum-Rich Aluminum-Copper Alloys during Solidification. *Metall. Trans. A*, 18A, 721-723.

Hammer, O. and G. Grunbaum (1974). Influence of Back Diffusion on Microsegregation during Solidification of Low-Alloy Steels. *Scand. J. Met.*, 3, 11-20.

Ho, C.Y., M.W. Ackerman, K.Y. Wu, S.G. Oh, and T.N. Havill (1978). Thermal Conductivity of Ten Selected Binary Alloy Systems. *J. Phys. Chem. Ref. Data*, 7, 959-1177.

Ho, C.Y., R.W. Powell, and P.E. Liley (1974). Thermal Conductivity of the Elements: A Comprehensive Review. *J. Phys. Chem. Ref. Data*, 3, supplement 1.

Howe, A. A. (1987). Development of a Computer Model of Dendritic Microsegregation for use with Multicomponent Steels. *Appl. Sci. Res.*, 44, 51-59.

Hultgren, R., P.D. Desai, D.T. Hawkins, M. Gleiser, K.K. Kelley, and D.D. Wagman (1973). *Selected Values of the Thermodynamic Properties of the Elements*, American Society for Metals, Metals Park, Ohio.

Hunsicker, H.Y.(1967). The Metallurgy of Heat Treatment. In K. R. Van Horn (Ed.), *Aluminum, Volume I: Properties, Physical Metallurgy, and Phase Diagram*, American Society for Metals, Metals Park, Ohio.

Kirkwood, D. H. (1984a). Microsegregation. *Mat. Sci. and Eng.*, 65, 101-109.

Kirkwood, D.H. (1984b). A Simple Model for Dendrite-Arm Coarsening During Solidification. *Mat. Sci. Eng.*, 73, L1-L4.

Kubaschewski O. and C.B. Alcock (1979). *Metallurgical Thermochemistry*, 5th edition, Pergamon Press, Elmsford, NY.

Loria, E.A. (1988). The Status and Prospects of Alloy 718. *J. Met.*, 40(7), 36-41.

Meyer, G.H. (1973). *Initial Value Methods for Boundary Value Problems*, Academic Press, New York.

Meyer, G.H. (1981). A Numerical Method for the Solidification of a Binary Alloy. *Int. J. Heat Mass Transfer*, 24, 778-781.

Meyer, G.H. (1981). The Method of Lines and Invariant Imbedding for Elliptic and Parabolic Free Boundary Problems. *SIAM J. Numer. Anal.*, 18, 150-164.

Murphy, J.B. (1961). Interdiffusion in Dilute Aluminum-Copper Solid Solutions. *Acta Met.*, 9, 563-569.

Murray, J.L. (1985). The Aluminum-Copper System. *Intl. Met. Rev.*, 30, 211-233.

Ogilvy, A.J.W. and D.H. Kirkwood (1987). A Model for the Numerical Computation of Microsegregation in Alloys. *Appl. Sci. Res.*, 44, 43-49.

Rappaz, M., W. Kurz, and M.E. Glicksman (1985). Statistical Model of Dendrite Arms Coarsening. Paper presented at 1985 Fall Meeting of TMS-AIME, Toronto.

Rickinson, B.A. and D.H. Kirkwood (1979). Microsegregation in Fe-Cr-C alloys solidified under steady-state conditions. In *Solidification and Casting*, Institute of Metals, London.

Roosz, A., E. Halder, and H.E. Exner (1986). Numerical Calculation of Microsegregation in Coarsened Dendritic Microstructures. *Mat. Sci. and Techn.*, 2, 1149-1155.

Sarreal, J.A. and G.J. Abbaschian (1986). The Effect of Solidification Rate on Microsegregation. *Metall. Trans. A*, 17A, 2063-2073.

Seshadri, S.K. and D.B. Downie (1979). High-Temperature Lattice Parameters of Copper-Aluminum Alloys. *Met. Sci.*, 13, 696-698.

Solari, M. and H. Biloni (1980). Microsegregation in Cellular and Cellular Dendritic Growth. *J. Cryst. Growth*, 49, 451-457.

Heat and Mass Transfer Simulations on a Personal Computer

M.J. Brown and F. Mucciardi
Department of Mining and Metallurgical Engineering
3450 University Street
McGill University, Montreal, Quebec

Abstract

A computer software package for the analysis of heat and mass transfer has been developed for the metallurgical industry. This interactive program, referred to as FASTP (Facility for the Analysis of Systems in Transport Phenomena) can be used to solve a variety of problems frequently encountered in the processing of metals and materials.

FASTP (ver.1.01) is completely menu driven and has extensive graphics capabilities. A description of these features will be highlighted. In addition, this paper will focus on the potential of using FASTP to analyze metallurgical processes such as continuous casting, ladle cycling, and scrap dissolution. Sample input and output computer sessions for some of the above examples will be presented.

Introduction

The absence of similar software products in the marketplace has prompted the development of a transient heat and mass transfer modelling package (FASTP) for the personal computer environment. The Facility for the Analysis of Systems in Transport Phenomena is based on an explicit finite difference algorithm that is capable of rapidly generating results for one, two, or three dimensional configurations in spherical, cylindrical, or rectangular coordinates.

FASTP (ver. 1.01) allows its user to track transient heat or mass transfer behaviour in simple shapes such as cylinders, slabs, and spheres. The concept is to use these simple shapes as building blocks in models of more complicated process engineering problems.

Both engineers and students may benefit from the use of such a tool. Engineers, already possessing an understanding of heat or mass transfer, will find that the package can perform a variety of calculations which until recently

were not easily carried out. FASTP provides an excellent means of evaluating
the relative merits of implementing or altering a process prior to actually making
the changes in the plant. In this capacity the package serves as a decision tool
– saving time, money, and effort. It has already, for example, been used in the
redesign of a spray zone for a continuous caster and in the prediction of heat
losses from a ladle as it cycles through a steel plant. Students, only beginning
to learn transport phenomena, will also undoubtedly find the package to be a
definite asset as an educational tool.

In order to illustrate how this package can be used, this paper will begin
with a description of the parameters which must be supplied to the program
when setting up a simulation. This will be followed by a description of the
package's features and its hardware requirements. Examples have been included
in the Appendices to demonstrate its viability as a process modelling tool.

Discussion

Configuring the model – available features

Reference was made in the Introduction to the building block approach
whereby solutions for the simple shapes available to FASTP are combined in
order to model more complicated process engineering problems. The first part
of the discussion will describe the parameters which are required and the options which may be specified when configuring a building block for a FASTP
simulation.

1. Physical setup

When initially setting up an analysis, the coordinate system in which the
user plans to work must be specified. The available options are the standard
coordinate systems – spherical, cylindrical, and rectangular. Depending on the
coordinate system, a transient analysis may be performed on up to a maximum
of one, two, or three dimensions (Table 1).

Table 1 – FASTP coordinates and maximum model dimensions

Coordinates system	Maximum number of dimensions
spherical	1
cylindrical	2
rectangular	3

1.1 Nodes

The explicit finite difference algorithm which has been written for FASTP
performs calculations on linearly arranged nodes (Fig. 1.a). The equations gov-
erning the interaction between the nodes are derived from Fourier's or Fick's
second laws. With FASTP, the user may specify any nodal progression within

the system's bounds. In the finite difference analysis the nodal points are viewed as discrete locations in a system at which values, temperatures or concentrations, are calculated on an iterative basis. The iterative process models the system's progress with time. During the course of the simulation the two end nodes, or surface nodes, are subject to applied boundary conditions resulting in a net gain or loss of energy / mass from the system.

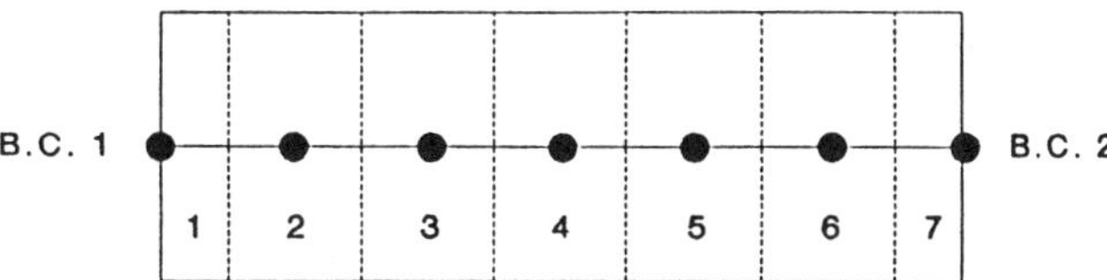

Fig. 1.a – 1-D Rectangular Coordinates Nodal Progression

1.2 Dimensions

Two and three dimensional analyses may be generated from these 1-D nodal arrangements (Figs. 1.b-1.c). The reasoning follows that of the classical approach of obtaining a product solution. Two types of nodes will arise from such a configuration, principal axis nodes and profile nodes. The principal axis nodes are those nodes upon which the iterative calculations are performed. Values for the profile nodes are derived from the principal axis nodes at times specified by the user. The scheme makes the algorithm amenable for use on a personal computer by reducing the overall number of calculations. Extensive testing has demonstrated that the accuracy of the solution is not significantly reduced provided ambient conditions are specified. FASTP allows the user to track the progress of either type of node with time.

A 2-D, rectangular coordinates configuration, for example, can be envisioned as a block having a rectangular shape. Boundary conditions are specified for each of the four sides (Fig. 2.a).

A 1-D, cylindrical coordinates setup can be visualized as a circle or annular ring with two specified boundary conditions applied uniformly to the inner and outer radii of the problem configuration.

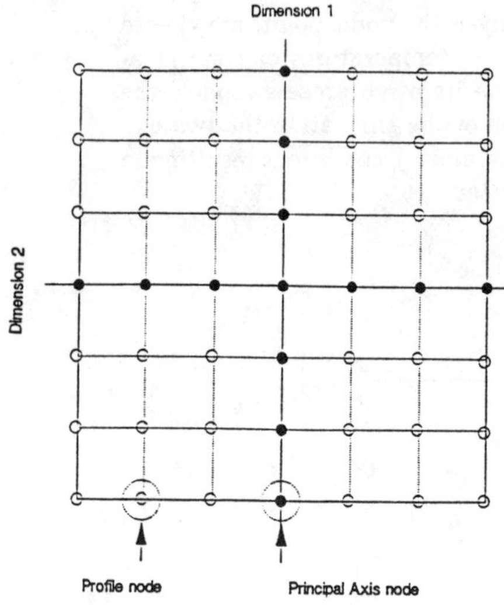

Figure 1.b 2-D Rect. Coords. Approximation
using two, 1-D solutions

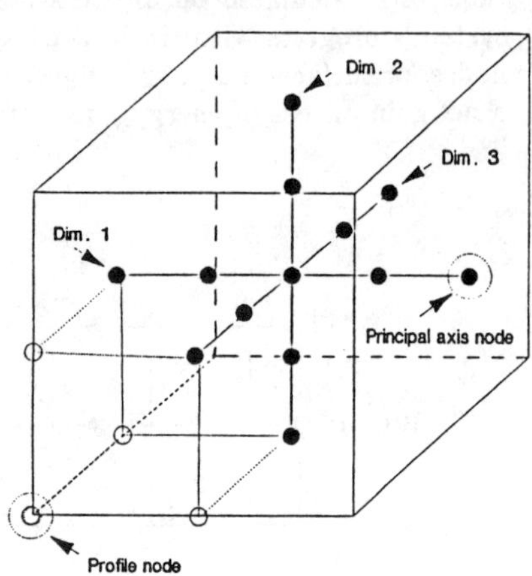

Figure 1.c 3-D Rect. Coords. Approximation
using three, 1-D solutions

1.3 Sections

For each dimension of the problem setup, up to five separate sections may
be specified. Consequently, a layered analysis may be configured in which each
section corresponds to a different material possessing its own set of material
properties (Table 2). The thickness and number of nodes assigned to each
section are parameters which are specified. As well, initial conditions (i.e. temperatures or concentrations) must be assigned to all nodes in each section and
dimension. The use of multiple sections in a problem configuration is demonstrated in the ladle cycling example (Appendix D).

Table 2 – Material properties required by FASTP

Heat Transfer	Mass Transfer
thermal conductivity	diffusion coefficient
density	
heat capacity	

2. Boundary conditions

The surface nodes, or end nodes, are subject to boundary conditions which
will cause a net loss or gain of energy or mass from the system during the course

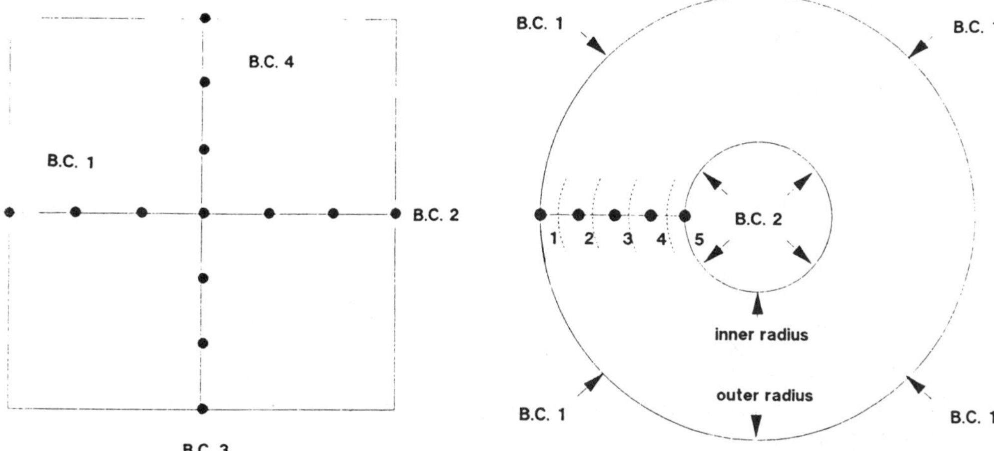

Fig. 2-a   2-D Rect. Coords. configuration
with Boundary Conditions applied

Fig. 2-b   1-D Cylind. Coords. configuration
with Boundary Conditions applied

of the analysis. Many types of boundary conditions are available for both heat
and mass transfer modelling (Table 3). A number of the boundary condition
options have been used in the examples included in the Appendices wherein the
various options are described.

Table 3 – Boundary Condition options

Heat transfer	Mass transfer
Fixed surface temperature	Fixed surface concentration
Convection	Convection
Radiation	——
Conv. & Rad.	——
Zero flux (adiabatic)	Zero flux
Surface flux	Surface flux
Surface flux	Surface flux
(with Conv. and / or Rad.)	(with Conv.)

An advanced feature, referred to as Batch Mode Operation, has been incorporated into the program's algorithm. This feature allows the user to program
times and/or surface conditions at which a specified boundary condition is to be
applied to a surface node. Consequently, analyses with dynamically changing
boundary conditions may be easily modelled. The feature offers tremendous
flexibility and was used to model the reheat and solidification behaviour of
a continuously cast steel billet as it progressed through the mold, secondary
cooling, and air cooled zone (Appendix F).

3. Other FASTP configuration features

The remaining configuration features described herein are optional when setting up a model. Wherever possible, reference to the use of these features in the appended examples has been made.

3.1 Interboundary resistances / partition coefficients

Interboundary resistances due to non-smooth contact between two surfaces may be modelled using FASTP. A resistance factor is specified between two designated sections of the problem configuration. During the transient analysis the algorithm will use this factor to maintain a temperature ratio between the two, contacting, end nodes of the designated sections while maintaining the energy balance. Thermal contact resistance is defined as the inverse of the heat transfer coefficient (i.e. $R_{th} = 1/h$).

The equivalent feature in mass transfer is the partition coefficient. Presently, only fixed partition coefficients may be applied. However, modification to the algorithm to accommodate floating point partition coefficients is underway. A floating point partition coefficient will change with surface concentration in order to reflect the solute's changing activity coefficient. This feature is useful in modelling the transfer of species at slag / metal interfaces.

3.2 Heat generation / mass accumulation

This feature can be used to model induction heating, electrical resistance heating, simulation of heat generation due to friction or, in mass transfer, to simulate the generation or depletion of a species by a chemical reaction. This feature enables the user to target specific nodes as sources of heat generation / mass accumulation.

3.3 Stirring

The stirring option allows the user to model the redistributive effect of mixing in a fluid system. Without any form of stirring specified, FASTP will assume that heat / mass transfer through a system will be due solely to diffusion, whether thermal or molecular. In other words the fluid is assumed to be stagnant. However, it is well known that fluid flow, especially turbulent flow, will have a tremendous effect on the redistribution of thermal energy / mass in the system.

FASTP will allow for the specification of a maximum of six stirred sections. The maximum is, of course, further limited by the number of sections in the given problem configuration.

A user-defined parameter, the mixing time, will affect the extent to which energy / mass redistribution can occur over the course of one node iteration. Thus, control over the degree of stirring is possible. An example of the application of the stirring feature has been included in Appendix F.

3.4 Phase changes

FASTP's algorithm has been designed to accommodate phase changes in a material. The thermal conductivity, heat capacity, and density of a material change with state. The package's database, described below, allows the user to specify three sets of material properties corresponding to three different states (Table 4).

Table 4 - State assignment for solid-liquid transformation

State 1	solid
State 2	transforming
State 3	liquid

When modelling solidification, the transforming state, also referred to as the 'mushy' state, corresponds to that region between the solidus and liquidus lines on a phase diagram (Fig. 3.a). This state exists over a defined temperature range. The range would most likely be chosen with prior knowledge of the cooling behaviour of the system being modelled (Fig. 3.b).

Materials which may be described as possessing three definite cooling trends (liquid, transforming, solid) may be modelled using FASTP. These would include pure metals, simple binary alloys, eutectics, and any other materials which approximate this behaviour. However, it is to be noted that solute redistribution is not taken into account. The examples in Appendices B and F demonstrate the use of the phase change feature in a model.

The three state property sets can also be used to model a changing thermal diffusivity (e.g. thermal conductivity) in a solid over a wide temperature range. Use of the phase change feature in this manner can find application in the modelling of heat transfer behaviour in ceramic materials because their properties can vary considerably with temperature.

4. Package facilities

The many package features of FASTP (ver.1.01) to be described below have been implemented to make the package a very easy and useful tool to work with.

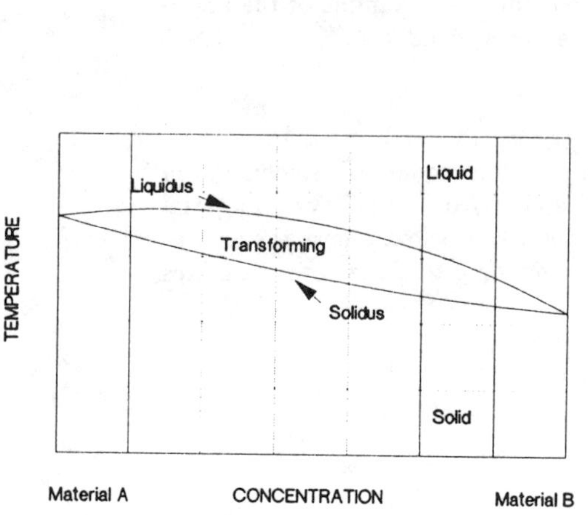

Figure 3.a Phase Diagram for simple binary alloy

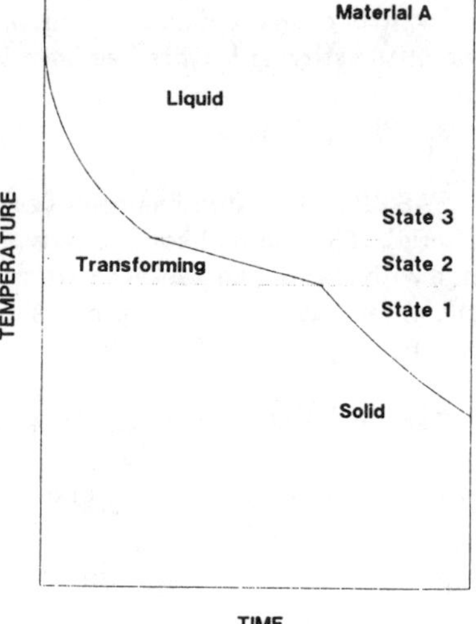

Fig. 3-b Cooling curve for simple binary alloy

FASTP operates on a screen by screen basis. Selection of options and data entry is simplified through the use of menus and function keys. Once all pertinent parameters have been entered and prior to execution of the simulation, all data may be stored to a user specified file. This feature, combined with an edit feature, allows the user to retrieve data at a later date, edit parts of it, and rerun the same simulation without having to reenter the entire configuration.

Also included is a database facility which allows for the storage of material property data. Data may be entered, viewed, edited, or deleted using this facility. Each data set is assigned a material code which is specified when entering data from the main program. The main program searches the database, retrieves, and installs all data associated with that code. This facility reduces time, effort, and the potential for making data entry errors.

FASTP output can be displayed via the package's graph generating facility. With the facility, the user may create Variable vs Time, Variable vs Section, and Isovariable plots where Variable can be temperature, energy content, or concentration. Results from the iterative calculations may also be sent directly to a printer.

The graphics display options allow the user tremendous flexibility in choosing which aspect of the simulation is to be observed. The Variable vs Time option allows the user to track up to three nodes, principal or profile, and plot their progress on one graph. The Variable vs Section and Isovariable options allow the user to track the progress of selected parts of the simulation on a frame by frame, time lapse, basis. Sample output is presented in Appendix F.

Although effective, graphics support for FASTP (ver.1.01) is still somewhat rudimentary. Efforts will likely be made to improve the facility in later versions. A file export facility presently enables the user to rewrite Variable vs Time data in a form (ASCII) which may be used by other commercially available graphics packages.

5. Package breakdown

FASTP (ver.1.01) is comprised of four (4) diskettes, a manual, and a copy protection, hardware device.

Table 5 – Package breakdown

Diskette	Function
1	Data entry
2	Calculations algorithm
3	Graphics package
4	Support files

Manual
Copy-protect hardware device

6. Hardware requirements

- IBM PC/XT/AT or IBM compatible
- 2 floppy disk drives (hard disk preferable)
- 80x87 Math Co-processor
- CGA, EGA, Olivetti graphics adaptor (Hercules not supported)
- Parallel port
- DOS 3.0 (or above)

Acknowledgements

The authors would like to thank A. Grandillo and K.E. O'Leary for their valuable input in compiling the many examples used in this paper. In addition, the authors are grateful to the Natural Sciences and Engineering Research Council (NSERC) for their financial support.

Summary

A heat / mass transfer modelling software package designed for the IBM PC environment has been described. FASTP (ver.1.01) allows the user to track transient heat or mass transfer behaviour in simple shapes which are used as the building blocks for modelling more complicated process engineering problems. To date, some of the problems which have been modelled include the continuous casting of steel billets, ladle cycling, wire feeding, and scrap dissolution. Results from FASTP have also been successfully correlated with a number of available exact solutions. Examples are included in the Appendices.

References

1. Carslaw, H.S. and Jaeger, J.C., <u>Conduction of Heat in Solids</u>, 2nd Ed., Oxford Univ. Press, London, 1959.

2. Grandillo, A., "A Temperature Control Strategy for Stelco McMaster Works", M.Eng. Thesis, McGill University, Montreal, 1988.

3. Mucciardi, F. and Guthrie, R., "Aluminum Wire Feeding in Steelmaking" Trans. Iron and Steel Soc., Vol III, Nov. 1983, pp. 53-59.

4. Guthrie, R.I.L. and Stubbs, P., "Kinetics of Scrap Melting in Baths of Molten Pig Iron", Cdn. Met. Quart., Vol. 12, No. 4, 1973, pp. 465-473.

5. Mucciardi, F. and Grandillo A.,"Thermal Cycling of Ladles at Stelco Mc-Master Works", Iron and Steel Engineer, Dec. 1987, pp.24-30.

6. Brimacombe J.K., Samarasekera I.V., and Lait J.E., <u>Continuous Casting</u>, Vol. II, "Heat Flow, Solidification and Crack Formation", ISS-AIME, 1984.

APPENDIX A

Heat Transfer Through a 1-D Rectangular Slab

This simple example serves to illustrate the computation of the transient thermal behaviour of a rectangular slab subjected to a convective environment. Shown in Fig. A1 are the properties and half-thickness of the active heat transfer axis of the slab. It was assumed that both faces of the slab are exposed to the same environment. Thus, symmetrical considerations allow for the analysis of half the thickness (denoted as L in Fig. A1).

The following equation represents the exact solution to this problem[1]:

$$\frac{T_{x,t} - T_\infty}{T_i - T_\infty} = 2 \sum_{n=1}^{\infty} exp(-\delta_n^2 Fo) \frac{\sin \delta_n \cos (\delta_n x/L)}{\delta_n + \sin \delta_n \cos \delta_n} \qquad (A1)$$

where $\cot \delta_n = \delta_n / Bi$

$$Bi = \frac{hL}{k} \qquad Fo = \frac{kt}{\rho C_p L^2}$$

Upon substituting the appropriate values for the required parameters, temperatures at x=0 (center) and x=L (surface) were calculated and are shown plotted for various times in Fig. A2. Also shown are some FASTP results. It will be noted that the exact and numerical solutions compare very favorably.

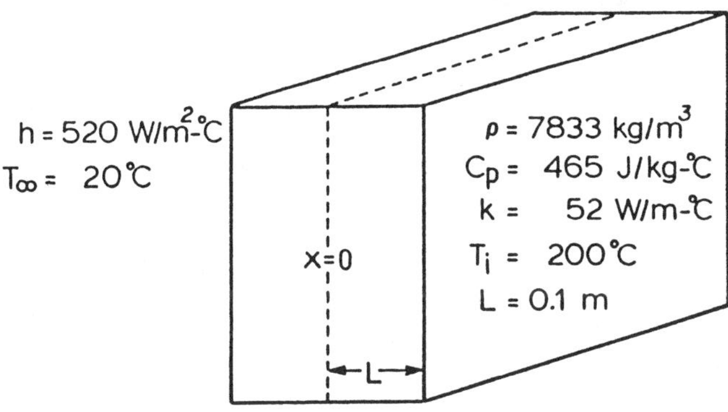

**Fig. A1 Schematic of a Rectangular Slab Subjected
to a Convective Environment**

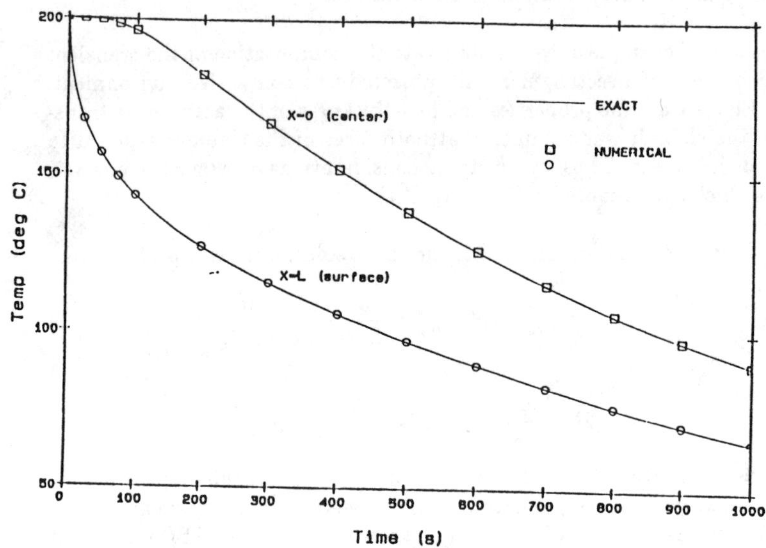

Fig. A2 Comparison of the Computed and FASTP Results
for the Surface and Center of a Cooling Slab.

APPENDIX B

Melting of Steel Cylinders in Molten Steel

Simulations were carried out in an effort to determine the kinetics of melting scrap steel rods in, for example, the tundish[2]. To successfully analyze a process one must first assess the possibilities and then decide on a course of action that best reflects reality. For this example, previous research efforts indicated that when a cold solid material contacts a liquid of low superheat, a layer of liquid may solidify on the surface of the solid[3]. Eventually, this solidified material (i.e. shell) will melt and expose the original material which will also melt if it has a low enough melting point.

From a mathematical point of view, this type of configuration requires an algorithm that can handle a moving interface to which is supplied a convective heat flux. This flux is, in turn, a function of the heat transfer coefficient and the temperature driving force of the system (i.e. bath temperature less the interfacial temperature). FASTP cannot directly handle convection at a moving interface, however one can devise a scheme that yields the equivalent and which fits within the FASTP framework. Since the heat transfer coefficient, h, can be expressed using the concept of the hypothetical stagnant boundary layer as k/δ where k is the state 2 thermal conductivity of the liquid and δ is the spacing between the surface nodal point and the first liquid (i.e. state 3) nodal point, one can use conduction to simulate convection. The only remaining item to be clarified is how to achieve the appropriate driving force given the liquid is subjected to conduction. The answer lies in the fact that FASTP interprets a fixed temperature boundary condition when applied to a stirred liquid phase as a signal to propagate this value throughout the liquid. Thus, by setting up the problem in this manner, one can simulate a moving boundary in a bath of fixed bulk temperature.

To simplify the analysis a number of assumptions were made and are summarized by the following statements. Both the bath and the cylinders are made of the same grade of steel having a solidus of 1490 °C and a liquidus of 1535 °C. The heat of fusion of 247,000 J/kg is dissipated over this temperature range. The bath temperature is assumed to be 1570 °C and the initial temperature of the cylinders is taken to be 25 °C.

Many simulations were carried out in an effort to determine a generalized set of results applicable to a variety of convective conditions and bar diameters. The results are presented in Fig. B1 where the Biot number is defined as:

$$Bi = \frac{h\,R_i}{k} \qquad (B1)$$

where, h = convective heat transfer coefficient

R_i = initial radius of the cylinder, and
k = thermal conductivity of the solid cylinder (i.e. state 1)

The Fourier number, Fo, in Fig. B1 is defined as:

$$Fo = \frac{\alpha t}{R_i^2} \tag{B2}$$

where, α = thermal diffusivity of the solid cylinder (i.e. state 1)
   t = time

On examining Fig. B1, it is seen that for lower values of Bi, R/R_i goes to values greater than one and then reverts to values less than one, ultimately going to zero. This implies that if conditions are such that Bi is sufficiently small, upon immersion of a cylinder, a shell forms (i.e. the cylinder radius increases beyond its initial value). After the shell has reached a maximum size it begins to melt back. This effect is very important in determining the total time required for the cylinder to completely melt (i.e. $R/R_i = 0$).

An expression relating the Biot number to the Fourier number was also derived for determining the total required melting time (i.e. time when $R/R_i = 0$). A regression of the results yielded the following expression:

$$Fo = 14.88 \; Bi^{-0.87} \tag{B3}$$

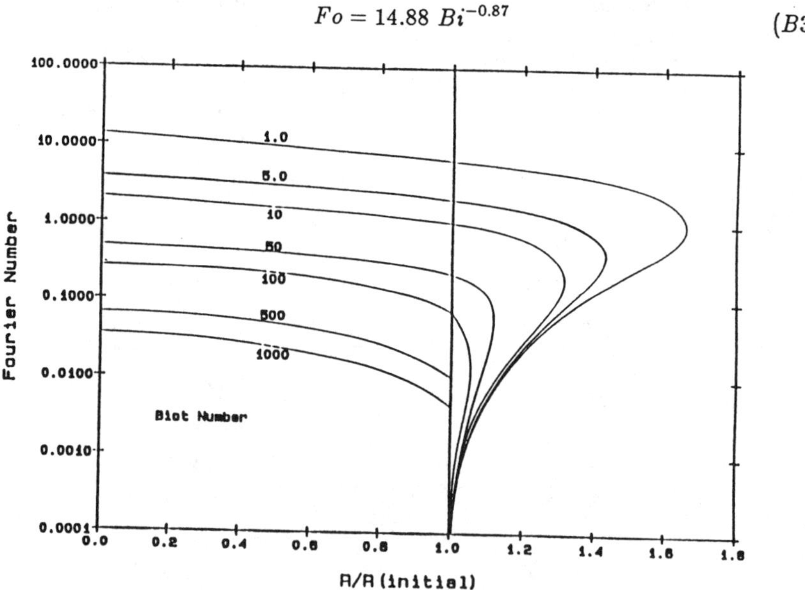

Fig. B 1 Correlations for the Melting of Steel Cylinders in Molten Steel.

APPENDIX C

Dissolution of Steel Cylinders in Pig Iron

A natural extension of the previous example (heat transfer) is the present one (mass transfer) wherein the steel cylinder does not melt but instead dissolves as a result of carbon diffusion from the bulk of the bath. In this case, heat transfer is not rate controlling. Because of its low carbon concentration, the steel has a high enough melting point that it does not melt until enough carbon has diffused into the surface layer to cause the liquidus of the cylinder material to drop below the bath temperature. As was the case in the previous example, a moving boundary condition will be utilized in essentially the same manner.

The analysis was simplified with the following assumptions. The bath and cylinder have approximately the same density. The bath is assumed to be of a uniform concentration (i.e. 4.74 %C at 1482 °C) while the cylinder has an initial carbon concentration of 0.1 wt %C. The solidus and liquidus concentrations at 1482 °C are taken to be 0.25% and 0.61%C respectively. A schematic illustrating the problem is shown in Fig. C1.

A number of simulations were carried out in an effort to determine dimensionless number plots applicable to a variety of convective conditions and bar diameters. The results are presented in Fig. C2 wherein the mass transfer Biot number is defined as:

$$Bi_M = \frac{k_M R_i}{D} \qquad (C1)$$

where, k_M = convective mass transfer coefficient
D = diffusion coefficient of carbon in solid (i.e. state 1) steel cylinder
R_i = initial radius of the cylinder

The mass transfer Fourier number in Fig. C2 is defined as:

$$Fo_M = \frac{D t}{R_i^2} \qquad (C2)$$

where, t = time

With Fig. C2 it is then possible to determine the melting time for a wide range of initial diameters, convective mass transfer coefficients or diffusion coefficients. A series of test results for this system have been reported by Guthrie and Stubbs[4] and are presented in Table C1. Also shown are the corresponding melting times computed with FASTP. The first 5 experimental results are in excellent agreement with the FASTP results, however the 3 extrapolated results advanced by Guthrie et al. are substantially different from the FASTP results. Upon careful examination of the data, the present authors found that the experimental data was extrapolated linearly when in fact, first principles indicate an exponential extrapolation is warranted. Thus, the validity of the FASTP analysis was confirmed once again.

Table C1   Comparison of Measured and Predicted "Melting" Times of Steel Cylinders

Experiment #	Measured Melting Time (min) *	Predicted Melting Time (FASTP) (min)
1	2.5	2.3
2	4.0	3.6
3	5.0	4.4
4	7.0	7.4
5	8.0	8.4
6	12.0 **	9.6
7	36.0 **	26.7
8	50.0 **	34.0

* Guthrie and Stubbs (1973)
** extrapolation

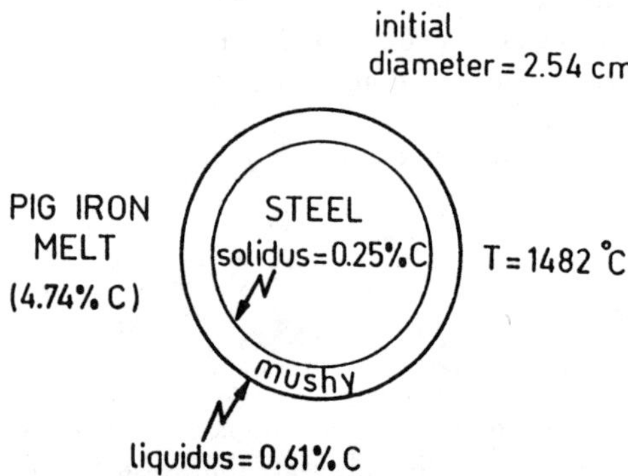

Fig. C1 Schematic of a Steel Rod Dissolving in Molten Pig Iron.

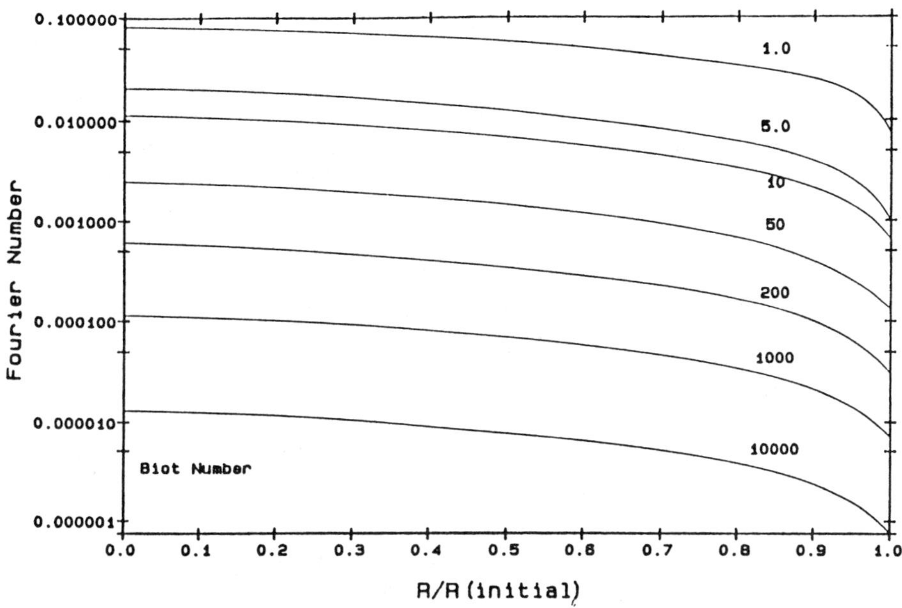

Fig. C2 Correlations for the Dissolution of Steel
Cylinders in Molten Pig Iron.

APPENDIX D

Ladle Cycling at Stelco's McMaster Works

Temperature control at every stage in the steelmaking operation can lead to improved product quality and productivity. To achieve temperature control, a basic understanding of the temperature losses in the system is required. To this end, ladle cycling simulations were carried out with FASTP in an effort to quantify the following objectives[5]:

1. Determine the maximum hot-spot temperature of the ladle shell.

2. Determine the effect of incorporating insulating tiles between the ladle shell and safety lining on the hot-spot temperature. This action was necessitated by the excessive shell temperatures inherent with normal shop operation.

3. Quantify liquid steel temperature losses both with and without insulating tiles.

An ideal ladle cycle based on a shop production rate of 14.5 heats/day was simulated (Fig. D1). Calculations were based on a 1-D system wherein energy is transferred radially. Fig. D2 shows schematically a comparison of two lining configurations. The only difference between the two is the incorporation of an insulating tile layer between the safety lining and the steel shell. Both configurations were analyzed with FASTP.

The steel was assumed to have an initial temperature in the ladle of 1650 °C. During casting, the rate of descent of the steel level in the ladle was relatively slow. Thus, the ladle was divided into five segments in the vertical direction. Each segment of 16.0 tonnes required 20 minutes to empty. Thus, the uppermost segment was assumed to have a liquid steel/lining contact time of 50 min. (30 min. hold for stirring and ladle metallurgy plus 20 min. emptying time) while the bottom segment had a contact time of 130 min. (30 min. plus 100 min. to empty the segment).

With FASTP, each step in the cycle was allowed to run for the corresponding amount of time specified in Fig. D1. At the end of each time period, execution stopped and a thermal profile was produced. Prior to execution of the next step, the appropriate boundary conditions (or other conditions) were changed and the last calculated thermal profile became the initial thermal profile for the next step in the cycle. This was continued until the thermal profiles from one cycle to another converged. It took 4 complete cycles to achieve a dynamic steady state temperature profile (i.e. working temperature). The predicted liquid steel tap temperature compensation for the first four heats on a new lining are shown in Table D1.

Fig. D3 presents the temperature profiles through the working and safety linings at various stages in the dynamic steady state cycling of a regular ladle. Equivalent temperature profiles for an insulated ladle are shown in Fig. D4. Both sets of profiles were computed for the bottom 1/5 section of the ladle where steel contact times are 130 min. Based on these results, one can see that the use of insulating tiles reduces the cold face temperature by 79 °C. Plant data correlated well with this value. A drop of 87 °C was reported.

The presence of insulating tiles reduces shell temperature, thus energy losses to the surroundings are reduced. FASTP calculations indicate that the incorporation of insulating tiles to the regular lining can result in an average decrease in liquid steel temperature loss of 12 °C. This implies that the tap temperature can be reduced by an equivalent amount which represents an approximate gain of 3% in shop productivity.

Table D1   Electric Furnace Tap Temperature Compensation
            Required for Heats Prior to Attaining Lining
            Working Temperature.

	Tap temperature increase, (°C)	
Heat number	Regular ladle	Insulated
1 (new lining)	37	39
2	8	10
3	3	4
4	0	2

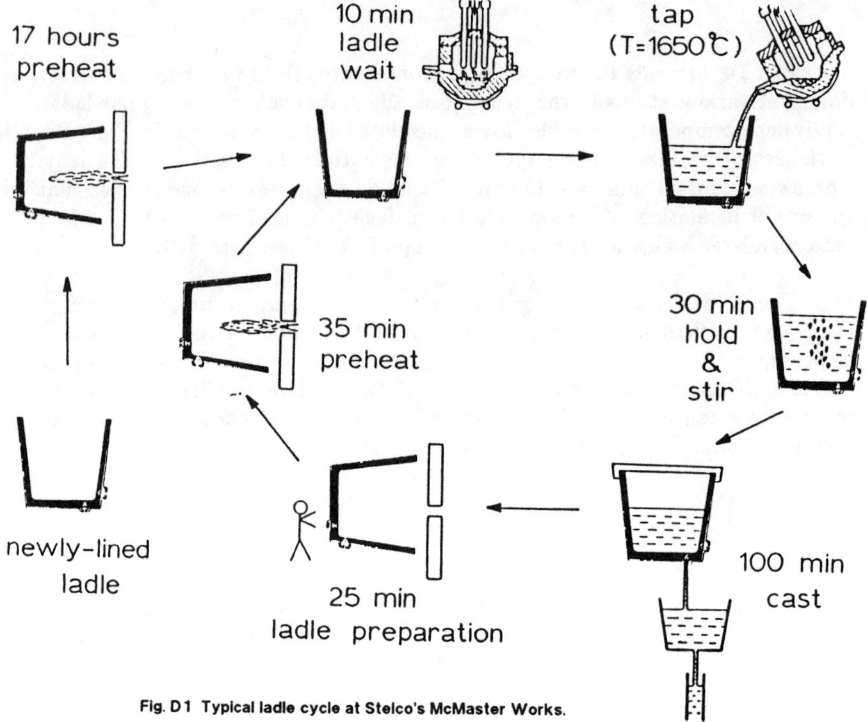

Fig. D1 Typical ladle cycle at Stelco's McMaster Works.

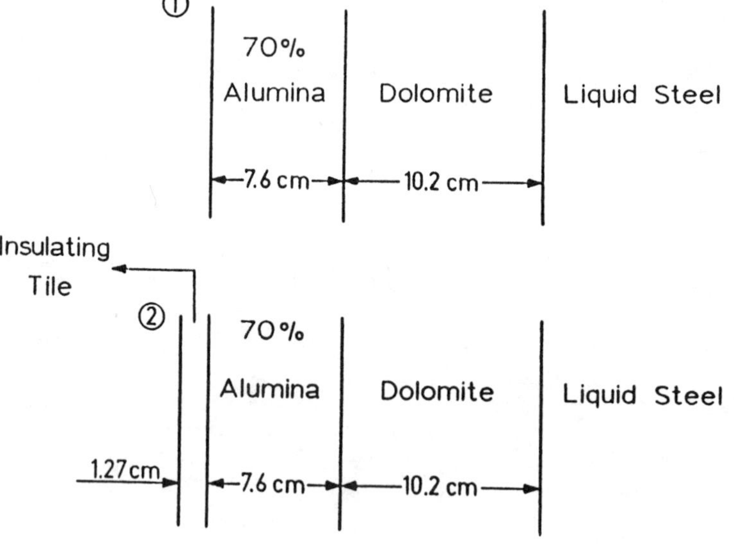

Fig. D2 Two Lining Configurations Analyzed with FASTP.

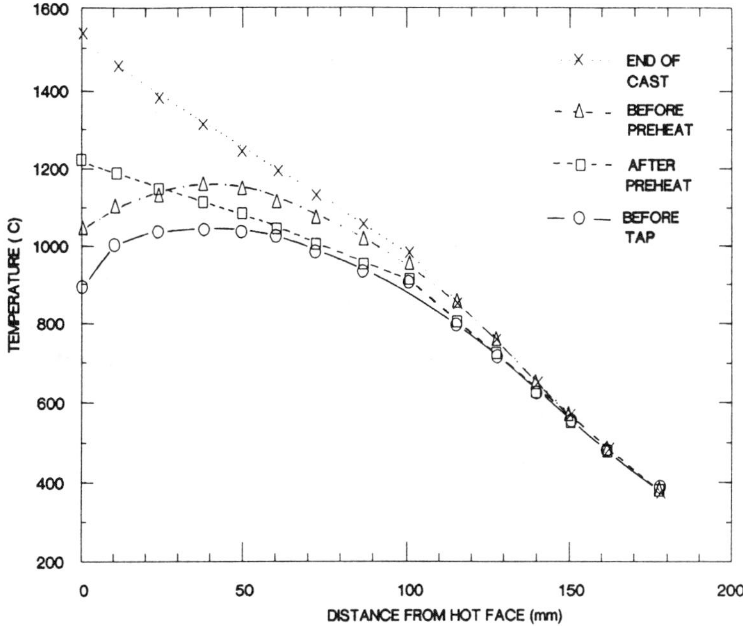

Fig. D3 Computed Steady State Temperature Profiles for a Regular Ladle.

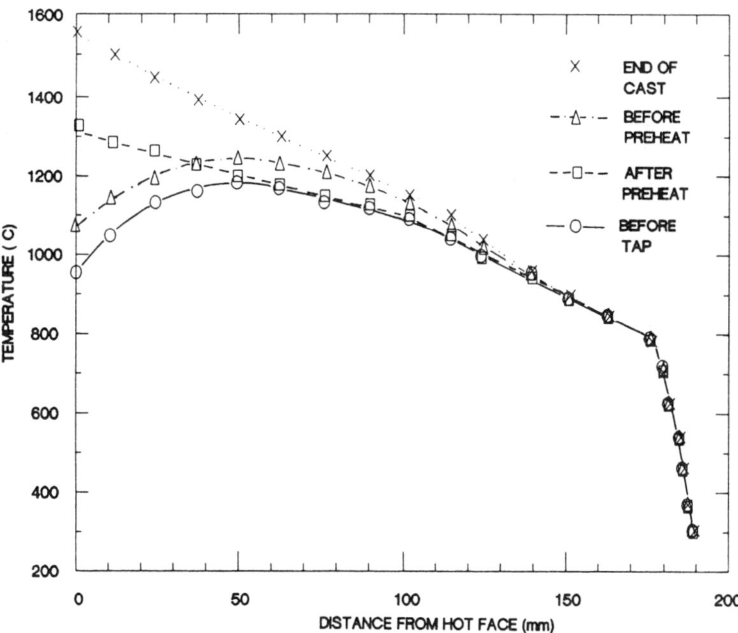

Fig. D4 Computed Steady State Temperature Profiles for an Insulated Ladle.

APPENDIX E

Transient Heat Transfer Analysis of Lance Injection

Injection of a reagent through a submerged lance is a common procedure in many pyrometallurgical processes. A schematic of a typical lance injection operation encountered in steelmaking is presented in Fig. E1. In general terms, a refractory lance of relatively low initial temperature is inserted into a melt while an inert gas flows through the lance. Once at the desired depth, the reagent is fed through the lance and into the melt. As the reagent normally starts off at room temperature, it can cause a substantial chilling effect at the nozzle as it contacts the melt - especially if it undergoes little heating as it moves through the lance. The result of insufficient heating is normally a blocked nozzle which can delay production.

To better understand lance injection, it was decided to carry out a complete simulation of the process with FASTP. A schematic of the cross-section of the lance is shown in Fig. E2. A number of basic assumptions were made and are summarized by the following statements. The lance is cylindrical and its body is made up of two materials (denoted as pipe and refractory in Fig. E2). The carrier gas and reagent form one uniform phase whose thermal and physical properties are assumed to be constant. In addition, the initial temperatures of the lance and reagent are equal.

In setting up this analysis, it is to be noted that this problem requires at least a complete 2-D grid, however FASTP cannot handle this type of configuration. Thus, it was decided that a 1-D grid would be used as the building block and analyzed repeatedly to produce results for a 2-D configuration. The approach adopted in the present study was to section the lance into 10 slices of thickness, Δz. Each slice was analyzed sequentially for periods of Δt time units. The energy transferred through the shell was absorbed by the fluid - thus causing its temperature to increase. This new fluid temperature was used as the boundary condition for the slice immediately below. The remaining slices were analyzed, the clock was advanced by Δt and the same procedure for all slices repeated.

To make the results more useful, it was decided that dimensionless parameters would be derived for this analysis. In order to do this, it was necessary to carry out a theoretical analysis to determine which ones are relevant. Since the simplest is the steady state case, the corresponding analysis was carried out and the following solution derived:

$$ln\,[\theta] = -\frac{Bi}{Gz} \tag{E1}$$

The above expression contains 3 dimensionless numbers: 1) θ, the dimensionless temperature, 2) Bi, the modified Biot number and 3) Gz, the modified

Graetz number. These are defined as:

$$\theta = \frac{\theta_z - \theta_b}{\theta_i - \theta_b} \qquad (E2)$$

$$\frac{1}{Bi} = ln(r_2/r_1) + \frac{k_1}{k_2}ln\frac{r_3}{r_2} + \frac{k_1}{h_1 r_1} + \frac{k_1}{h_3 r_3}$$

$$Gz = \frac{(\rho C u)_{gas} r_1^2}{2 k_1 z}$$

where, r_1 = inside radius of lance opening
r_2 = outer radius of material comprising the first radial section of the lance
r_3 = outer radius of lance in contact with the melt
k_1 = thermal conductivity of inner material of lance
k_2 = thermal conductivity of outer material of lance
h_1 = effective heat transfer coefficient at reagent/lance interface
h_2 = effective heat transfer coefficient at lance/melt interface
z = vertical displacement from the melt surface
θ_i = initial temperature of reagent and lance
θ_b = melt temperature
θ_z = temperature of reagent at position, z
ρ = density of reagent
C = heat capacity of reagent
u = velocity

For the transient case, the dimensionless Fo number is also required. The following form for Fo was derived and used in this study:

$$Fo = \frac{k_1 k_2}{(r_3 - r_2)k_1 + (r_2 - r_1)k_2} * \frac{(r_3^2 - r_1^2)}{\rho_1 C_1(r_2^2 - r_1^2) + \rho_2 C_2(r_3^2 - r_2^2)} * \frac{t}{(r_3 - r_1)} \quad (E3)$$

Results for the case where $1/Bi = 50.3$ are presented in Fig. E3. It will be noted that the dimensionless temperature attains a constant value at relatively small Fo numbers. This actual value of θ represents a steady state result and is as such an asymptotic solution. Fig. E4 presents results for $1/Bi = 62.6$ while Fig. E5 contains the results for $1/Bi = 176.3$. The latter results clearly indicate that steady state is achieved at higher Fo numbers (i.e. larger times).

The accuracy of these plots was checked by comparing computed steady state dimensionless temperatures to the corresponding asymptotic values obtained from Figs. E3-E5. The results are shown plotted in Fig. E6. It is apparent that the agreement is excellent.

In summary, FASTP was used to generate a relatively complex set of solutions for lance injection systems.

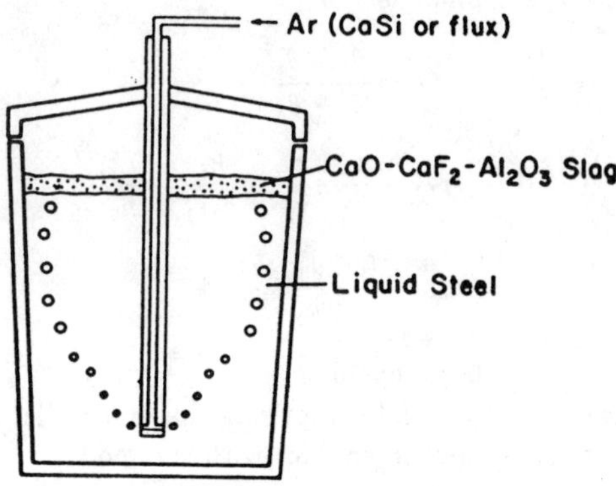

Fig. E1   Typical Lance Injection System for the
Treatment of Molten Steel.

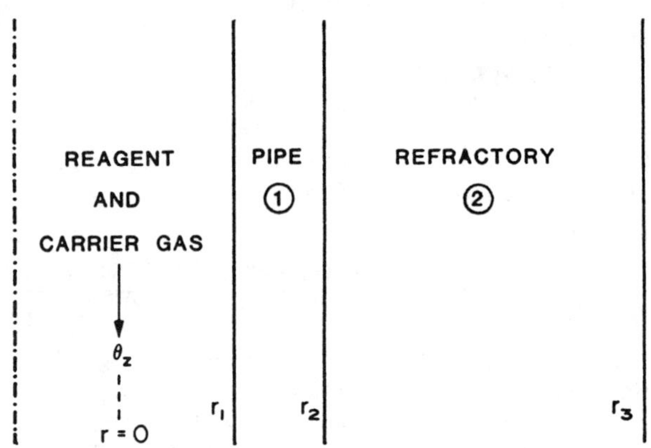

Fig. E2   Schematic of Lance Cross Section.

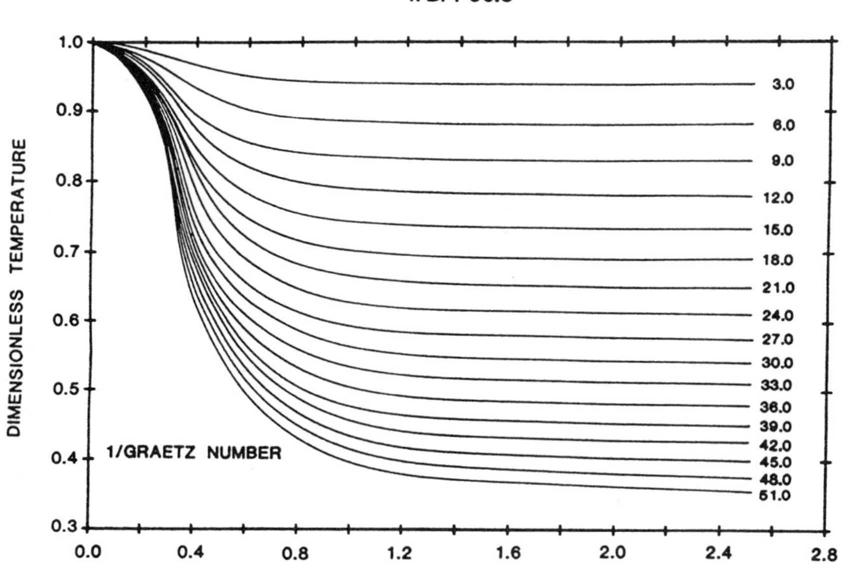

Fig. E3 Dimensionless Correlations for Lance Injection
when 1/Bi = 50.3.

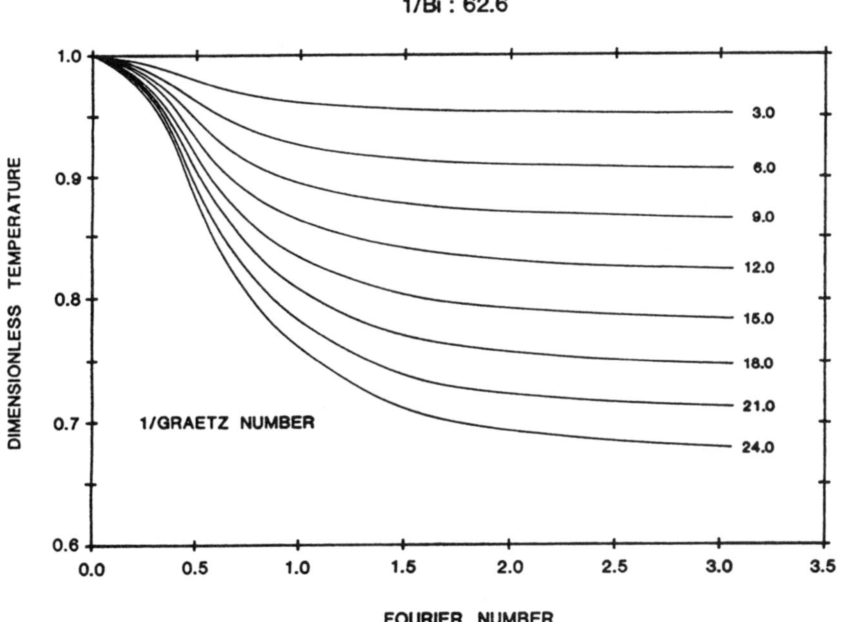

Fig. E4 Dimensionless Correlations for Lance Injection
when 1/Bi = 62.6

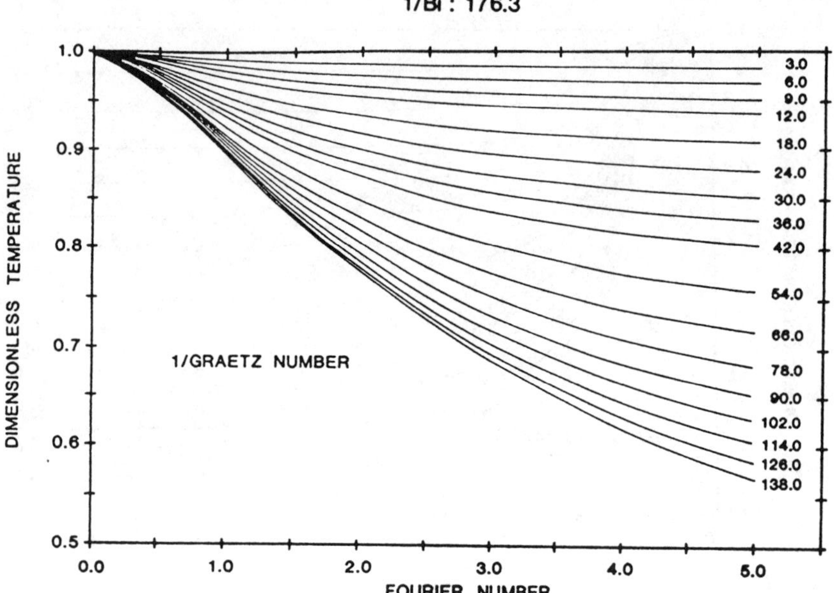

Fig. E5   Dimensionless Correlations for Lance Injection
when 1/Bi = 176.3.

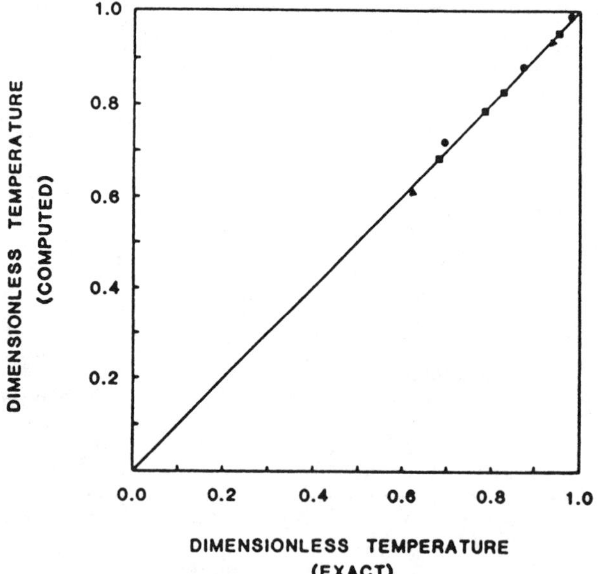

Fig. E6   Comparison of FASTP and Exact Solution
Results at Steady State.

APPENDIX F

Continuous Casting of Steel Billets

The following is an extract of some work which had been undertaken for a major Canadian steel producer. The work was carried out in order to evaluate the effectiveness of a proposed spray zone redesign. The design changes were intended to reduce billet reheat and the possibility of liquid metal breakout.

Objectives

The principal objectives of this work were to use FASTP to:

1. Model the cooling behaviour of a 152mm x 152mm transverse cross section of a continuously cast steel billet.

2. Predict the reheat behaviour of the above described billet at the billet's corner and midface.

3. Generate temperature isotherms of the same cross section at specific times during the simulation.

With FASTP, a 2-dimensional nodal configuration was used in order to simulate the cooling behaviour of a transverse cross section of the billet as it passed progressively through the three principal zones of the process:

1. copper mold,
2. secondary cooling zone,
3. air cooled zone.

FASTP generates an approximate solution to a 2-D problem by generating a product solution from two 1-D analyses (Fig. F1).

Assuming biaxial symmetric heat loss behaviour from the four faces of the billet throughout the entire casting process, only a quarter of the transverse section was actually modelled. The remaining three quarters, by symmetry were assumed to behave in a similar manner.

A geometric nodal progression, with a higher concentration of nodes near the outer surface, was chosen for the analysis. It was thought that the higher concentration would lead to a more accurate analysis of the surface behaviour.

Material property data was supplied to model the solidification behaviour of the system from a macro balance perspective. The model was not designed to take into account every phase transformation which would occur in the system in real life.

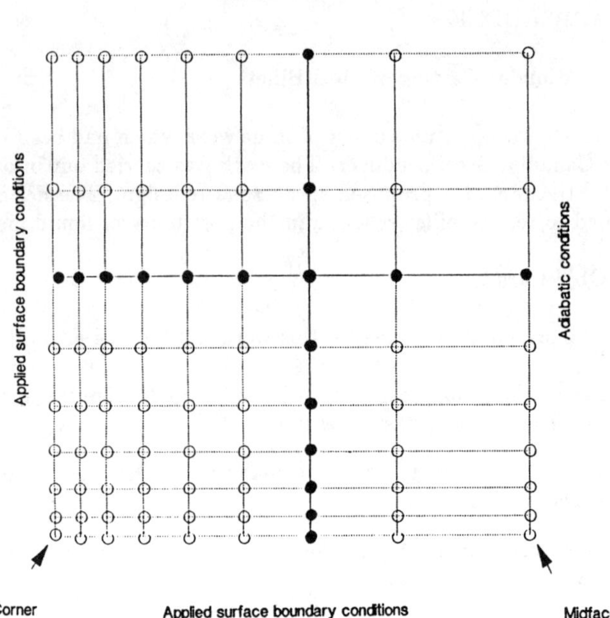

**Fig. F1 Nodal Distribution Across a Quarter Section
of a Continuously Cast Steel Billet.**

Mixing times were assigned in order to model the energy redistribution due to the fluid motion within the remaining liquid portion of the billet.

A feature referred to as Batch Mode Operation was used which greatly reduced the amount of time required to carry out an analysis. Batch operation allows the user to program the simulation to alter the applied boundary conditions at a preset time or when a surface node has reached a preset temperature. Using Batch Operation, the applied boundary conditions can be programmed to reflect the dynamic changes which occur as the transverse cross section passes through the three zones.

For modelling heat losses within the mold a negative heat flux was applied to the boundaries of the configuration corresponding to the billet surface. Using Batch programming the flux value was reduced in a stepwise fashion with time to model air gap formation within the mold as suggested by Brimacombe et al.[6].

Secondary cooling within the spray subzones was treated in a similar manner using Batch Operation by programming the simulation to alter the applied heat transfer coefficient, h, pending a preset time (to model entry into a new subzone) or a preset temperature (to model the effect of billet surface temperature on h).

Post-secondary cooling was modelled by applying the appropriate combination of convective and radiative boundary conditions to the billet surface.

A series of analyses were performed in which the starting conditions, casting speed, mold design, and spray zone design were altered in order to determine an optimum combination producing the least drastic reheat behaviour. Sample results from an analysis are presented in Figs. F2 & F3. These were generated with FASTP's graphics utility. Fig. F2 shows the predicted reheat behaviour at a corner and midface. Conditions modelling the spray subzones were applied at approximately the 19.0 second mark and continued until 120.0 seconds. Reheat is predicted to occur throughout. Fig. 3 is an example of the isothermal plots generated at selected times during the simulation. The feature enables the user to observe the development of the temperature profile with time.

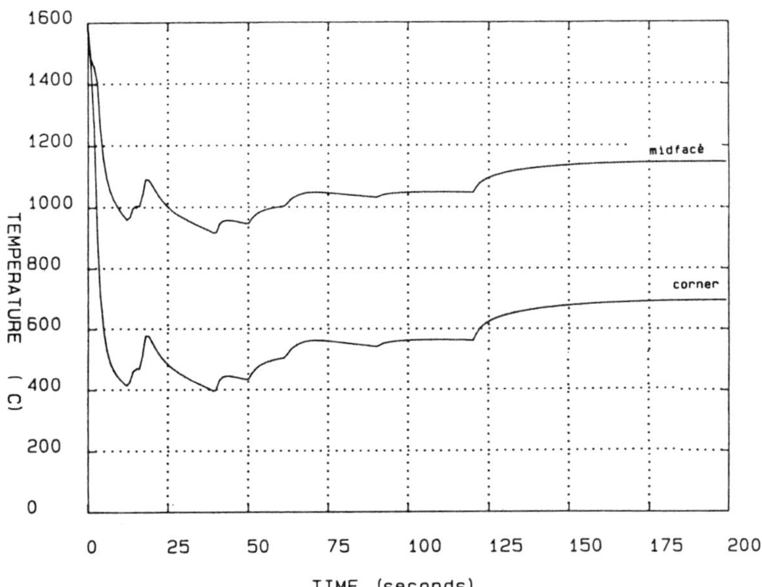

**Fig. F2 Predicted Cooling Curves for the Corner
and Midface of a Steel Billet During
Continuous Casting.**

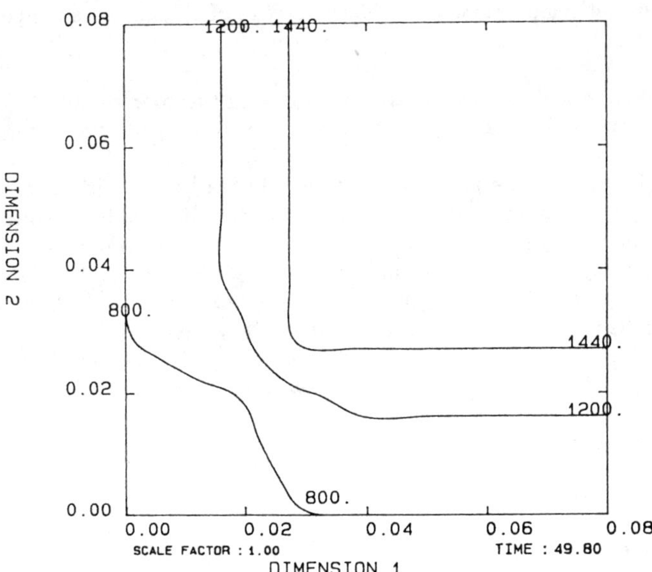

**Fig. F3 Three Computed Isotherms in a Quarter
Section of Billet.
Note : 1440°C corresponds to Solidus.**

Heat Transfer Tutor and Caster for
Educational Computing and Simulation

H. Henein, Department of Metallurgical Engineering and
Materials Science, Carnegie-Mellon University, Pittsburgh, PA 15213

Abstract

This paper presents two types of software for solving heat transfer problems: **Heat Transfer Tutor (HTT)** and **Caster**. The HTT is currently intended for the novice on the subject. It guides the user through the methodology of solving the Energy Equation for Conduction in cartesian, cylindrical and spherical co-ordinates. The proper choice of boundary and initial conditions and of assumptions are stressed throughout the tutorial with ample use of graphics. Steady state and transient Lumped Capacitance problems can currently be handled by the Tutor. This computerized instructional approach is particularly effective for students learning the material for the first time or for practicing engineers who are unfamiliar with the subject. HTT is programmed in the C-MU Tutor language.

Caster on the other hand is a fully flexible software package for modelling the heat flow and optimization in continuous casting of rectangular cross-sections for any alloy, caster geometry and boundary conditions. Several heat flow models may be used including finite difference explicit, ADI and ADI with iterations in time. SQP is used for optimization modelling with three objective functions, maximum and minimum casting rate, and maximum strand enthalpy content. Quality constraints are also specified but their definition is flexible. Full use of computer graphics is made to view the temperature profile in the strand. Caster runs on a MicroVax II in a VMS operating environment.

Keywords: heat transfer, CAD, computer software, educational computing, computerized instruction, modelling.

Introduction

Computers continue to play an increasingly important role in our field of materials science and engineering. Their primary functions in research and education are in number crunshing, word processing and data acquisition. There is a growing trend toward the decentralization of computer facilities through the use of networked PC's and workstations. There is also a dramatic increase in the useage of computers and software without it being nessecary for the user to have any knowledge of programming languages. These trends have collectively contributed to an unprecedented change in the way computers are used and has increased the access of the computer to 'untraditional' users and to users at a younger age. For example, engineering students at C-MU begin to use PC's and workstations prior to their traditional computer programming course. In fact many of the engineering students that are entering the University are already computer literate.

These dramatic changes have an important and an inevitable effect and influence on our educational approach and curriculum, not to mention the pedigree of the young graduate. One of these changes is the use of the computer as an instructional tool in teaching undergraduates and graduate students. This paper will describe two software packages designed for instruction: Heat Transfer Tutor (HTT) and Caster. While Caster was developed as a result of a research program, both software developments can be used for simulation in problem solving and design. The description in this paper will include an outline of the structure of the software packages, their capabilities, hardware requirements and some of our experiences in using them.

Heat Transfer Tutor

HTT was first conceived as an undergraduate project in 1984. It has since gone through three major revisions where it has expanded considerably in scope. The first version of HTT was written in PASCAL and ran on an HP9836 computer, because of its superior color graphics capabilities. Currently HTT is written in CMU Tutor[1] and runs on an IBM-RT using the Andrew operating system. The CMU Tutor language is a computer language especially designed for writing computer-based educational materials. The reasons for the change are due to the future portability of CMU Tutor to Macintoshes and IBM-PC, and to increased graphics resolution and capability of PC's. Furthermore, CMU Tutor was adopted because of its capabilities to compile code line by line and the ease by which the software can be developed incrementally.

Transport is one of the more difficult subjects taught to undergraduates in the materials field. The theoretical concepts and their mathematical representation is quite new to students. They must learn to integrate conceptual, phenomenological and mathematical knowledge to solve Transport problems. Illustrating problem solutions in a classroom is constrained by time and the medium of the blackboard. When seeing a solution to a problem presented in class, the average student is often more confused and fails to see the rigour and methodology in the problem solving approach. Other students show anger at their self-perceived lack of abilities and begin to lose confidence. These and other issues build-up over the duration of the course into a very high level of frustration and stress, and in a historical stigmatization of the course. Students often carry these perceptions and confusions about a specific subject matter into their professional career. Thus, many of the querries and difficulties would be more efficiently and effectively dealt with at the time the problem is initially being attempted by the student. Well designed interactive computer software thus presents unique capabilities for educational useage. These were some of the goals and motivation for developing HTT.

The approach of HTT is to prompt and guide the user through the methodology of solving a transport problem. This could otherwise be referred to as a type of user-friendly interface. The user must define the following issues in a prescribed order:

- Co-Ordinate system

- Assumptions

- Boundary Conditions

- Property values

This prescribed order is chosen to reinforce to the user good methodology in problem solving skills. HTT then computes the solution and illustrates it to the user. The tutor has been designed to determine if the choices made by the user are self-consistent, but it does not have the solution to the specific problem pre-programmed. This maintains the generic features of HTT and demands that the user retains the initiative in problem solving. This is particularly important in educational applications. User input in HTT is varied between specific key board entries, and mouse useage on screen graphics. The following is a more detailed description of the operation of HTT.

Co-Ordinate System

On activating HTT, the user must choose in which co-ordinate system the problem will be solved: cartesian, cylindrical or spherical. On making a selection, HTT displays the Heat Conduction partial differential equation of the respective co-ordinate system. Figure 1 shows a screen dump of this part of HTT run with the Andrew system on an IBM-RT. A sample geometry, wall, sphere or cylinder, is also displayed with a co-ordinate axis overlayed. This clearly helps the user define the relationship between the mathematics and the graphical representation of the object in the problem.

Assumptions

This part of HTT deals with making simplifications to the Heat Conduction Equation which is fully displayed on the monitor screen (see Figure 2). A menu of the possible assumptions is offered to the user. Currently HTT can only solve one-dimensional problems with constant thermophysical properties (i.e. thermal conductivity), and with or without heat generation. For transient problems, only the Lumped Capacitance Method can be solved at the present time. A full explanation is provided on-line for each of the possible choices of assumptions.

After choosing an assumption, HTT displays the effect of the assumption both graphically and mathematically (Figure 3). Thus, at the completion of this step which is defined by the user the final differential equation is clearly visible on the monitor. HTT now directs the user's attention to the choice of Boundary Conditions.

Boundary Conditions

Most boundary conditions for heat transfer problems have the characteristics of one of the following four types:

- Constant surface temperature
- Constant heat flux
- Adiabatic surface
- Convective heat flux

These choices are available in HTT which is also aware of how many boundary conditions are required. For lumped capacitance problems, this section of HTT is not encountered. A full on-line documentation of the meaning and significance of each of these boundary conditions is available (Figure 4).[1]

---

[1]Figure 4 and all subsequent figures are screen dumps from a Macintosh II running CMU Tutor - CT, version 1.0

Property Values

On the basis of the choices made by the user to this point, HTT prompts the user for data. This includes an initial condition if a transient problem is to be solved, thermophysical properties, size of geometry anlyzed, etc.

Solution

The differential equation is then displayed along with the boundary conditions , initial conditions and other relevant parameters (Figures 5 to 7). The integrated form of the equation is illustrated followed by some of the relevant numerical values imputted for the problem. HTT also displays a graphical representation of the geometry analyzed with a clear identification of where the boundary conditions are applied and a temperature profile is plotted (Figure 8). The solution may be reviewed by the user as many times as needed and there is an option to change some of the variables to study their effect on the solution.

Experience

HTT has been in use for three years in a semester long course on Transport and Kinetics. A quantitative and rigorous study of the useage of HTT has not been conducted. The results reported here are based on my observations of the students in class and on comments from the students. A quarter of the way into the semester, the first topic of transport is introduced - heat transfer. This follows an introduction to chemical kinetics and reactor design. While there are analogies between kinetics and heat transfer (eg. rate equations, rate constants, etc.), students still found the transition difficult to make.

Following a couple of sessions with HTT, students are more confident with the subject and the questions they ask tend to be at a more advanced level than those that occurred previously. The students have found that they can get help when they need it, whatever time of day. Their time is utilized more efficiently and they develop a better discipline in problem solving. Students quickly realize the potential of HTT and offer suggestions for future implementations and developments based on the portions of the course that they find more difficult. For example, after the first year of implementation the class requested that HTT be expanded to deal with transient problems. After further development of HTT, the next class clearly indicated that further addition to the transient solutions were not needed at that time. A more pressing need was the expansion of the treatment of convection.

The most critical comments made by the students concern the current limitations of HTT. They would like to see it further expanded to tutor them through the more complex issues in heat transfer. They see it as a strong potential aid for problem solving in their homework and as a pivotal tool for mastering the course content.

Thus it is clear that the goals in developping HTT have been met and that the approach is sound and productive. The potential for further development of HTT is great. I am confident that the future of this approach will be to include tutorial software with textbooks, to focus class discussion, to simplify continuing education and make it more accessible. Furthermore, with the addition of more complex solutions, such as say numerical methods, HTT can be quite a useful tool to professionals and researchers for problem solving in design and research without the need of the user to do any programming.

CASTER

Caster differs from HTT in a number of key ways. Caster at this stage of development is not intended for users who are not familiar with any computer operating systems or programming languages. HTT on the other hand can be used easily by such users. Furthermore, Caster is a dedicated software package for solving numerical heat flow problems in continuous casting. In contrast HTT is a much more generic package for solving heat transfer problems. In instructional applications, Caster is intended for graduate or advanced undergraduate students; while HTT is only helpful to the novice in heat transfer at this time. In future, by building a user friendly interface on top of the current command level structure in Caster, it is quite feasible for novice users to run Caster. A more interesting approach and the goal of my software development is to incorporate Caster and other numerical solutions into the structure of HTT, thus making HTT a fully flexible, integrated, and comprihensive software package for instructional, design and research purposes.

The philosophy in developing the current version of Caster is to provide the user with the means to run process models without having to deal with detailed equation derivation and computer code debugging. This is accomplished without compromising the integrity of the modelling process. For design and research applications, the user must still be aware of convergence and stability issues which in turn allow Caster to be used as an effective tool in instruction of modelling methodology. For routine or production useage, Caster allows the user to preset model parameters. The user then focuses on graphical results more quickly and with confidence.

Caster is the product of a research study in optimization modelling of continuously cast steel.[2] It runs on a MicroVax II with VMS operating system and an IRIS graphics workstation. Written in FORTRAN and C, it is designed to meet instructional, design and research goals. The functions of Caster are to edit, excecute, and control I/O operations. Different from HTT, Caster is a command driven software package. The commands were designed to build on many VMS features of the MicroVax operating system.

Editing allows changes to be made to the caster geometry, metal properties and cooling boundary conditions. Only square and rectangular sections can be modelled at this time by Caster. Constraints and Objectives functions may also be chosen or changed using the Edit mode of Caster.

Several heat flow models are available for excecution. All are based on the finite difference method, explicit and implicit formulations. Both the Kirchoff transformation and time iteration are available. The user chooses the grid size and time steps desired. Depending on the model algorithm and the discretization chosen, model excecution can be completed in 3 hours to 2 minutes. The result will be a complete thermal profile for the cross-section of the strand in the entire length of the caster specified. This data can be scanned numerically or more simply viewed on an IRIS workstation graphics monitor.

Other Excecute functions of Caster include the running of a Squared Quadratic Programming Optimization model. Objective functions include maximum and minimum casting speeds and maximum strand enthalpy. Constraints have been explained in detail in previous publications.[2,3] Further changes to constraints and objective functions can be made with the Edit function of Caster.

The I/O function of Caster is quite versatile. Temperature distribution in the strand can be accessed in numerical form or in graphical form. Graphical representation can be visible for any transverse cross section at any strand length down the caster. Temperatures at any longitudinal cross section are also available at any transverse cross section. The length of the strand and the cross-sectional views are user defined by mouse on the graphics screen. The color codes and shades used to define temperature increments are user defined. Temperature gradients rather than temperatures may also be defined by the user.

Closure

It is clear from the development of HTT and Caster that computers are a versatile tool for instructional and research activities in the materials field. Many of the uses and applications of the computer are yet to be defined and explored. These two software packages are examples which clearly demonstrate the potential of these activities. Development of well designed software is however a slow and tedious task, hampered by severe shortage of resources.

References

1. B.A. Sherwood and J.N. Sherwood: The CMU Tutor Language, Stipes Publishing Co., Chapaign, Ill, 1986.

2. B. Lally, H. Henein and L. Biegler: Proceedings of the Mathematical Modelling of Materials Processing Operations, AIME-TMS, Warrendale, PA, 1987, pp 1055 - 1069.

3. B. Lally, H. Henein and L. Biegler: Proceedings of the Modelling of Casting and Welding Processes, AIME - TMS, Warrendale, PA, 1988, in press.

Acknowledgments

The efforts of J. Melia, C. Abel, F. Coffman, D. Sundo and J. Osterle in programming HTT is especially recognized. Caster is a product of B. Lally's PhD thesis. Also acknowledged are the painstaking efforts of D. Holic to program the various software elements into Caster. Funding from the Center for the Development of Educational Computing, and the Department of Metallurgical Engineering and Materials Science, C-MU, is acknowledged for HTT. The Center for Iron and Steelmaking Research and National Science Foundation (grant 84-21112) are acknowledged for funding Caster. Finally and above all the students in Transport and Kinetics who were willing participants of the HTT experiment deserve more than my appreciation for their invaluable contributions.

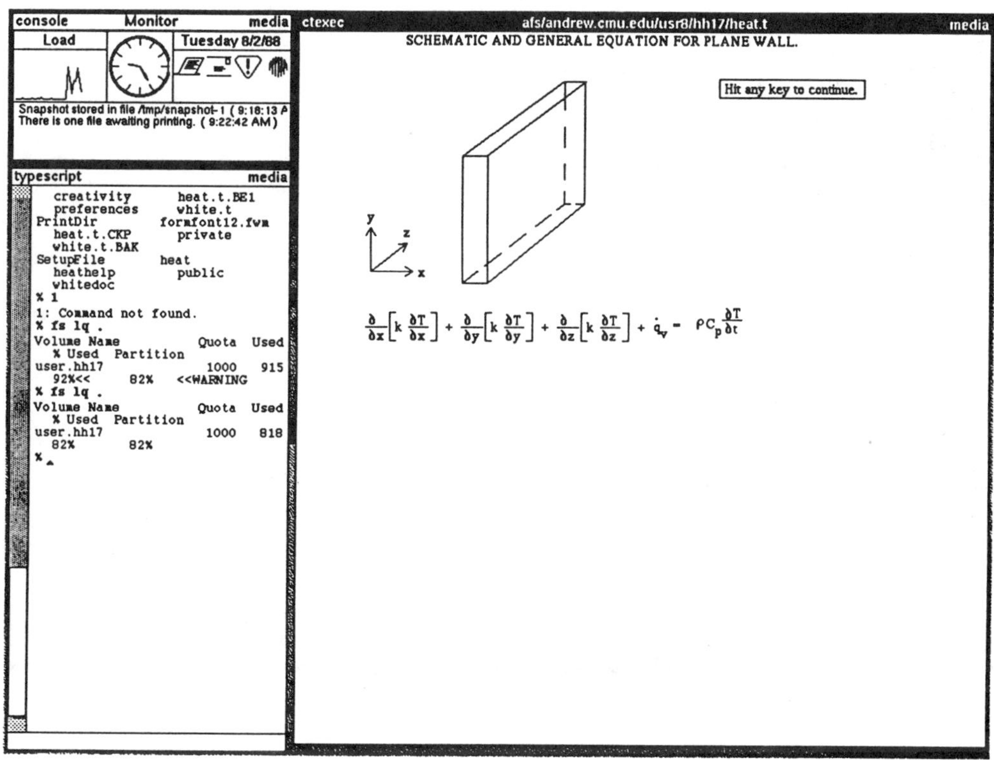

$$\frac{\partial}{\partial x}\left[k\,\frac{\partial T}{\partial x}\right] + \frac{\partial}{\partial y}\left[k\,\frac{\partial T}{\partial y}\right] + \frac{\partial}{\partial z}\left[k\,\frac{\partial T}{\partial z}\right] + \dot{q}_v = \rho C_p \frac{\partial T}{\partial t}$$

Figure 1: Screen Dump from IBM-RT running Andrew operating system. Screen graphics from HTT showing cartesian co-ordinates and corresponding Heat Conduction Equation.

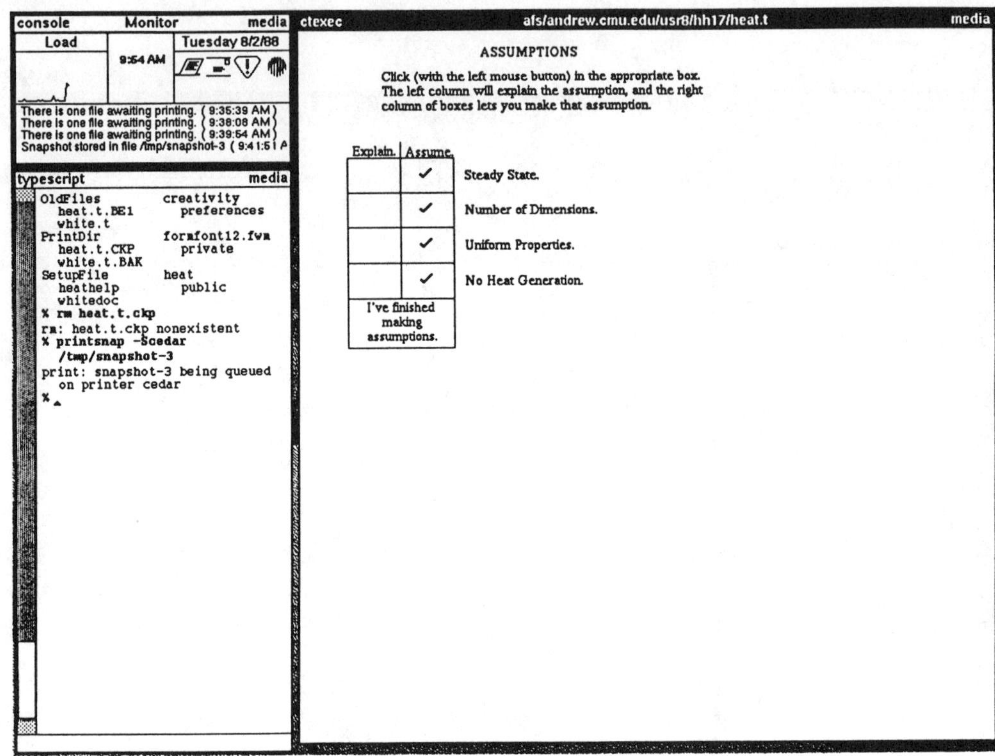

Figure 2: List of Assumptions screen from HTT running on IBM-RT.

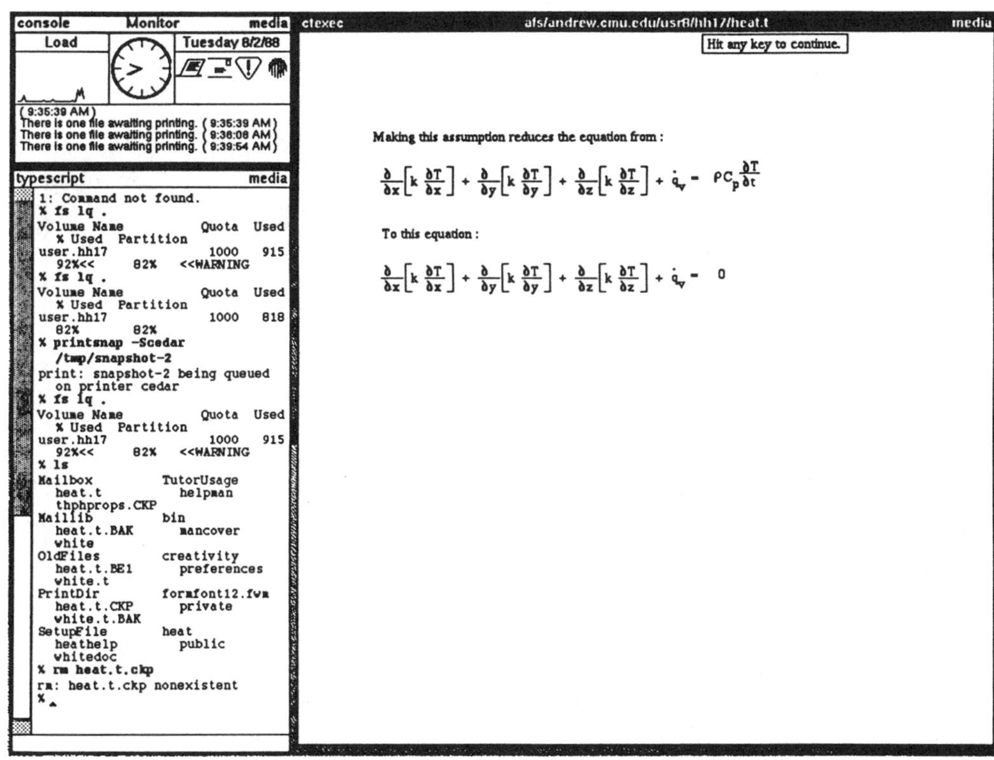

Making this assumption reduces the equation from :

$$\frac{\partial}{\partial x}\left[k\,\frac{\partial T}{\partial x}\right] + \frac{\partial}{\partial y}\left[k\,\frac{\partial T}{\partial y}\right] + \frac{\partial}{\partial z}\left[k\,\frac{\partial T}{\partial z}\right] + \dot{q}_v = \rho C_p \frac{\partial T}{\partial t}$$

To this equation :

$$\frac{\partial}{\partial x}\left[k\,\frac{\partial T}{\partial x}\right] + \frac{\partial}{\partial y}\left[k\,\frac{\partial T}{\partial y}\right] + \frac{\partial}{\partial z}\left[k\,\frac{\partial T}{\partial z}\right] + \dot{q}_v = 0$$

Figure 3: IBM-RT monitor displaying the effect of assumptions on the Heat Conduction Equation.

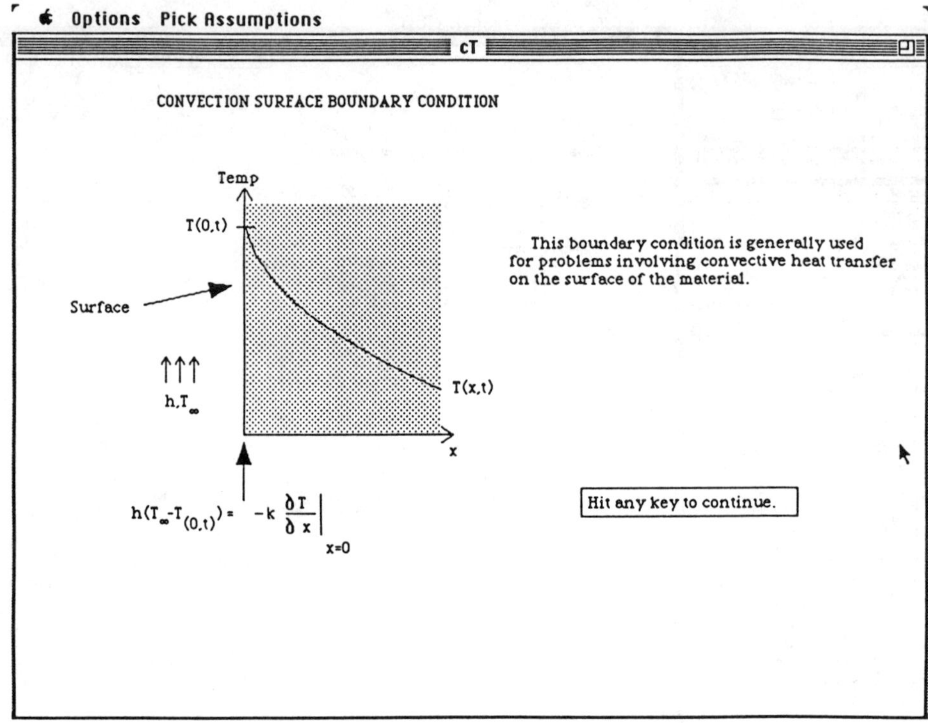

Figure 4: Screen dump from Macintosh II running CT, and displaying an explanation for one of the Boundary Conditions.

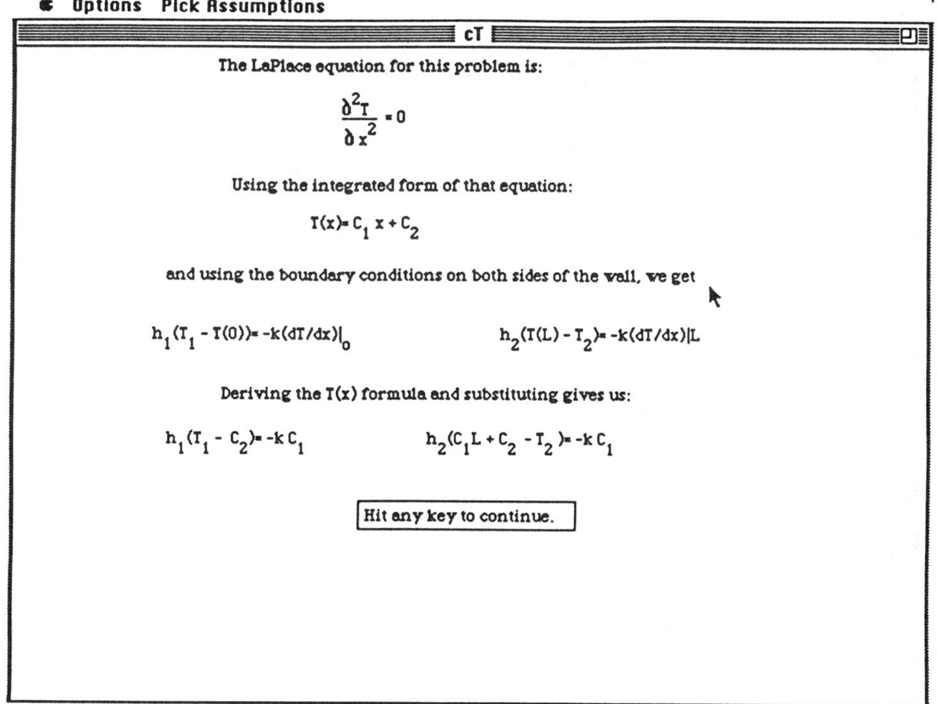

Figure 5: Screen dump from Macintosh II running CT, and displaying the first of four solution screens.

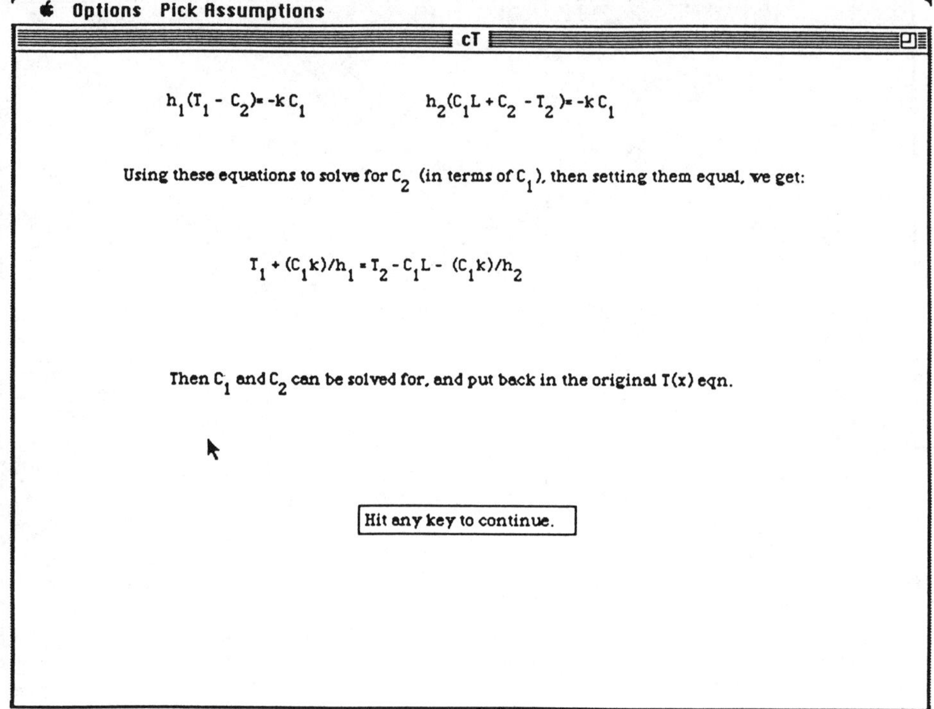

Figure 6: Screen dump from Macintosh II running CT, and displaying the second of four solution screens.

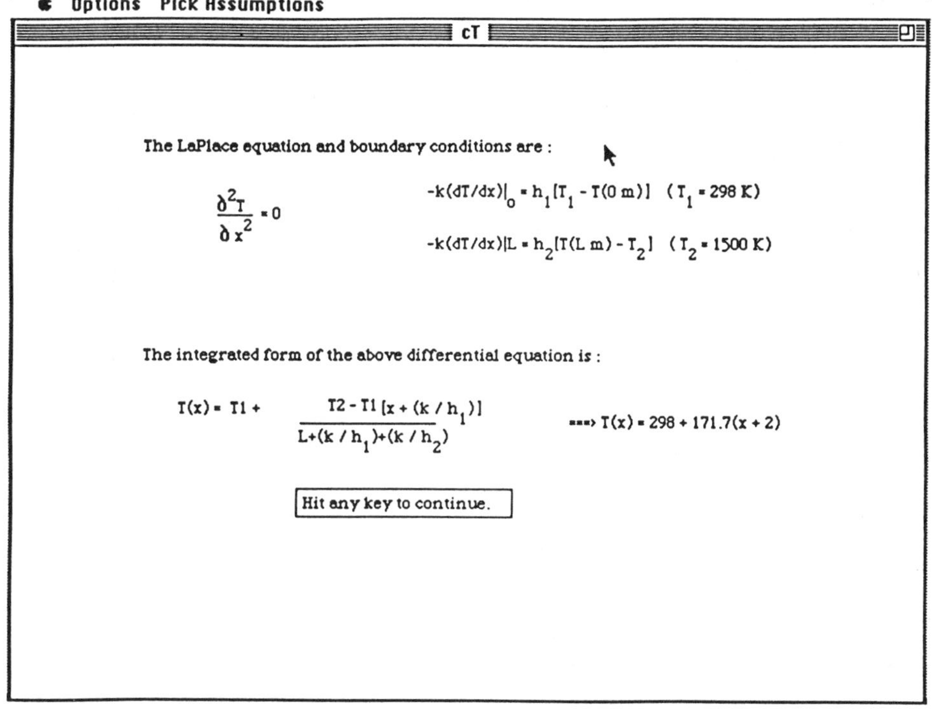

$$\frac{\delta^2 T}{\delta x^2} = 0$$

$$T(x) = T1 + \frac{T2 - T1\,[x + (k / h_1)]}{L + (k / h_1) + (k / h_2)} \quad \text{---> } T(x) = 298 + 171.7(x + 2)$$

Figure 7: Screen dump from Macintosh II running CT, and displaying the third of four solution screens.

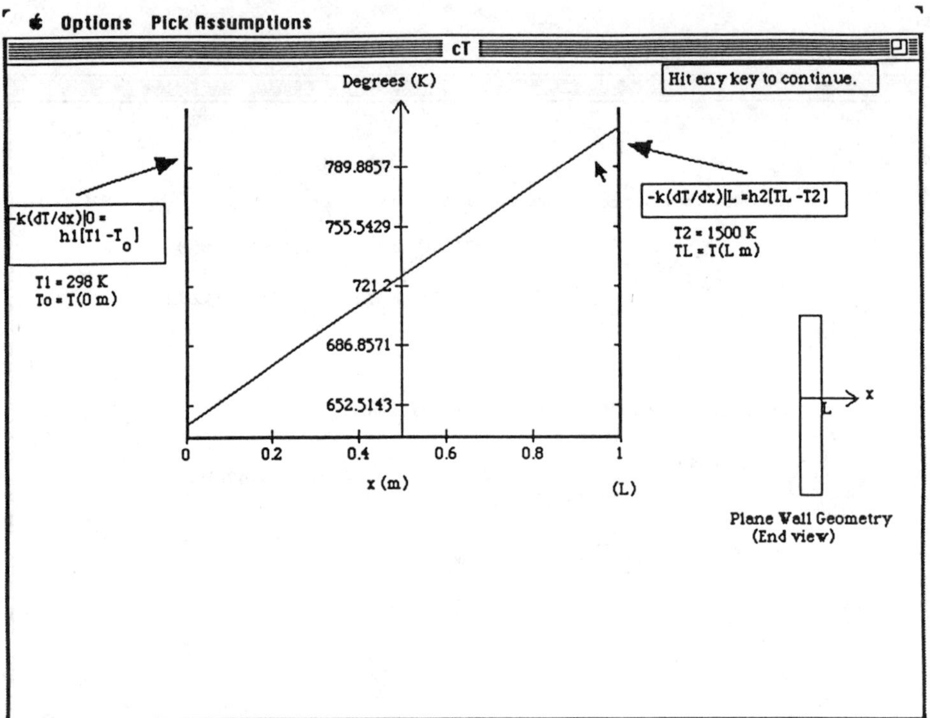

Figure 8: Screen dump from Macintosh II running CT, and displaying the fourth of four solution screens.

MATHEMATICAL MODELLING OF ZINC PROCESSES

G.G. Richards, D. Dreisinger, E. Peters and J.K. Brimacombe

The Centre for Metallurgical Process Engineering
The University of British Columbia
Vancouver, B.C.   V6T 1W5
Canada

ABSTRACT

A number of mathematical models have been developed at the Centre to study various zinc processes. In each case the work has been directed at using the models to understand the basic mechanisms of the process as a preliminary step in optimization and design. Three models are presented in this paper: zinc slag fuming, zinc pressure leaching and direct-chill casting of zinc jumbos. Zinc fuming is the reductive treatment of lead blast furnace slag with a mixture of coal and air. A model of the process was developed which accounts for the direct reduction of slag by entrained coal particles and oxidation phenomena in the tuyere gas stream. Analysis and subsequent in-plant experiments have shown that increasing coal entrainment in the slag results in increased fuming efficiency. The oxygen pressure leach of zinc sulphide was modelled on the basis of a series of reaction kinetics and mass transfer steps. Account was taken of the size distribution of the input feed. Results indicate that the ferric ion leaching step becomes increasingly important in the later stages of the leach. Preliminary studies suggest that if a uniformly high ferric ion leaching rate could be maintained throughout the autoclave, the process could be run with higher feed rates and zinc recoveries. A three-dimensional heat-flow model was developed to study the direct-chill (D.-C.) casting of zinc jumbos and to eliminate internal cracks in the cast product. The alternating-direction, implicit finite-difference method was employed to solve for the temperature distribution and sump profile. The model was validated by comparison with D.-C. casting experiments on zinc jumbos. The model led to the elimination of cracks through a redesign of the sub-mould spray cooling system.

KEYWORDS

Mathematical modelling; zinc processes; zinc slag fuming; oxygen pressure leach; zinc casting.

OVERVIEW

Mathematical modelling has been an integral part of our approach to the analysis of metallurgical processes. It is a very powerful tool when combined with industrial measurements in the elucidation of process kinetics, troubleshooting and optimization. This paper covers the application of modelling to three processes involved at various stages in the extraction and production of zinc: zinc slag fuming, zinc pressure leaching and the direct-chill casting of zinc jumbos. In each case modelling and industrial measurements have been brought together to improve our understanding of an existing process. Each case study, however, illustrates a different modelling technique which has been chosen to suit the particular problem under examination.

PART 1. ZINC SLAG FUMING

Introduction

The zinc slag fuming process was developed in the early part of this century to recover zinc from lead blast furnace slag (Richards, 1983). The first plant was commissioned in 1930 and the process has remained virtually unchanged since. In the process a 45 - 90 tonne batch of liquid slag is treated with coal and air to recover dissolved zinc oxide. This is done in a furnace with a rectangular cross-section (typically 2.4 m wide by 6.4 m long) made of individual water jackets. Two sets of submerged tuyeres run down the length of each side of the furnace just above the bottom. Coal is conveyed to the tuyere with primary air and enters the tuyere at an angle from above. The main blast, or secondary air, enters the tuyere behind the coal stream. Coal flow rates are of the order of 1-1.5 kg/s in a combined blast of 6-8 Nm^3/s. A schematic diagram of the process is shown in Fig. 1.

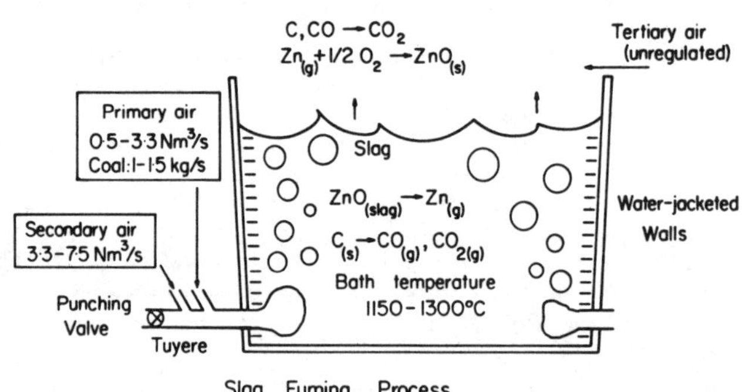

Fig. 1. Schematic diagram of fuming furnace cross-section showing construction and reactions.

Within the bath the coal-air mixture reduces the zinc oxide from the slag producing metallic zinc vapour. In the process the carbon is oxidized to CO and CO_2. Above the bath the zinc, CO and any unburnt coal are combusted with leakage or 'tertiary' air. The zinc oxide fume produced this way is captured in a baghouse for subsequent leaching. The process is normally applied to lead blast furnace slag which may contain from 11 to 18 wt.% Zn. Two operations, however, treat slag from a reverberatory furnace and start at significantly lower zinc levels, about 8% (Richards, 1983). Fuming is carried out until the zinc concentration reaches 1.5 to 2.5 wt.% depending on smelter schedules and recovery economics. Typical cycles last three hours which includes a total of one hour for tapping and charging. The overall efficiency varies from 1 to 2 kg coal per kg zinc. In the process, virtually all the lead in the slag is recovered.

The process has operated virtually unchanged since 1930. Major developments such as blast preheat (Blaskett, 1970), continuous fuming with fuel oil (Abrashev, 1972) and fuming with natural gas (Evdokimenko and others, 1977) have been confined to individual plants and not generally adopted. There has been historical debate over the fundamental mechanism of the process (Kellogg, 1957). One position has been that the process operates at thermodynamic equilibrium: the injected coal and air react and come to thermal and chemical equilibrium with the slag before leaving the surface. The rate of zinc fuming is thus dictated by thermodynamic variables such as temperature and the activity of zinc oxide. This assumption provided the basis for a series of mathematical modelling studies of the process (Bell, 1955; Grant, 1975; Kellogg, 1967). The alternate position held that fuming took place on entrained coal particles in the slag. Although this is not necessarily at variance with the thermodynamic view, it does suggest that kinetic factors such as injection phenomena are important. Following this line of reasoning, a major evaluation of the process was undertaken by two of the authors (G.G. Richards and J.K. Brimacombe) starting in 1980. A number of crucial pieces of evidence show that the process is in fact kinetically controlled (Richards, 1985a), and that the fuming rate is a function of both kinetic and thermodynamic factors (Richards, 1985b, 1985c; Cockcroft, 1988). Recent work has translated these ideas into significant process improvements (Cockcroft, 1987).

Modelling

Work on the process began with a series of measurements on several operating furnaces. In addition to slag composition-versus-time profiles, a number of other types of data were collected. The following conclusions were drawn (Richards, 1985a, 1985b, 1985c):

a) the slag contains from 0.1 to 1 percent carbon in the form of entrained particles,
b) concentration gradients of Zn and Fe are found adjacent to some pores in the slag,
c) about 10 percent of the injected coal passes through the bath unreacted, and
d) the blast air enters the bath in the form of bubbles which rise close to the furnace wall.

These findings indicate that factors like injection dynamics and diffusion of ZnO in the slag are important process parameters and therefore kinetics are a very important aspect of furnace operation. Furthermore, the picture which emerges is that, in its most simple form, the process can be divided into two main reaction zones: reduction in the slag by entrained coal particles, and oxidation of coal and ferrous iron in the slag by oxygen in the column of gas discharging from the tuyeres. This conception of the process is summarized in Fig. 2.

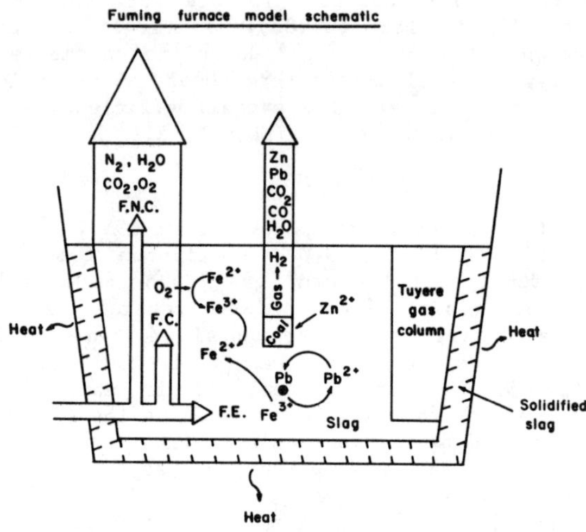

Fig. 2.   Basic conception of reaction processes in zinc slag fuming.

In order to model this complex system it was necessary to break it down into several interconnected modules:

1) a coal-slag reaction model which predicts the extent of reaction of injected carbon with ZnO, FeO and Fe$_3$O$_4$ as a function of time (Fig. 3(a)),
2) a lead prill-slag reaction model which calculates the rate of. liquid lead oxidation by magnetite,
3) an entrained coal particle residence time model, which for a given furnace charge and blowing condition, predicts the total length of time the coal reacts with the slag before it reaches the surface and exits to the freeboard (Fig. 3(b)),
4) a model of the injection process which determines the fraction of coal entrained in the slag, the fraction which remains in the tuyere gas column and combusts, and the fraction which short-circuits the bath,
5) the air-slag reaction model which predicts the rate of ferrous iron oxidation to magnetite,
6) the wall model which determines the amount of slag that melts from, or freezes onto, the wall as a function of changes in bath temperature (Fig. 3(c)),
7) the overall slag bath mass balance on each species of interest (including ZnO, FeO, Fe$_3$O$_4$, CaO and SiO$_2$), and
8) the overall slag bath heat balance.

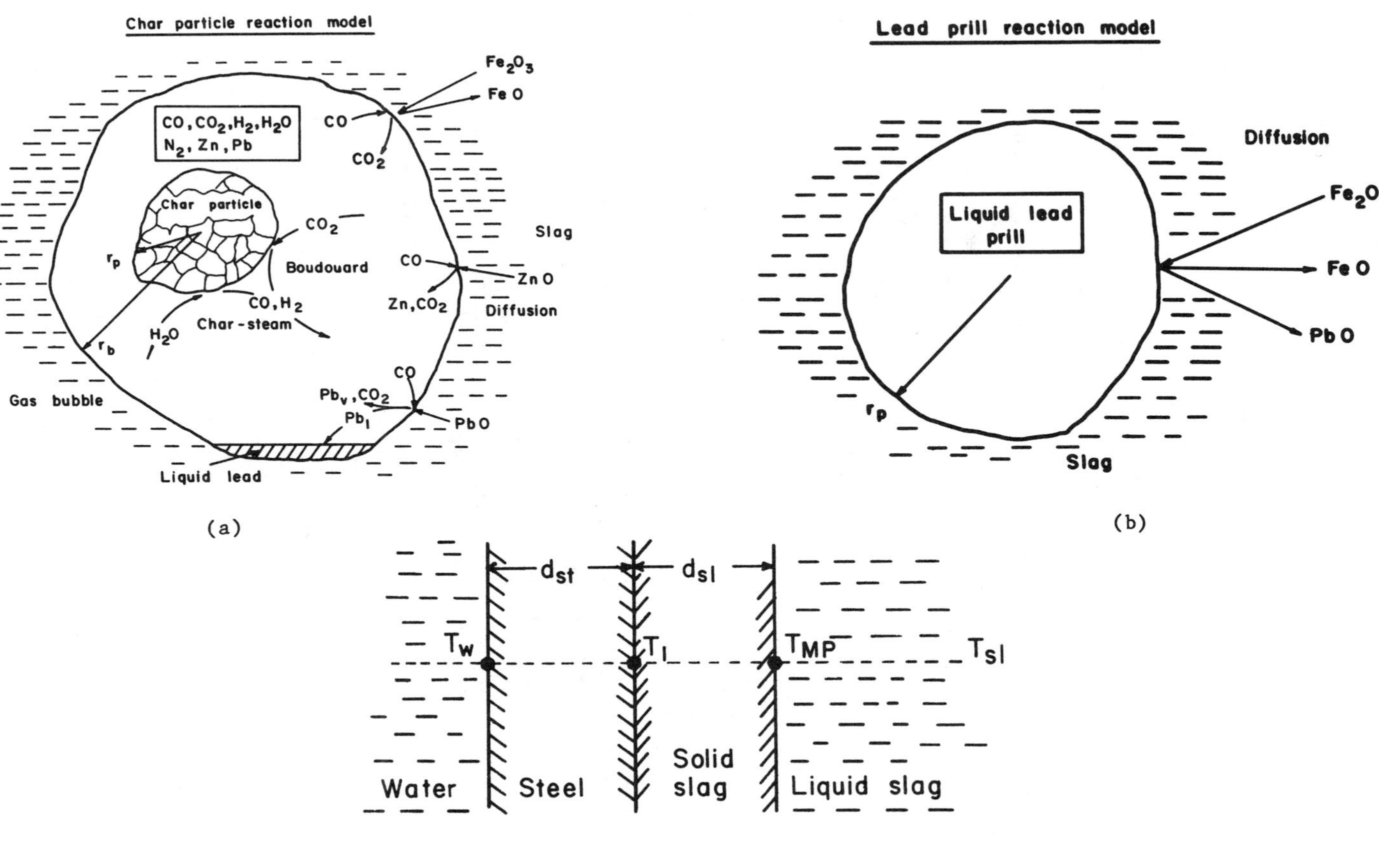

Fig. 3. Schematic diagrams of submodel systems in the fuming model: (a) slag-char particle reaction, (b) entrained coal particle residence time, and (c) furnace wall freeze layer.

A detailed description of these models is beyond the scope of this paper and the interested reader is referred to Richards (1985b) and Cockcroft (1988). Table 1 summarizes the basic details of each. The FORTRAN programme of this model is 3362 lines long including comments.

TABLE 1   Summary of Mathematical Models: Zinc Slag Fuming

	SYSTEM	COMMENTS
1	Slag – Char Particle	14 ordinary differential equations. Diffusion of ZnO, FeO, Fe_2O_3 and Boudouard and Char–Steam reactions considered rate controlling.
2	Lead Prill – Slag	Rate of lead oxidation calculated assuming a constant surface area of liquid lead in each time interval. Diffusion of PbO, FeO and Fe_2O_3 assumed rate controlling.
3	Residence Time	Calculated on the basis of the height of foamed slag in the furnace, an assumed 'bath circulation velocity' and the calculated tuyere bubble diameter. Used to determine total reaction time of model #1.
4	Coal Partition	$$F_{CE} + F_{CC} + F_{SC} = 1$$ The fraction of coal entrained, F_{CE}, and the fraction of coal combusted, F_{CC}, used to fit the model to industrial data. (F_{SC} calculated by difference).
5	Air – Slag Reaction	$$F_{OXY}$$ The fraction of oxygen not used in combustion which oxidizes FeO to magnetite. Used as a fitting parameter.
6	Wall	The thickness of slag on the wall was assumed to follow temperature changes in the bath instantaneously. The model calculated a steady-state thickness at each increment in bath temperature.
7	Bath Mass Balances	One balance for each species. Carried out over a 1 or 2 minute time step assuming a constant fuming rate. Input from models #1, #2, #4, #5 and #6.
8	Bath Heat Balance	Heat balance includes reaction enthalpies (from #1, #2, #4 and #5), heat losses through the wall as well as the latent heat release (#6), and the sensible heat load of coal and blast air. All products assumed to leave at bath temperature.

It is important to note that models #3 and #4 were used to fit the industrial data from five different operations. There was simply insufficient fundamental information in the literature from which to calculate the degree of coal entrainment, coal combustion or the rate of ferrous oxide oxidation under fuming conditions. The results of the fitting for one case are shown in Fig. 4 and the findings from all industrial trials are summarized in Table 2. A reasonable degree of consistency is seen, especially when account is taken of differences in bath depth and tuyere gas velocity.

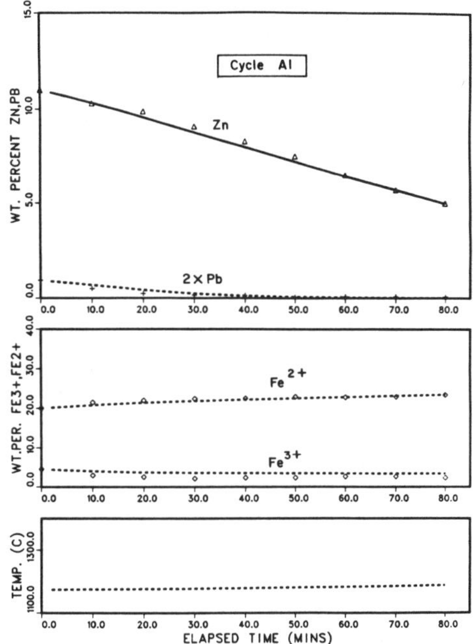

Fig. 4. Industrial data and model fit for fuming Cycle Al.

This pattern can be summarized fairly simply: the process operates in a dynamic equilibrium which balances ferrous oxidation with ferric reduction. If ferrous oxidation is high, ferric reduction rates must be high in balance and the rate of reduction of zinc oxide suffers. If the rate of ferrous oxidation is low, the rate of ferric reduction is correspondingly low and there is more reductant available for zinc oxide. Two routes are then available for process improvement: increase the amount of reductant (coal) entrained in the slag or decrease the rate of ferrous oxide oxidation. The former factor is the most easily varied and was the subject of a series of trials at Cominco's Trail Smelter (Cockcroft, 1987). In these tests a high-pressure coal injection unit was installed and operated to increase the fraction of coal entrained in the slag. In the actual test the total coal and air flow were kept constant and for some period the regular coal was cut back by about 10-20 kg/min while this amount was injected through one high-pressure tuyere. The results for one trial are shown in Fig. 5 and the improvements seen in several trials are reported in Table 3. In general, significant increases (of the order of 1.5-2 times) in fuming rate and fuming efficiency (wt. Zn/wt. coal) were seen.

TABLE 2 Model Parameters for Zinc Fuming Simulation

Cycle #	F_{CE}	F_{CC}	F_{OXY}	Predicted Oxygen Utilization	Calculated Bath Depth (m)	Nominal Tuyere Exit Velocity (m/s)
A1**	0.40	0.45	0.15	0.72	1.7	41
B1**	0.47,*0.20	0.45,0.50	0.20,0.22	0.85	2.0	31
C1	0.27	0.45,0.54	0.02,0.05	0.55,0.75	1.0	21
C3	0.30,*0.20	0.37,0.70	0.02,0.03	0.6–0.8	1.0	18
D1	0.37,*0.15	0.43	0.25	0.80	2.1	38
E1**	0.35	0.48	0.10	0.72	2.3	27

\# Letters refer to different companies
* Short-term change in entrainment factor
** No temperature data available

Summary

The mathematical modelling analysis of the zinc slag fuming process has yielded a clear insight into the kinetics of the process. It essentially provided a way of organizing a large amount of industrial data according to basic metallurgical principles and the subsequent distillation of the basic rate controlling steps on both microscopic and macroscopic levels. This new understanding of the process was verified in a number of in-plant tests suggested by the model which have led to significant improvements in fuming rates and coal utilization efficiency. The use of a computer model was a key aspect of this development.

TABLE 3 Fuming Rates and Efficiencies Determined from High-Pressure Injection Trials

	Fuming Rate Based on Assays (% Zn/min)	Fuming Efficiency Based on Assays (kg Zn/kg coal)	Increase in Fuming Rate	Increase in Fuming Efficiency	Predicted Fuming Efficiency From Model Analysis
Run 1 - 50:50 Hot-Cold (Steady State High-Pressure Injection)	0.10	0.635			
			1.69 X	1.38 X	0.92
Plant Average for 50:50 Hot-Cold Charges*	0.059	0.460			
Run 2 - 70:30 Hot-Cold (Steady State High-Pressure Injection)	0.124	1.02			
			1.85 X	1.90 X	1.1
Plant Average for 70:30 Hot-Cold Charges*	0.067	0.537			
Run 3 - 100% Hot (Steady State High-Pressure Operation)	0.150	1.27			
			1.83 X	1.99 X	1.25
Plant Average for 100% Hot Charges*	0.082	0.639			

*Low-Pressure Operation

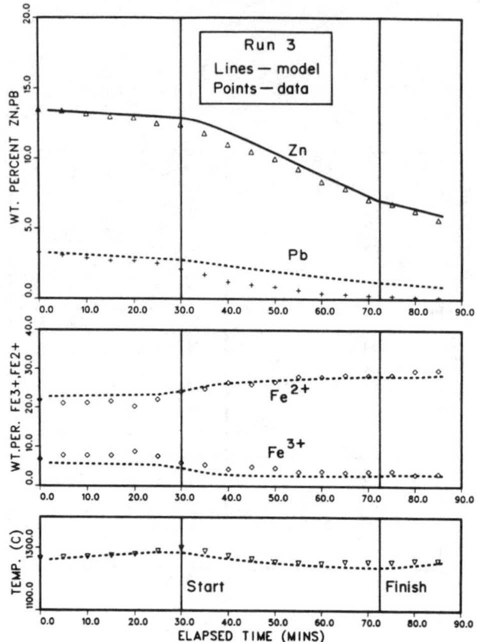

Fig. 5.   Industrial data and model fit for fuming Run 3
          (vertical lines indicate start and finish of
          high-pressure coal injection).

PART 2.   THE ZINC PRESSURE LEACH MODEL

<u>Introduction</u>

The direct oxygen pressure leaching of zinc concentrates at temperatures above
the melting point of sulphur (119°C) was demonstrated by Sherritt-Gordon Mines
Ltd. and Cominco Ltd. in a pilot plant operation in 1977.   The direct leach
process was installed at Trail, B.C. alongside Cominco's conventional
roast-leach plant in 1981.   The Cominco autoclave has a rated capacity of
190 tonnes per day of zinc concentrate.

Martin and Jankola (1985) have published a description of the plant and
included operating data for leaching at 97% and 127% of rated capacity.   The
autoclave is currently being operated at rates as high as 200% of rated
capacity (Parker and McKay, 1987).   For the purposes of this paper, however,
only the published data at 127% of capacity will be used for modelling.
Figure 6 shows a schematic flowsheet for the pressure leach while Table 4
presents the operating data abstracted from Martin and Jankola (1985).

The autoclave installed at Trail, B.C. contains four compartments, representing
the equivalent of four stirred tanks in series.   Oxygen is introduced to the
solution by spargers in the first two compartments.   A dual agitator

configuration is used in each compartment with the agitators dispersing gas bubbles, mixing the slurry and pumping additional gas into the slurry from the plenum space. A zinc extraction of 83.5% is achieved in the first compartment of the autoclave in 27.8 minutes of mean residence time at 127% of rated capacity. This is equivalent to an oxygen demand in the first stage of about 22.4 moles.m^{-3}min^{-1} (0.50 std m^3 of oxygen gas per m^3 of slurry per minute at 150°C).

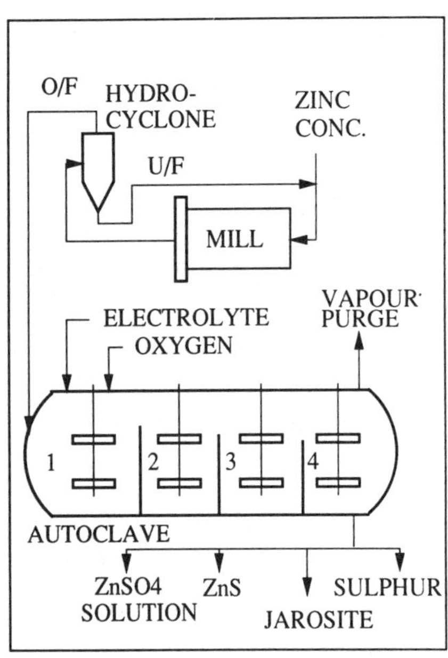

Fig. 6. The Cominco zinc pressure leach flowsheet.

TABLE 4 Selected Operating Data for the Cominco Autoclave

	Assay g/1 or %					Distribution %		
	Zn	Pb	Fe	S	H_2SO_4	Zn	Pb	Fe
Leach Input								
Zinc Outputs	48.6	5.7	11.6	32	–	100	100	100
Feed Acid	50	–	–	–	165	–	–	–
Leach Outputs								
$ZnSO_4$ Soln.	115	–	5.0	–	30	98	–	33
Flot. Conc.	1.8	0.6	0.8	95.2	–	1	1	2
Flot. Tail.	1.4	16.0	21.3	16.5	–	1	99	65

The leach chemistry is summarized by Eqs.(1)-(3) below. It should be noted that most of the sulphur reports in elemental form in the leach residue and that much of the iron reports as a plumbojarosite solid material.

$$ZnS + H_2SO_4 + 0.5O_2 \longrightarrow ZnSO_4 + H_2O + S \tag{1}$$

$$PbS + H_2SO_4 + 0.5O_2 \longrightarrow PbSO_4(s) + H_2O + S \tag{2}$$

$$FeS + 0.5H_2SO_4 + 0.75O_2 + 0.167PbSO_4 + 0.5H_2O$$

$$\longrightarrow 0.333\ Pb_{0.5}Fe_3(OH)_6(SO_4)_2 + S \tag{3}$$

Model Development

Two interdependent models were developed to simulate the zinc pressure leach process. The first is referred to as the "macromodel" and involves the integration of the various physicochemical processes of leaching. These processes include oxygen absorption, homogeneous ferrous oxidation and the ferric leaching of sulphide minerals. The second model is referred to as the "micromodel". The micromodel accounts for the leaching of a concentrate containing particles of various sizes under linear kinetics conditions for variable residence times in each stage of the autoclave. Both the macromodel and the micromodel are described in detail below.

The macromodel. Traditionally it has been assumed that heavily agitated leaching systems are kinetically controlled at the mineral-solution interface, even when a gaseous reagent is involved. However with gas absorption rates as high as 20-40 mole m^{-3} min^{-1}, the resistance to gas-liquid mass transfer is probably significant. There is also compelling evidence that the presence of ferric ions in solution is necessary for high leaching rates. Dissolved iron acts as a catalyst, being oxidized homogeneously by dissolved oxygen and subsequently reduced by reaction with a sulphide mineral as leaching proceeds. There are, therefore, at least three steps in the leaching process involving reactions taking place at different sites. Equations (4)-(6) provide mathematical statements for the rate at which each step proceeds.

Step 1 $O_2(g) \longrightarrow O_2(aq)$

$$R_1 = R(O_2) = k_g([O_2]^* - [O_2]) \tag{4}$$

Step 2 $4Fe^{2+} + 4H^+ + O_2 \longrightarrow 4Fe^{3+} + 2H_2O$

$$R_2 = 4R(O_2) = k_2[Fe^{2+}]^2[O_2] \tag{5}$$

Step 3 $2Fe^{3+} + ZnS \longrightarrow 2Fe^{2+} + Zn^{2+} + S$

$$R_3 = 2R(O_2) = k_z A_{ZnS}[Fe^{3+}] \tag{6}$$

The rate expressions for each of the three steps are based on assumed mechanisms. Since the first compartment is normally operated at steady state, the rate of reaction for each of the three steps (reported as $R(O_2)$, the oxygen consumption rate) must be constant.

Equations (1)-(3) are a good representation of the total pressure leach chemistry but are inadequate in describing the first stage leach condition. The first compartment is usually more acidic and reducing than the other three compartments. This inhibits the formation of jarosite. The model therefore assumes that no jarosite forms in the first compartment and that all iron leached remains in solution as either ferrous or ferric ion.

Since iron (present in the concentrates as marmatite) and lead (present as galena) and some sulphur are oxidized (to ferrous, ferric, plumbous and sulphate), it is necessary to adjust the total oxygen demand of the system by a factor of 1.454 above that required solely for zinc leaching. This factor has been used in all subsequent calculations.

To solve Eqs.(4)-(6), values of the rate constants k_g (gas absorption), k_2 (a third order homogeneous rate constant for ferrous oxidation), k_z (a heterogeneous rate constant for ferric attack of the sulphide mineral) and A_{ZnS}(the steady state sulphide mineral surface area) must be chosen. These equations contain three internal variables, $[O_2]$, $[Fe^{2+}]$ and $[Fe^{3+}]$, which are at steady state in each compartment. In addition, for the first compartment, the sum of the ferric and ferrous concentrations in solution must equal C_F, the total concentration of iron leached.

$$[Fe^{2+}] + [Fe^{3+}] = C_F \qquad (7)$$

Values for the rate constants k_g, k_2 and k_z are discussed below. The computation of A_{ZnS} will be discussed in the micromodel.

The value of k_g for oxygen mass transfer. k_g is a constant that combines the gas-liquid interfacial area with the mass-transfer coefficient, i.e.

$$k_g = k_{g/1} \cdot A_{g/1}, \ (\min^{-1}) \qquad (8)$$

Unforunately, no reliable correlations for predicting values of k_g under zinc pressure leach conditions have been published. A starting value of k_g must, therefore, be extracted from the operating data. The first two compartments have dual agitators and gas spargers. The value of k_g for these compartments is probably similar. The third and fourth compartments do not have spargers and therefore rely on oxygen pumping by the agitators. The oxygen demand in these stages is quite low because most of the zinc extraction occurs in the first two compartments. For the purposes of modelling, it will therefore be assumed that the slurry in the final two compartments is saturated in oxygen.

The value of k_2 for ferrous oxidation by dissolved oxygen. The rate of ferrous oxidation by dissolved oxygen has been studied by numerous workers (Chmielewski, 1984; McKay, 1958; Cornelius, 1958; Mathews, 1972; Bielpolskij, 1948; Pound, 1939) who have usually presented their results in the form of a termolecular rate law (Eq.(5)). A recent study (Dreisinger and Peters, 1988) determined an average value for $k_2 = 2.54(10^{-3})$moles^{-1} m^3 min^{-1} atm^{-1} for typical zinc pressure leach conditions. This value will be used for all calculations in this paper.

The value of k_z for ferric oxidation of ZnS. The value of k_z has been

estimated based on a calculated average leaching rate of 0.4 μm/min (from Martin and Jankola's data and the micromodel). k_z represents the rate constant for either a direct chemical attack of the ZnS and other sulphides or the electrochemical attack (via the Fe^{3+}/Fe^{2+} couple) of the sulphides. A discussion of other possibilities (e.g. mass-transfer control, H_2S intermediate reaction, etc.) has been published elsewhere (Dreisinger and Peters, 1987). Table 5 summarizes the calculated values of k_z under different extraction conditions.

TABLE 5   Calculated Values of k_z for Various Amounts of Zinc

Extraction

Assume that half the iron is in the ferric state and the mineral dissolves at the rate of 0.4 μm/min. Reaction at 150°C.

Zinc Extraction	70	80	90
C_F, moles/m^3	198	227	255
$[Fe^{3+}]$, mole/m^3	99	114	128
k_z, m/min	3.79×10^{-4}	3.31×10^{-4}	2.94×10^{-4}
$[Fe^{3+}] \cdot k_z$, moles/m^2/min	0.03752	0.03752	0.03752

The micromodel. The micromodel simulates the leaching of concentrate particles having a range of initial sizes that have been exposed to widely varying residence times in the autoclave compartments.

The following procedure was adopted to develop the micromodel.

1. The feed to the autoclave is reground zinc concentrate that is nominally 95% −325 mesh (44 μm). Particle size data were obtained from Cominco (Parker and McKay, 1987) for a typical feed material to the autoclave. The Rosin-Rammler-Bennett equation (Rosin and Rammler, 1933) was fitted to this data using non-linear, least-squares estimation. Equation (9) shows the general form of the function while Eq.(10) gives the fitted form for the feed material to the autoclave. Figure 7 illustrates the fit of the function to the data.

$$Y = 1 - \exp(-(u/u')^n) \tag{9}$$

$$Y = 1.026 - \exp(-(u_o/22.76)^{1.563}) \tag{10}$$

where the subscript 'o' is added to refer to unleached particles; Y is the cumulative fraction of material found below a size u. It is also possible to differentiate Eq.(10) to give dY/du_o (or $\Psi(u_o)$), the probability of finding a given differential unit of material in a size range between u_o and $u_o + du_o$.

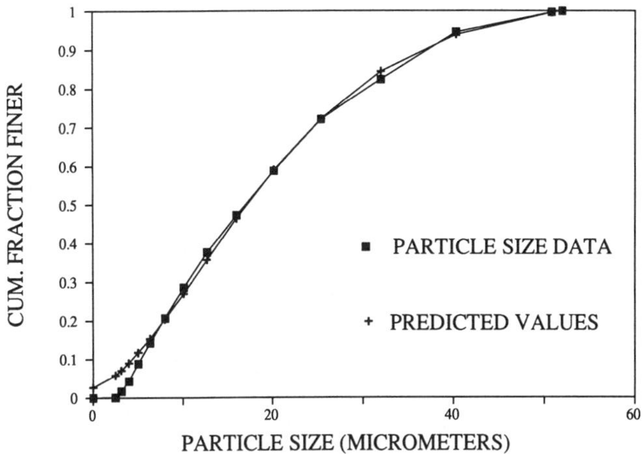

Fig. 7. Particle size distribution data for feed to the pressure leach. Predicted values from the Rosin-Rammler-Bennett distribution function.

$$dY/du_o = \Psi(u_o) = 1.181(10^{-2})u_o^{0.563} \exp(-(u_o/22.76)^{1.563}) \qquad (11)$$

2. The probability of a residence time, t, in an ideal continuously stirred reactor is given by:

$$P(t) = \frac{1}{t_{ave}} \exp(t/t_{ave}) \qquad (12)$$

where t_{ave} is the mean residence time in the reactor.

3. For the purposes of the model it has been assumed that the particles leach at a linear rate, k_1. The particle size after residence time t (initial size u_o) is given as,

$$u = u_o - 2k_1 t \qquad (13)$$

The volume fraction (or weight fraction) of a particle remaining unleached in time t is therefore:

$$\frac{V}{V_o} = \frac{W}{W_o} = (\frac{u}{u_o})^3 = (1 - \frac{2k_1 t}{u_o})^3 \qquad (14)$$

When the equations for particle size, residence time and linear leaching rate are assembled, an overall equation describing the fraction of all particles unleached at time t results.

$$\frac{V}{V_o} = \int_{2k_1 t}^{u_o(\text{max})} \int_0^{\frac{u_o(\text{max})}{2k_1}} (1 - \frac{2k_1 t}{u_o}) \ \Psi(u_o) \ \frac{\exp(- \frac{t}{t_{ave}})}{t_{ave}} \ dt du_o \qquad (15)$$

The value of A_{ZnS}. The concentrate particles have been assumed to be spherical. Based on this assumption and the initial particle size distribution fitted to the Rosin-Rammler-Bennett function, an effective surface area for the feed material of 144 m^2/kg of concentrate or $1.96(10^4)m^2/m^3$ of slurry was calculated. In 27.8 min of residence time in compartment 1 at a linear multiplication leach rate of 0.418 μm/min, the area at steady state is 16.9 m^2/kg of feed or 2306 m^2/m^3 of slurry. The zinc extraction is 83.5% but the surface area depletion is about 88%. This is, of course, due to the disappearance of much of the fine material in stage 1 of the leach.

Modelling of stages 1-4 of the zinc pressure leach. The first stage of the autoclave deserves special attention because more than 80% of the leaching is typically carried out in this compartment. In order to model the first stage, the macromodel equations (4), (16)-(18) must be solved for $R(O_2)$, $[Fe^{2+}]$, $[Fe^{3+}]$ and $[O_2]$. Also a value of k_1 must be chosen for the micromodel to produce the required 83.5% zinc extraction reported by Martin and Jankola (1985). The micromodel then permits a calculation of A_{ZnS} as outlined above.

$$R(O_2) = 0.25k_2[Fe^{2+}]^2[O_2] \qquad (16)$$

$$R(O_2) = 0.5k_z A_{ZnS}[Fe^{3+}] \qquad (17)$$

$$C_F = [Fe^{2+}] + [Fe^{3+}] \qquad (18)$$

Equations (4) and (16)-(18) have been rearranged to give one equation in $[Fe^{2+}]$. Equation (19) can be solved numerically following the bisection root finding method.

$$0.25k_2[Fe^{2+}]^2 \ ([O_2]^* - \frac{0.5(C_F - [Fe^{2+}])A_{ZnS}k_z}{k_g}) - 0.5(C_F - [Fe^{2+}])A_{ZnS}k_z = 0$$

$$(19)$$

The modelling of stages 2-4 has been carried out in the same manner as the modelling of stage 1. The output particle size information from the micromodel leach in each stage was used as input to the subsequent stage (this was done in the form of a $\Psi(u)$ data file). Values of k_1 were determined with the micromodel for stages 2-4 to match the extraction results given by Martin and Jankola (1985). An additional equation has been introduced to take into account the precipitation of jarosite in stages 2-4.

$$3Fe^{3+} + 0.5PbSO_4 + 1.5SO_4^= + 6H_2O \longrightarrow Pb_{0.5}Fe_3(OH)_6SO_4 + 6H^+ \qquad (20)$$

At equilibrium for reaction (20),

$$\log[Fe^{3+}] = 0.333\log K - 2pH - 0.5\log[SO_4] \qquad (21)$$

The usual exit solution from the autoclave is approximately 30 gpl H_2SO_4 and 5 gpl Fe^{3+}. Assuming that these values are close to equilibrium, Eq.(21) can be rewritten as:

$$\log[Fe^{3+}] = 0.621 - 2pH \qquad (22)$$

For modelling purposes we have assumed that Eq.(22) adequately represents the jarosite precipitation in stages 2-4. That is, we are using a pseudo-equilibrium expression to describe a process that is probably occurring relatively close to equilibrium.

Table 6 contains a summary of the model calculations for stages 1-4. The value of k_g for stage 2 was set equal to that of stage 1 because of the similarity between stages. The slurry in stages 3 and 4 was considered to be saturated in oxygen due to the low oxygen consumption rates in these stages. k_2 was kept constant at $2.54 \times (10^{-3})\,mole^{-1}\,m^3min^{-1}\,atm^{-1}$. Values of k_z were calculated from Eq.(17).

The results of the micromodel are presented in Fig. 8. This figure illustrates the normalized Ψ distributions for the input feed and the output from stages 1-4. It is apparent that the fine material in the feed disappears rather quickly in the leach.

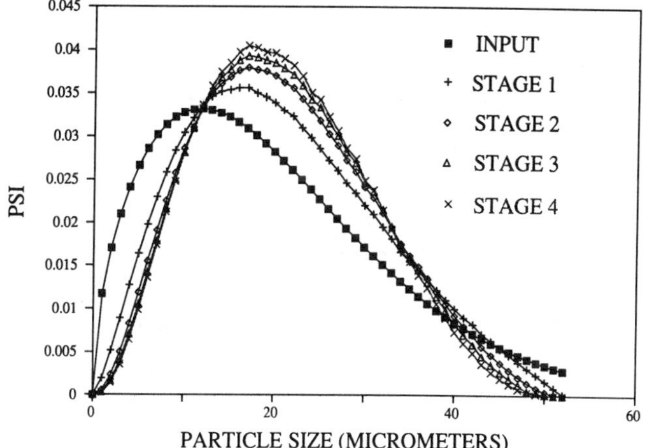

Fig. 8. Model predictions of normalized Ψ distributions for the four stages of the zinc pressure leach.

TABLE 6   Computed Steady-State Dynamic Properties in a Four-Compartment Autoclave during Zinc Pressure Leaching

$[O_2]^* = 7.5$ atm.

Compartment	1	2	3	4
Zinc Extraction, %	83.5	95.3	97.0	97.8
Residence Time, min	27.8	19.6	19.1	20.4
$R(O_2)$, moles $O_2 \cdot m^{-3} \cdot min^{-1}$	23.4	4.69	0.69	0.30
k_g, min^{-1}	3.70	3.70	–	–
k_2, $mole^{-1} \cdot m^3 \cdot min^{-1} \cdot atm^{-1}$	2.54×10^{-3}	2.54×10^{-3}	2.54×10^{-3}	2.54×10^{-3}
k_z, $m \cdot min^{-1}$	3.42×10^{-4}	1.03×10^{-4}	2.40×10^{-5}	1.54×10^{-5}
k_1, $\mu m \cdot min^{-1}$	0.418	0.336	0.0782	0.047
A_{ZnS}, m^{-1}	2306	594	371	270
$[O_2]$, atm	1.17	6.23	7.5	7.5
$[Fe^{2+}]$, $moles \cdot m^{-3}$	177.3	34	12	8
$[Fe^{3+}]$, $moles \cdot m^{-3}$	59.4	153	150	147
Jarosite Ppted., $moles \cdot m^{-3}$	0	83	113	123
Total Fe Leached, $moles \cdot m^{-3}$	237	270	275	277
$[H_2SO_4]$, $moles \cdot m^{-3}$	550	401	395	392

Table 6 has the following interesting features.

1. The value of k_1, the linear leaching rate constant, decreases from stage 1 to stage 4 by about one order of magnitude. This is reflected in a similar decrease in k_z. The leaching rate should be increasing as the feed moves through the autoclave because the particles are encountering progressively more oxidizing conditions. The decreasing rate may be explained by (a) the presence of refractory zinc minerals which leach rather slowly, (b) the loss of activity of the detergent (lignin sulphonate) used to prevent liquid sulphur from coating unreacted ZnS in the later stages of the leach and/or, (c) the particle size distribution data used for calculations may be in error, i.e. there may have been more coarse material present than reported (a problem in Coulter Counter analysis).

2. The first compartment seems to be operating in an oxygen deficient condition. In fact, calculations show that the first compartment rate would benefit significantly from enhanced oxygen mass transfer and to a lesser extent by raising the rate constant for ferrous oxidation and the

rate constant for ferric leaching of zinc sulphide. The first compartment appears to be controlled kinetically by a mixed mechanism involving all three reactions. Rate control of stages 2-4 of the leach appears to shift progressively to ferric ion leaching of zinc sulphide.

3. The bulk of the leaching work is carried out in stage 1 with almost no work being done in stage 4. Stage 4 may be most useful for optimizing jarosite precipitation and sulphur coalescence.

4. Most of the jarosite precipitates in compartment 2 of the autoclave. This is in agreement with the industrial data.

5. The predicted ferric-to-ferrous ratio increases dramatically from less than 1.0 in stage 1 to over 10 in stage 4, in agreement with the industrial data. This highly oxidizing condition could account for the destruction of detergent in stages 3 and 4 and hence the lower leaching rates.

<u>Summary</u>

A preliminary mathematical model of the zinc pressure leach has been presented. The model is comprised of two components, a macromodel which describes the bulk chemistry of the process and a micromodel which follows the leaching of individual feed particles as they move through the autoclave. Based on this preliminary concept, a number of experimental studies have been initiated to refine the model.

1. A more precise determination of the particle size distribution of the feed material to the autoclave must be made. Small errors in the distribution, especially of the coarse particle sizes, could result in large errors in the fraction unleached. Particle size measurements by Cominco and image analysis work at UBC are being pursued to provide better data in this area.
2. The study of gas-liquid mass transfer remains a compelling area of investigation. Degraaf (1984) has carried out studies at UBC related to finding the optimum agitator design and sparging rates for the first stage of the Cominco autoclave and these studies are continuing.
3. The rate of ferric ion leaching of ZnS and the role of the detergent must be explained. Leaching experiments and interfacial tension measurements in the sulphur - leach solution - zinc sulphide system are underway in the UBC laboratories.
4. A comparison between model predictions and actual compartment samples taken by Cominco from the autoclave has been initiated. Particle size data and solution chemistry profiles are being compared.

The mathematical modelling of the zinc pressure leach has provided a framework for ongoing study and possible process enhancement. The authors look forward to providing important updates on the progress of the studies outlined above.

PART 3. DIRECT-CHILL CASTING OF ZINC JUMBOS

<u>Introduction</u>

Direct-chill (D.-C.) casting was developed fifty years ago and has proven to be an economical, reliable process for solidifying non-ferrous metals such as aluminum, copper, magnesium and zinc. The process consists of a short, stationary, water-cooled mould into which liquid metal is fed continuously at a controlled rate to maintain constant metal level. Normally the mould cooling water is directed onto the surface of the solidifying metal below the mould to provide a chilling action (hence the name "direct-chill") but in some operations, such as in zinc casting, the secondary cooling is achieved with a bank of water sprays. The solidifying shape rests on a stool which is inserted into the mould at the beginning of casting and is subsequently lowered at constant speed. In contrast to the continuous casting of steel, the withdrawal

rates in D.-C. casting operations are about an order of magnitude lower; the liquid pool, or sump, depths in the latter process are also considerably smaller.

At the Trail Smelter of Cominco, the D.-C. process is employed for the production of zinc "jumbos" which are sold as remelting stock for operations such as galvanizing. In cross-section, the jumbos are rectangular (508 x 546 mm) with notches near the top of the short sides to facilitate handling by fork-lift vehicles. Figure 9 shows a photograph of a jumbo section.

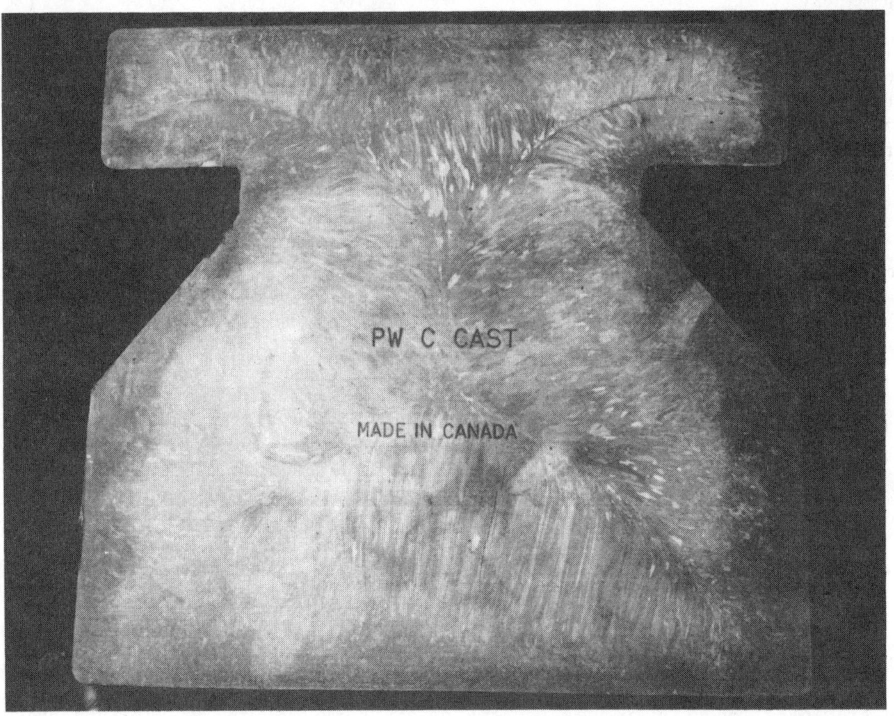

Fig. 9. Transverse section of Prime Western Grade zinc jumbo.

The study described in this part of the paper was undertaken (Venkateswaran, 1980; Venkateswaran and Brimacombe, 1983) because internal cracks appeared in jumbos cast from Prime Western Grade zinc (containing approximately 1 wt.% lead). The cracks normally were found to be running orthogonally from the bottom and notched faces of these jumbos. The cracks typically would start about 25 mm from each face and would extend inward 100 mm or more. In severe cases the cracks were several mm wide. For simple remelting applications, the cracks should not have posed a problem; however because the jumbos frequently were stored or handled in exposed areas, water (from rain or snow) could accumulate in the defects with the potential of causing explosions when contacted with molten zinc during remelting. The objective of the study, therefore, was to eliminate the internal cracks.

At the outset, it was suspected that the tensile stresses causing the cracks were thermal in origin. Earlier studies on similar crack problems encountered in the continuous casting of steel billets (Van Drunen, Brimacombe and Weinberg, 1975; Brimacombe, 1976; Grill, Brimacombe and Weinberg; 1976) had revealed that surface reheating of the solid shell due to a sharp reduction in cooling below the spray zone caused internal cracks to form in a zone of extremely low ductility close to the solidification front. The cracks could be eliminated by proper design of the spray cooling system to minimize the surface reheating.

Consequently for the present study, it was decided to model the D.-C. casting process thermally and, combined with plant measurements on an operating caster, to ascertain the origin of, and solution to, the internal cracking problem.

Mathematical Model

Formulation.   The mathematical model is based on three-dimensional heat conduction in the Cartesian coordinate system as follows

$$k \left(\frac{\partial^2 T}{\partial x^2} + \frac{\partial^2 T}{\partial y^2} + \frac{\partial^2 T}{\partial z^2} \right) = \rho C_p \frac{\partial T}{\partial t} \qquad (23)$$

Several assumptions were made in formulating the model:

1) A plane of symmetry exists running vertically through the jumbo section. Thus calculations need only be performed for one-half of the jumbo as shown in Fig. 10;
2) Mixing in the sump is neglected;
3) The thermal conductivity of both solid and liquid zinc has been assigned the same constant value, independent of temperature;
4) The latent heat is released linearly between the liquidus and solidus temperatures by adjusting the specific heat.

The initial and boundary conditions applied were the following:

Initial condition (when the stool has been raised into the mould):

    $t = 0$ (instantaneous filling of mould)
    $T = T_p$ (throughout calculation domain)

Boundary conditions.   At the _liquid metal level_, the pouring temperature is held constant, $T = T_p$. At the _bottom of the jumbo_ in contact with the stool, $-k \frac{\partial T}{\partial z} = h(T - T_b)$. The heat-transfer coefficient, h, which was assigned a value of 0.209 kW/m^{2}K, had no influence on the calculations for steady-state operations. On the _faces_ of the jumbo in contact with the mould and sprays, $-k \frac{\partial T}{\partial x} = h(z)(T - T_w)$; $-k \frac{\partial T}{\partial y} = h(z)(T - T_w)$, where values of h were estimated from plant measurements involving a tracer and cast-in thermocouples. At the _centre plane_ the jumbo, about which symmetry has been assumed,

$$-k \frac{\partial T}{\partial x} = 0; \quad -k \frac{\partial T}{\partial y} = 0$$

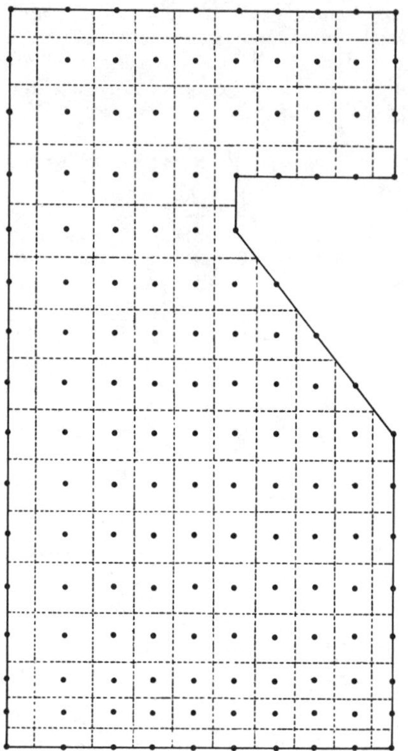

Fig. 10. Domain of calculation and nodal arrangement in
transverse section of D.-C. cast zinc jumbo.

<u>Solution</u>. Equation (23) was solved, subject to the initial and boundary
conditions, by the alternating-direction, implicit (ADI) finite-difference
method. Discretization of the transverse (x-y) plane into nodes for the
numerical solution is shown in Fig. 10. To simplify the calculation, the
radiative component of the external heat transfer was linearized by introducing
a temperature-dependent radiative heat-transfer coefficient. The growth of the
jumbo was simulated through periodic addition of a layer of nodes at the
pouring temperature to the top of the sump (liquid level). Checks were made at
each time step to ensure that a newly solidified nodal volume had released all
its latent heat. Checks on internal consistency and maximum node size also
were made by comparing model predictions to analytical solutions for simplified
cases. The thermophysical properties employed in the model are presented in
Table 7.

<u>Validation</u>. The ability of the model to simulate the Cominco D.-C. caster was
assessed by performing three types of trials in-plant.

1) <u>Thermocouple measurements</u>: Thermocouples were frozen into the solidifying
  shell, close to the surface of the jumbo (~ 20 mm), during steady-state
  operation, and were allowed to descend with the metal. The exact position
  of the thermocouple tip was located subsequently by sectioning the jumbo.

TABLE 7   Thermophysical Properties of Zinc Used in Mathematical
Model of D.-C. Casting

Property	Value
Liquid density, kg/m^3	6620
Solid density, kg/m^3	6981
Liquid specific heat, J/gK	0.480
Solid specific heat, J/gK	$0.343 + 0.154(10^{-3})T$
Thermal conductivity, W/mK	113
Latent heat of fusion, J/g	113
Liquidus temperature, °C	420
Solidus temperature, °C	410

From the measured temperature-time plots, the surface heat-transfer coefficient, h(z), was back-calculated by a trial-and-error procedure. The values obtained are presented in Table 8. This, of course, is not a validation of the model but is a necessary procedure to "tune" it to the process owing to the uncertainty of the surface heat-transfer coefficients.

2) <u>Tracer additions</u>: The sump profile was delineated at a casting speed of 1.27 mm/s by adding an alloy of 10% copper (in zinc) to the mould after steady-state operation had been reached. After making the addition, the pool was stirred with a paddle driven by a power drill. The jumbo was later saw-cut longitudinally along the mid-plane perpendicular to the two unnotched faces. The exposed face was polished carefully to remove the saw-cut marks then was etched with 4% Nital solution to delineate the copper-rich region (liquid at the time of addition). The etched section (sump region darker) is shown in Fig. 11 together with the model prediction of the sump profile, given as a dashed line. Agreement is seen to be very close including the asymmetry of the sump due to the presence of the notches.

3) <u>Pool depth sounding</u>: The development of the liquid pool from the beginning of the cast through to steady-state was followed at two casting speeds, 1.27 and 1.69 mm/s, by a simple sounding technique. A steel rod was dipped into the liquid pool to the bottom periodically throughout the casting runs. The measurements made at the slower speed are shown as open circles in Fig. 12 and are compared to the model-predicted sump depths (solid line) as a function of time. Agreement is excellent. It should be noted that a solid shell does not commence growth at the bottom effectively until after about 500-mm length of jumbo has been cast because heat extraction downward through the steel is very slow. Also it is evident that steady state is not reached until about 10 min have elapsed.

TABLE 8 Heat-Transfer Coefficients Back-Calculated from
Industrial Trials on D.-C. Jumbo Caster

casting speed: 1.69 mm/s

Distance from top of mould (mm)	Heat-transfer coefficient W/m^2K
0-212	20,934
212-216	4187
216-1450	Exponential drop from 4187 to 125
1450-1454	125

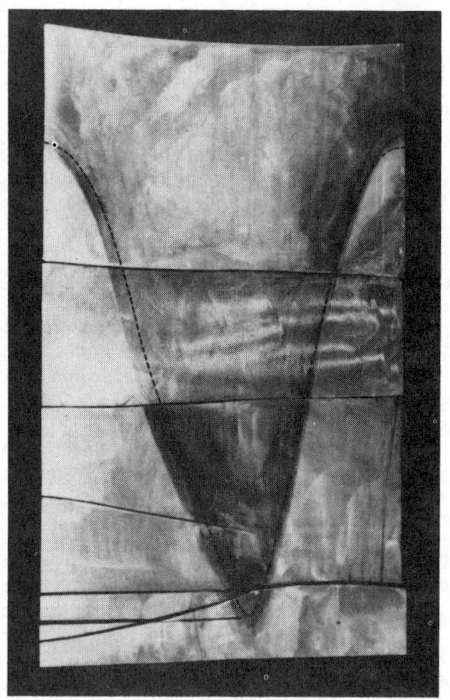

Fig. 11. Sump profile delineated by copper tracer in a
zinc jumbo cast at 1.27 mm/s as compared to
model prediction (dashed line).

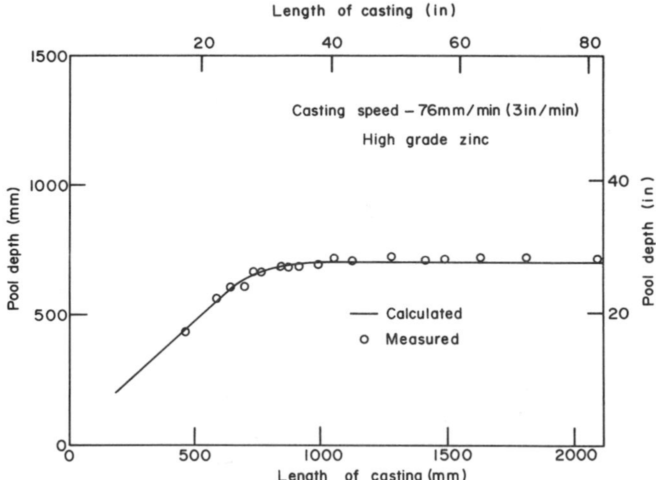

Fig. 12. Comparison between measured and calculated
sump depths as a function of time from the
start of casting a zinc jumbo at 1.27 mm/s.

Application of the Model to Formulation of Internal Crack Mechanism

To assess the role of surface reheating in internal crack formation, the
mathematical model was run for casting conditions corresponding to plant trials
in which jumbo sections containing cracks were obtained for study. An example
of the predicted transient temperature profiles at the surface and in two
interior locations of a zinc jumbo cast at 1.69 mm/s is shown in Fig. 13. Thus
it is evident that the surface of the jumbo indeed experiences reheating, as
originally suspected, beyond 120 s. Translating time into distance, this
corresponds to the bottom of the spray cooling zone where heat extraction drops
sharply. The situation then is an exact parallel to the steel billet casting
system.

If the internal cracks were generated by the surface reheating, they would have
to form close to the solidification front because this is the location where
the resulting thermal stresses are tensile in the transverse plane.
Metallographic analysis of the interior surface of the cracks confirmed that
this was indeed the case. The cracks clearly had formed interdendritically and
exhibited a smooth topography characteristic of hot tearing close to the
solidification front. Analysis of the crack surface with a scanning electron
microscope further revealed the presence of a lead-rich phase which was
evidently liquid when the cracks formed. Thus the hot tearing mechanism
involves the separation of dendrites surrounded by a liquid lead-rich phase at
temperatures in the solid shell below the solidus temperature of zinc.

A stronger link between the surface reheating and cracking events can be made
from consideration of the predicted solidification front and knowledge that the
cracks have formed close to it. Then the location of the cracks (inner-most
tip if the cracks open from the solidification front toward the adjacent face)
observed in the jumbo sections should correspond roughly to the shell thickness
at the time of crack formation. The distance below the liquid level at which
the shell in the crack location has reached this thickness can be determined

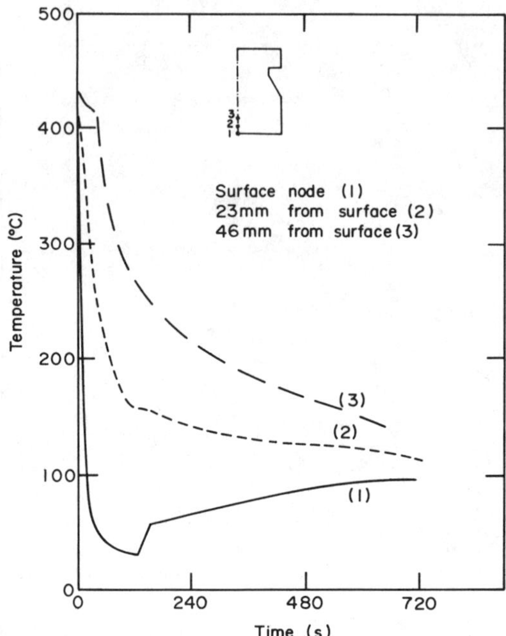

Fig. 13. Model-predicted transient temperatures at the
surface and two interior locations of a zinc
jumbo cast at 1.69 mm/s.

easily from the model predictions of sump profile. Figure 14 shows the
predicted level at which observed cracks formed as a vertical line adjacent to
the sump profile calculated for a jumbo cast at 1.69 mm/s. It is seen that all
the cracks examined formed during the reheating event.

Solution to the Cracking Problem

Having ascertained that surface reheating below the sprays was the cause of
cracks in Prime Western Grade zinc jumbos, the design of a new spray system
which would minimize this phenomenon, particularly over the length of the sump,
was undertaken. A number of runs were made with the mathematical model to
determine the influence of the spray heat-transfer coefficient on the magnitude
of the surface reheating. Thus it became clear that the reheat could be
minimized by decreasing the intensity of the existing sprays and by extending
the spray zone to below the sump. The arrangement of the sprays in the newly
designed cooling assembly is shown in Fig. 15 for the bottom unnotched face.
The sprays for the other faces were of similar design.

The new spray assembly was tested in-plant by making four production runs. In
the first run, high-grade zinc, which was not crack susceptible, was cast at
1.27 mm/s to check out the assembly. In the ensuing three runs, Prime Western
Grade zinc was cast at 1.27 and 1.48 mm/s. Twenty-two sections were cut from
two strands in two of the runs and were inspected for cracks. None were
found. The spray design is now part of the operating casting system.

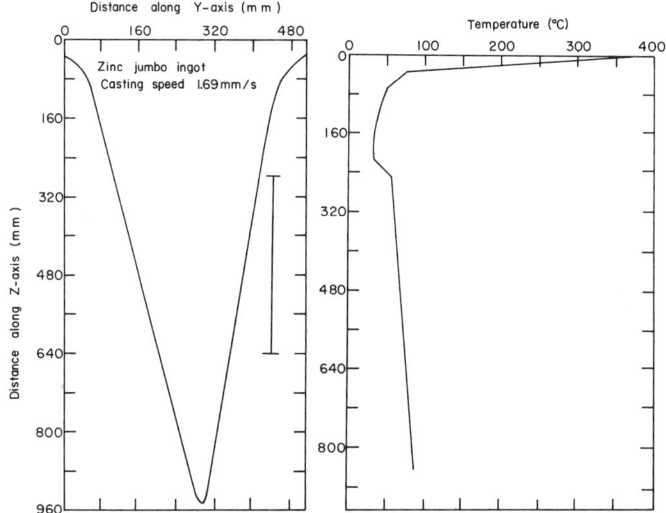

Fig. 14. Model predictions of sump profile in the
mid-plane passing through the unnotched faces
(left) and of bottom mid-face temperature
profile (right) for a zinc jumbo cast at 1.69
mm/s. Note vertical line indicating location
of crack formation corresponding to surface
reheating.

Summary

A mathematical model, based on the principles of three-dimensional heat
conduction, was developed to simulate the operation of a D.-C. casting machine
for the production of zinc jumbos. The specific objective of the modelling
exercise was to eliminate the presence of internal cracks in jumbos cast from
Prime Western Grade zinc. Numerous plant trials were performed to validate the
mathematical model and to obtain crack samples under known conditions for
metallographic analysis. Examination of the crack surfaces, combined with
model predictions, showed that the stresses causing the cracks were generated
by surface reheating below the secondary cooling (sprays) zone. The model then
was applied to the redesign of the spray system which, when implemented,
eliminated the cracking problem. The strength of modelling, in combination
with plant and laboratory studies, is again apparent.

Acknowledgements

The authors owe a considerable debt of gratitude to the research and operating
personnel of Cominco at Trail, B.C. who have provided much needed financial
support, gave us access to their operating processes, supplied us with data and
shared their knowledge and experience. C.A. Sutherland and G.W. Toop deserve
special mention. Without such cooperation and the abiding support of the
Natural Sciences and Engineering Research Council of Canada and B.C. Science
Council, the work described in this paper could not have been done.

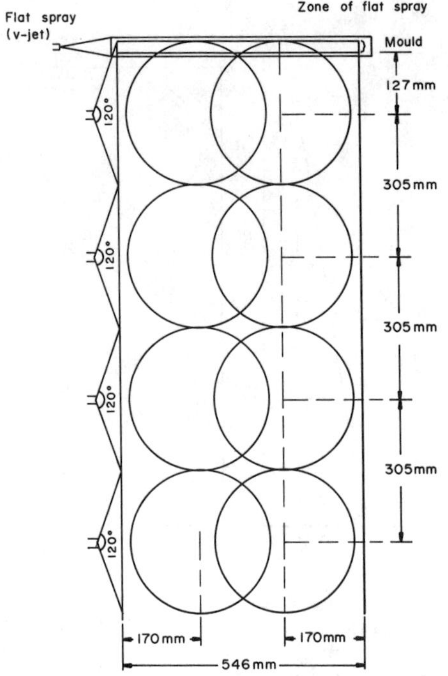

Fig. 15. New spray system to prevent formation of
internal cracks in zinc jumbos.

TABLE OF NOMENCLATURE

Part 1

F_{CC} Fraction of coal combusted.
F_{CE} Fraction of coal entrained in the slag.
F_{OXY} Fraction of oxygen not used in combustion which oxidizes FeO to
            Fe_3O_4.
F_{SC} Fraction of coal which short-circuits the bath.

Part 2

$A_{g/1}$ Gas-liquid interfacial area (m^2).
A_{ZnS} Steady-state area of sulphide mineral surface per unit volume
            (m^2/m^3).
C_F Total concentration of iron leached ($mole/m^3$).
k_g Oxygen gas absorption rate constant (min^{-1}).
$k_{g/1}$ Oxygen mass transfer coefficient in solution (per unit volume)
            ($min^{-1}\ m^{-2}$).
k_z Heterogeneous rate constant for attack of sulphide minerals by
            ferric ions (m/min).

k_1	Linear particle leaching rate (m/min).
k_2	Third order homogeneous rate constant for the oxidation of ferrous ions (m^3/mole/min/atm).
K	Equilibrium constant for Equation (20).
P(t)	Probability of a given residence time 't' in an autoclave compartment.
R_i	Rate of reaction of step i, Equations (4)-(6) (moles/m^3/min).
$R(O_2)$	Rate of oxygen consumption (moles O_2/m^3/min).
t	time (min).
t_{ave}	Mean residence time of slurry in autoclave compartment (min).
u	Particle size (microns).
u'	Fitting constant in Equation (9) (microns).
V	Volume (m^3).
W	Weight (kg).
Y	Cumulative fraction of material found below a given size u.
$\Psi(u_o)$	Probability of finding a differential unit of material in a size range between u_o and $u_o + du_o$.
[]	Concentration (moles/m^3).

Superscript:

*	interfacial value.

Subscript:

o	unleached particles.

Part 3

C_p	Specific heat at constant pressure (J/g K)
h,h(Z)	Heat transfer coefficient (W/m^2 K)
k	Thermal conductivity of zinc (W/m K)
t	time (s).
T	temperature (K).
x,y,z	Cartesian coordinates (m)
ρ	Density (kg/m^3)

Subscripts:

b	bottom of casting
p	condition at pouring
w	water spray

REFERENCES

Abrashev, G. (1972). In M.J. Jones (Ed.), Advances in Extractive Metallurgy and Refining, IMM, London, pp. 317-26.

Bell, R.C., G.H. Turner and E. Peters (1955). Trans. TMS-AIME, 203, 472-7.

Bielpolskij, I. and N. Urusov (1948). Zh. Prikl. Khim.. 21, 903.

Blaskett, D.R. (1970). In M.J. Jones (Ed.), Ninth Commonwealth Mining and Metallurgical Congress 1969, Vol. 3, Mineral Processing and Extractive Metallurgy, IMM, London, pp. 879-89.

Brimacombe, J.K. (1976). Can. Metall. Q., 15, 163-75.

Chmielewski, T. and W.A. Charewicz (1984). Hydrometallurgy, 12, 21-30.

Cockcroft, S.L., G.G. Richards and J.K. Brimacombe (1987). Accepted for publication, Metall. Trans. B.

Cockcroft, S.L., G.G. Richards and J.K. Brimacombe (1988). Can. Metall. Q., 27, 27-40.

Cornelius, R.J. and J.T. Woodcock (1958). Proc. Australas. Inst. Min. Metal., 185, 65.

Evdokimenko, A.I., R.Zh. Khobdabergenov, A.P. Sosnin, S.P. Golger, and N.A. Kolesnikov (1977). Soviet J. of Non-Ferrous Metals, 18, No. 11, 62-5.

Degraaf, K.B. (1984). M.A.Sc. Thesis, The University of British Columbia.

Dreisinger, D.B. and E. Peters (1987). In J. Szekely and others (Ed.), Mathematical Modelling of Materials Processing Operations, TMS-AIME, Warrendale, 347-369.

Dreisinger, D.B. and E. Peters (1988). Accepted for publication, Hydrometallurgy.

Grant, R.J. and L.J. Barnett (1975). In South Australia Conference 1975, Australas. I.M.M., Melbourne, 247-65.

Grill, A., J.K. Brimacombe and F. Weinberg (1976). Ironmaking and Steelmaking, 3, 38-47.

Kellogg, H.H. (1957). Eng. Mining J., 158, No. 3, 90-2.

Kellogg, H.H. (1967). Trans. TMS-AIME, 239, 1439-49.

Martin, M.T. and W.A. Jankola (1985). CIM Bulletin, 78, No. 876, 77-81.

Mathews, C.T. and R.G. Robins (1972). Proc. Australas. Inst. Min. Metal., 242, 47.

McKay, D.R. and J. Halpern (1958). Trans. TMS-AIME, 212, 301.

Parker, E.G. and D.R. McKay (1987). Private communication with D. Dreisinger and E. Peters.

Pound, J.R. (1939). J. Phys. Chem., 43, 955-967.

Richards, G.G. (1983). Ph.D. Thesis, The University of British Columbia.

Richards, G.G., J.K. Brimacombe and G.W. Toop (1985a). Metall. Trans. B, 16B, 513-27.

Richards, G.G. and J.K. Brimacombe (1985b). Metall. Trans. B, 16B, 529-40.

Richards, G.G. and J.K. Brimacombe (1985c). Metall. Trans. B, 16B, 541-9.

Rosin, P. and E. Rammler (1933). J. Inst. Fuels, 7, 29-36.

Van Drunen, G., J.K. Brimacombe and F. Weinberg (1975). Ironmaking and Steelmaking, 2, 125-133.

Venkateswaran, V. (1980). Ph.D. Thesis, The University of British Columbia.

Venkateswaran, V. and J.K. Brimacombe (1983). Proc. Modeling of Casting and Welding Processes, eds. J.A. Dantzig and J.T. Berry, TMS-AIME, 365-8.

OVERALL ENERGY AND MATERIAL BALANCES
USING THE OUTOKUMPU-HSC AND THE LOTUS 123 PROGRAMS

Li Wu[1], Antti Roine[2], and N.J. Themelis[1]

The Outokumpu - HSC software is an interactive program which
provides enthalpy and free energy data for a large number of
chemical compounds of metallurgical interest. The LOTUS 123
program allows for matrix-algebra calculations to be carried out
on an array of numbers entered on a spreadsheet. The capabilities
of these two programs were combined and utilized to calculate the
overall material and energy balances in a high temperature
reactor where zinc calcine particles are reduced by carbonaceous
material. This is a quick and convenient method for flowsheet
development and for examining the effect of process parameters on
the flowsheet.

1. Henry Krumb School of Mines, Columbia University,
   New York, N.Y. 10027.
2. Outokumpu OY, P.O. Box 60, SF-28101, Pori, Finland

INTRODUCTION

Research and plant engineers are often faced with the rather
tedious job of carrying out calculations of process mass and
heat balances. Such work usually encompasses two different
aspects: search of the respective thermochemical and thermodynamic data, and construction and computation of the appropriate
conservative equations. The search of published data is time
consuming and dull. Also, the mass and heat balance equations for
an industrial process may involve multiple variables and unit
operations.

Computers have been used to "automate" this type of work for
a number of years. However, their use often requires the
construction of a program in Fortran, Basic, or other computer
languages. That, in itself, can be very time-consuming and, in
some cases, not justified economically. It is therefore desirable
to develop programming techniques which make full use of existing
general-use, "user-friendly" programs so that the amount of "new"
programming can be kept to a minimum.

An example illustrative of this approach is presented in
this paper. The authors developed the material and energy
balances for a novel process for producing zinc by utilizing two
commercially available computer programs: The HSC program (1),
developed by Outokumpu Company of Finland, and the well-known
Lotus 123 computerized spreadsheet, developed by Lotus Corp (2).

The HSC software is an interactive program which contains
thermochemical and thermodynamic data (3) for over two thousand
chemical compounds and elements of interest to extractive and
process metallurgists; the program can carry out simple heat and
material balance calculations on the basis of a given chemical
equation. However, problems of material and heat balances
encountered in process design and analysis are usually very
complex and require simultaneous solution of a number of
equations. In this respect, Lotus 123 is a very convenient tool
because it provides for the use of matrix algebra, as will be
described later, and for getting answers to the "what-if"
questions that must be asked during flowsheet and process design.

In addition to other important features, the HSC program provides for saving the resulting files in a special extension that is directly readable by Lotus 123. This feature enhances the compatability and utility of these two programs in the proposed application.

Process description

The example used in this paper is the development of the flowsheet for a novel, and as yet untried process, for producing zinc by direct smelting and reduction of zinc calcines in the presence of carbonaceous material and oxygen. As envisaged, the process block diagram (Figure 1) first involves roasting of the concentrates in a fluid bed reactor. The zinc calcines are then introduced in a high-temperature reactor with carbonaceous material and bulk oxygen.

In the reaction chamber, partial combustion of carbon with oxygen produces carbon monoxide which reacts simultaneously with zinc oxides to produce zinc vapor and carbon dioxide; some of the latter reacts with carbon to regenerate carbon monoxide which is used further for zinc reduction.

The carbon combustion and the carbon monoxide regeneration reactions are as follows:

$$C + O_2 \quad = \quad CO_2 \tag{1}$$

$$C + CO_2 \quad = \quad 2CO \tag{2}$$

The principal reaction is the reduction of zinc oxide by carbon monoxide:

$$ZnO + CO = Zn(g) + CO_2 \tag{3}$$

Hydrogen present in the fuel will also reduce zinc oxide according to the equation

$$ZnO + H_2 \quad = \quad Zn(g) + H_2O \tag{4}$$

With reference to the above ZnO reactions (3) and (4), at constant pressure of 1 atmosphere the respective equilibrium constants can be expressed as follows:

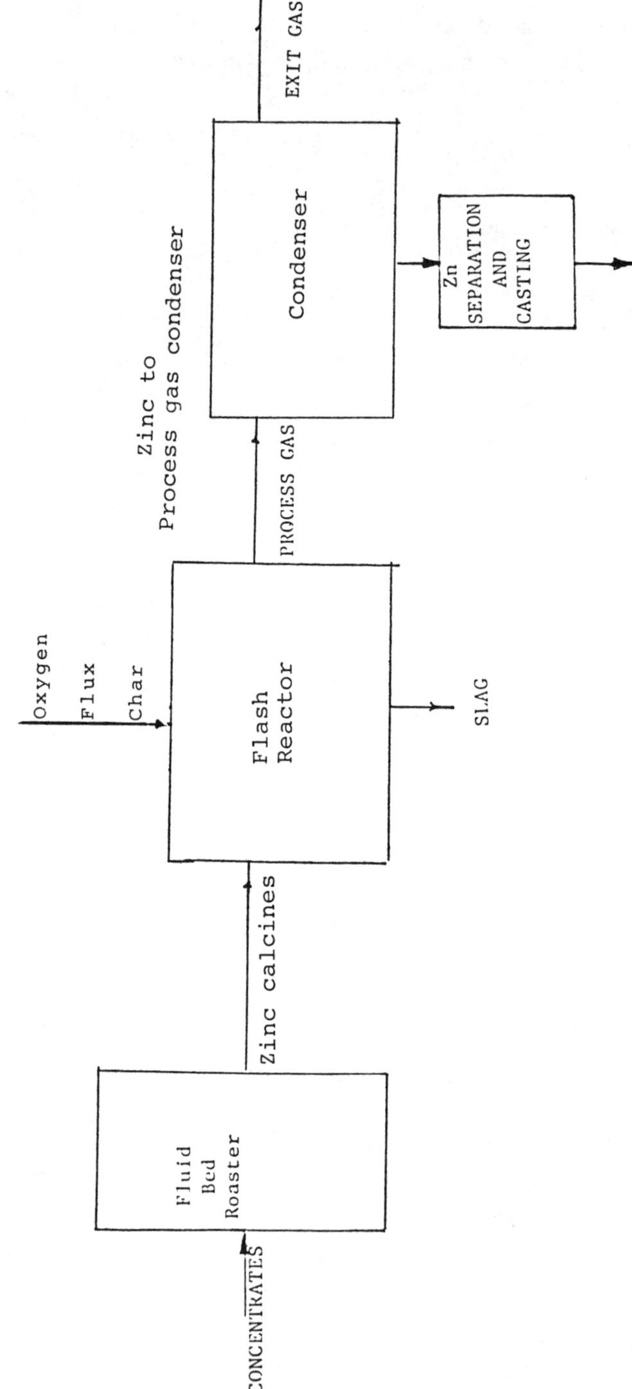

Figure 1. Flowsheet of Zinc Reduction Process

$$K_1 = f_1(T) = P_{Zn}P_{CO2}/P_{CO} \tag{5}$$

and

$$K_4 = f_1(T) = P_{Zn}P_{H2O}/P_{H2} \tag{6}$$

The exit gas from the reactor contains carbon monoxide, carbon dioxide and zinc vapor; the latter is condensed to liquid metal in a condenser and subjected to rectification to produce marketable zinc.

Feed materials and basic assumptions

The material and heat balances are based on processing a nominal 20 tonnes of zinc calcines per hour. The chemical analysis of the zinc calcine (Table 1) is based on a typical composition of zinc calcines produced in the conventional zinc roast-leach-electrowing process (4); the assumed compositions of silica flux, carbonaceous material and industrial oxygen are shown in Table 1.

The chemical composition of the slag produced is based on the desirable silica to iron ratio to provide for a relatively low zinc concentartion in the slag loss and was estimated on the basis of published information on zinc slags produced under reducing atmospheres. This simplification is made for the sake of brevity of presentation of this method of calculation; the HSC program can be readily expanded to include the required activity data in the slag phase so that the zinc concentration in the slag can be part of the overall calculation.

The following additional assumptions have been made to simplify the calculation:

i) All iron contained in the feed materials ends in the slag and exists in the form of FeO

ii) the exit gas from the flash furnace contains only $Zn(g)$, CO, CO_2 and N_2

iii) the carbon in the feed is assumed to be burned into either CO or CO_2.

TABLE 1.

Composition of Feed Materials

CALCINE:

ZnO	Fe_2O_3	$PbSO_4$	$ZnSO_4$	SiO_2	Al_2O_3	CaO
71.14	16.84	2.61	3.69	2.29	2.29	1.14

COKE: Oxygen: Flux:

C	SiO_2	Al_2O_3	O	N	SiO_2	CaO	Al_2O_3
91.8	6.1	2.1	98	2	85	7.5	7.5

The first assumption is justified because the oxygen potential in the furnace is controlled at such a level that reduction of FeO to metallic iron or oxidation to magnetite are thermodynamically unfavorable.

The second assumption cannot be made if the feed materials used contain an appreciable amount of hydrogen or water vapor. However, the system of equations developed below can be readily expanded to include hydrogen and water vapor balances.

The third assumption above represents a first approximation to the desired solution; as with other factors, the proposed technique can be used to indicate how the heat and material balances will change if a certain amount of carbon is not utilized in the reactor properly and is collected as residual carbon.

Development of system of equations

It is now necessary to calculate the mass flowrates of reagents, products and by-products corresponding to the given mass flowrate of feed material, i.e., zinc calcines, to be processed.

Thus, the only known variable is the feed rate of zinc calcines; the coke rate, W_{coke}, oxygen flowrate, V_{oxy}, and flux rate, W_{flux}, are unknown variables. Other unknown variables are the rate of slag production, W_{slag}, the total volume of process gas exiting the reactor (V_{total}) and the concentration of zinc vapor, CO, CO_2, and N_2 in this gas (X_{Zn}, etc.). For convenience, we shall consider the unknown product $X_{Zn} \times V_{total}$ to be represented by the zinc "flowrate", V_{Zn} and do the same for the other three components of the process gas, i.e. V_{CO}, V_{CO2} and V_{N2}.

Accordingly, there are, in all, eight dependent variables, or unknowns. Their solution requires eight equations.

Material Balances

Six equations can be constructed easily on the basis of the material balances for the six principal components of the system, namely zinc, oxygen, carbon, silica, iron, and nitrogen. For example, the zinc balance can be described as follows:

Zinc in calcines = zinc vapor in exit gas + zinc in slag

$$\%Zn_{calcine}\, W_{calcine}/100 = \%Zn_{slag}\, W_{slag}/100 + \\ + V_{zinc}\, M_{Zn}/22.4 \qquad (7)$$

where W symbolizes the respective solid and liquid flowrates in kg/h and V_{zinc} is the rate of zinc vapor production, in Nm^3/h. At this point, it is necessary to make an estimate of the projected zinc concentration in the slag; this can be done on the basis of the "designed" slag composition and of prior experimental and operating evidence from comparable systems.

Similarly, the carbon balance is expressed as follows:
$$\%C\, W_{coke}/100 = V_{CO}\, M_{CO}/22.4 + V_{CO2}\, M_{CO2}/22.4 \qquad (8)$$

The oxygen balance is also obtained by equating all input sources of oxygen to the sum of the outputs:

$$X_{O2}\, V_{O2}/22.4 + X_{O,red}\, W_{calcine}/M_{O2} = (0.5V_{CO} + V_{CO2})/22.4 \qquad (9)$$

where X_{O2} is the fraction of oxygen in the industrial oxygen used (taken to be 0.98); V_{O2} is the input flowrate of industrial oxygen, and V_{CO} and V_{CO2} represent the respective flowrates of carbon monoxide and dioxide in the process gas. The term $X_{O,red}$ represents the fractional concentration of oxygen in the calcines that must be removed during the reduction reaction; it is calculated from the stoichiometry of the calcines and on the assumption that all higher iron oxides will be reduced to FeO.

The same method is used to develop the silica, iron, and nitrogen balances and, thus, provide a total of six algebraic equations. It should be noted that during this part of the work, the HSC program can be used to facilitate the calculation of molecular weights and the fractional composition of a particular chemical compound.

Thermodynamics and driving force for reduction

The seventh equation is derived on the basis of the thermodynamic requirements of the principal reaction, i.e., the reduction of zinc oxide by carbon monoxide, in the following manner.

First, the HSC program is requested to provide the free energy of formation for reactions (1-4) shown earlier and the respective equilibrium constants, at an assumed reaction temperature of 1300°C. The response for the principal reduction reaction is shown in Table 2. It can be seen that the reaction is reversible; e.g., at an assumed zinc vapor concentration of 20% ($X_{Zn} = 0.2$) in the process gas from the reactor, the corresponding equilibrium ratio of CO/CO_2 is 0.23. This means that at CO concentrations below this ratio, zinc vapor will be oxidized back to ZnO.

It should be noted that the above ratio also affects the condensation efficiency and metal recovery in the condenser after the reactor.

TABLE 2.

REACTION EQUATION OR ONE CHEMICAL FORMULA ?   (ENTER = BACK TO THE
ZnO+CO(g)=Zn(g)+CO2(g)   MENU)

TEMPERATURE(S) / C ?   (XXX-XXXX = Temperature interval !)
1050-1400

GIVE STEP FOR TEMPERATURE INTERVAL ? (ENTER = 100 DEGREES)
50

ZnO+CO(g)=Zn(g)+CO2(g)

TEMPERATURE		ENTHALPY	FREE ENERGY	EQUILIBRIUM CONSTANT
T	T	CHANGE H	CHANGE G	OF THE REACTION K
K	C	kJ/mol	kJ/mol	
1323.15	1050.00	183.010	30.418	6.308E-02
1373.15	1100.00	182.377	24.664	1.154E-01
1423.15	1150.00	181.744	18.932	2.021E-01
1473.15	1200.00	181.110	13.223	3.399E-01
1523.15	1250.00	180.476	7.536	5.517E-01
1573.15	1300.00	179.841	1.869	8.669E-01
1623.15	1350.00	179.205	-3.777	1.323E+00
1673.15	1400.00	178.569	-9.404	1.965E+00

Since the reduction rate phenomena within the reactor are
driven by the difference between actual and equilibrium
concentration, one must also ensure at this point that the
exit gas is greater than the equilibrium concentration so that
reductionoccurs throughout the length of the reactor.

Of course, the desirable CO/CO_2 ratio in the exit gas will
determine the respective feedrates of carbonaceous material
and oxygen into the reactor, as well as the rate of carbon
consumption per kilogram of zinc produced.

In view of the importance of this factor, it is convenient
to examine the effect of selecting a range of CO/CO_2 ratios on
the overall energy and material balances for the process. Each
value of this ratio provides the seventh equation needed,
e.g.:

$$P_{CO}/P_{CO2} = 2 = V_{CO}/V_{CO2} \qquad (10)$$

Heat balance

The eighth equation is provided by the overall thermal
energy balance for the reactor, over the operating period of
one minute:

Heat of combustion of C in coke to CO + heat of combustion of CO to CO_2 = endothermic heat of reduction of ZnO to Zn + latent and sensible heat in process gas and in slag + heat loss   (11)

The calculation can be simplified by assuming that all feed materials and the injected oxygen are at 25°C. This assumption can be readily modified if any of the materials, e.g., the zinc calcines are preheated.

The heat balance for a process is, of course, largely determined by the reaction temperature. On both thermodynamic and kinetic counts, it is known that high temperatures favor the reduction of zinc oxide. On the other hand, however, higher gas and slag temperatures entail greater heat carryover in the product and by-product streams and higher heat losses from the furnace. For the purposes of this calculation, it was assumed that the exit gas will be at 1300°C while the slag will be tapped at 1200°C.

The heat losses from the reactor must be estimated from heat transfer considerations and the proposed geometry of the reactor. However, the reactor design must await the calculation of the overall heat and material balances. It is therefore necessary to proceed by a reiterative method: first make an "intelligent guess" as to the rate of heat loss and, after solving the balance equations, design the reactor dimensions and calculate the corresponding heat loss; then rework the material and energy balances. For the purposes of this calculation, the heat losses from a reactor processing 20 tph of zinc calcines were assumed to be 17.6 GJ/h.

The above overall heat balance (equation 11) can be written as follows:

$$-\frac{\%C}{100 M_c} W_{coke} \Delta H_1^\circ - V_{CO_2} \Delta H_2^\circ = [\left(H_{1300}^\circ - H_{25}^\circ\right)_{CO} V_{CO} + \left(H_{1300}^\circ - H_{25}^\circ\right)_{CO_2} V_{CO_2} + \left(H_{1300}^\circ - H_{25}^\circ\right)_{N_2} V_{N_2}$$

$$+ \left(H_{1300}^\circ - H_{25}^\circ\right)_{Zn} V_{Zn}]/22.4 + \Delta H_3^\circ V_{Zn} + \Delta H_4^\circ W_{slag} + \Delta H_{loss}$$

where ΔH_1°, ΔH_2°, ΔH_3°, ΔH_4° are, respectively, enthalpy changes for carbon combustion to form CO, oxidation of CO further to CO_2, ZnO reduction to Zn(g) and slag formation from materials at 25 °C, in kJ/kmole; $\left(H_{1300}^\circ - H_{25}^\circ\right)_{CO}$ etc., represents enthalpy increase for CO, etc., from 25 to 1300 °C, in kJ/kmole.

The enthalpy changes for carbon and CO combustions, for zinc reduction and for each gaseous species from 25 to 1300 $^{\circ}$C can be obtained from the HSC program. Table 3 shows the procedures and the results of calculating the enthalpy changes for carbon combustion, zinc reduction and CO from 25 to 1300 $^{\circ}$C. The enthalpy changes for other physical and chemical reactions in Equation 11 were obtained by following the same procedure as shown in Table 3; finally, Equation 11 can be rewritten as:

$$8.45W_{coke} + 12.634V_{CO2} = 22.557V_{zinc} + 1.831V_{CO} + 2.892V_{CO2}$$

$$+1.815V_{N2} + 2.5085W_{slag} +292.88 \quad (12)$$

where the coefficient of each term in this equation represents heat gain or loss per unit material (in kg for condensed, and in Nm^3 for gaseous species). E.g., the coefficient of the first term, 8.45, is the amount of heat released from burning one kilogram of coke to form CO; the third coefficient, 22.557, represents the heat required for reducing ZnO at 25°C and producing one normal cubic meter of Zn(g) at 1300 $^{\circ}$C.

TABLE 3.

```
*********************** HEAT BALANCE ****************************

INPUT:
=====
COMPONENT                TEMPERATURE   AMOUNT         kg      HEAT CONTENT
                               C        kmol         Nm3              MJ

C                          25.00        1.00        12.01            0.00
O2(g)                      25.00        0.50        11.21            0.00
EXTRA INPUT HEAT OF THE PROCESS                                      0.00

OUTPUT:
======
COMPONENT                TEMPERATURE   AMOUNT         kg      HEAT CONTENT
                               C        kmol         Nm3              MJ

CO(g)                      25.00        1.00        22.41         -110.62
HEAT LOSESS                                                          0.00


________________________________________________________________________

HEAT IS RELEASED / MJ                                            -110.62
```

```
*********************** HEAT BALANCE ****************************

INPUT:
=====
COMPONENT              TEMPERATURE    AMOUNT        kg      HEAT CONTENT
                          C           kmol         Nm3              MJ

CO(g)                    25.00        1.00        22.41        -110.62
ZnO                      25.00        1.00        81.38        -348.34
EXTRA INPUT HEAT OF THE PROCESS                                   0.00

OUTPUT:
======
COMPONENT              TEMPERATURE    AMOUNT        kg      HEAT CONTENT
                          C           kmol         Nm3              MJ

CO2(g)                   25.00        1.00        22.41        -393.77
Zn(g)                    25.00        1.00        22.41         130.49
HEAT LOSESS                                                      0.00
_______________________________    _____________________________________

MORE HEAT IS NEEDED / MJ                                         195.67

*********************** HEAT BALANCE ****************************

INPUT:
=====
COMPONENT              TEMPERATURE    AMOUNT        kg      HEAT CONTENT
                          C           kmol         Nm3              MJ

CO(g)                    25.00        0.04         1.00          -4.94
EXTRA INPUT HEAT OF THE PROCESS                                   0.00

OUTPUT:
=== ==
COMPONENT              TEMPERATURE    AMOUNT        kg      HEAT CONTENT
                          C           kmol         Nm3              MJ

CO(g)                  1300.00        0.04         1.00          -3.11
HEAT LOSESS                                                      0.00
_______________________________    _____________________________________

MORE HEAT IS NEEDED / MJ                                          1.83
```

Use of Lotus 123 to solve system of equations

In line with the above discussion and after introducing the known numerical values and simplifying, the resulting eight equations are written as follows:

$$AX = B \tag{21}$$

where **A**: matrix coefficient

    X: dependent variable vector

    B: constant vector

The solution vector, **X**, can be obtained by inverting the coefficient matrix, **A**, and multiplying by the constant vector, **B**:

$$X = BA^{-1} \tag{21}$$

The matrix algebra operations are performed readily on the LOTUS 123 program by entering the matrix coefficient matrix **A** and the constant vector, B, on the spreadsheet. The system of linear equations is then solved by entering on the keyboard

> Data
>
> Matrix
>
> Inversion
>
> Range (input & output)
>
> Intercept (computer)
>
> Go

which results in the inversion of matrix **A**, followed by multiplication of the inversion of **A** by the constant vector, B, to yield the values of the vector **X**.

The results of the simultaneous solution of the heat and material balances for a p_{CO}/p_{CO2} ratio of 2 are shown in Table 5.

Similar calculations can be made at different p_{CO} to p_{CO2} ratios by only changing the coefficient of the right hand term in Equation 19 and by repeating the procedure of matrix inversion and multiplication.

The effect of changes in other operating conditions, e.g., the composition of calcines and carbonaceous material, can be studied readily by making the appropriate changes in the matrix equation. Also, the results of such changes can be compared in tabular form or plotted directly, using the PRINTGRAPH subroutines of LOTUS 123. For example, Table 6 and Figure 2. compare the effect of the exit CO/CO_2 ratio on the consumption of carbonaceous material per tonne of zinc produced.

Table 5. Calculated results showing how to use Lotus 123 for solving linear equations of material and heat balances

Matrix A

Balan Eqn	Qoxy	Wcoke	Wflux	Wslag	Qzinc	QCO	QCO2	QN2	B	Solution
Oxygen	0.98	0	0	0	0	-0.5	-1	0	-42.348	155.7032
Carbon	0	0.918	0	0	0	-0.5357	-0.5357	0	0	170.6337
Silica	0	0.06	0.85	-0.26	0	0	0	0	-9.767	5.355995
Zinc	0	0	0	-0.0376	-2.919	0	0	0	-197.53	94.45238
Nitrogen	0.02	0	0	0	0	0	0	-1	0	66.45378
Iron	0	0	0	-0.42	0	0	0	0	-39.67	194.9371
CO/CO2	0	0	0	0	0	0.5	-1	0	0	97.46857
Heat	0	8.45	0	-2.5085	-22.557	-1.831	9.742	-1.815	292.88	3.114064

Inversion of matrix A

1.024042	-0.90318	0	-0.75824	-0.17808	-0.51815	0.415689	0.098121
0.003117	0.314554	0	-0.65043	-0.15276	-0.44448	0.648362	0.084170
-0.00022	-0.02220	1.176470	0.045913	0.010783	-0.69691	-0.04576	-0.00594
0	0	0	3.4E-15	0	-2.38095	0	0
0	0	0	-0.34258	0	0.030669	0	0
0.003561	-0.88512	0	-0.74308	-0.17452	-0.50779	1.407376	0.096158
0.001780	-0.44256	0	-0.37154	-0.08726	-0.25389	-0.29631	0.048079
0.020480	-0.01806	0	-0.01516	-1.00356	-0.01036	0.008313	0.001962

Table 6. Solutions of the involved material and heat balance equations at different partial pressure ratios of CO to CO2

CO2/CO	Qoxy Nm3/min	Wcoke kg/min	Wflux kg/min	Wslag kg/min	Qzinc Nm3/min	QCO Nm3/min	QCO2 Nm3/min	QN2 Nm3/min
0.5	155.7	170.63	5.36	94.45	66.45	194.94	97.47	3.114
0.75	140.72	147.26	7.01	94.45	66.45	144.2	108.15	2.81
1	131.92	133.54	7.97	94.45	66.45	114.42	114.42	2.64
1.5	122.04	118.13	9.06	94.45	66.45	80.97	121.46	2.44
2.5	113.22	104.37	10.03	94.45	66.45	51.1	127.75	2.26

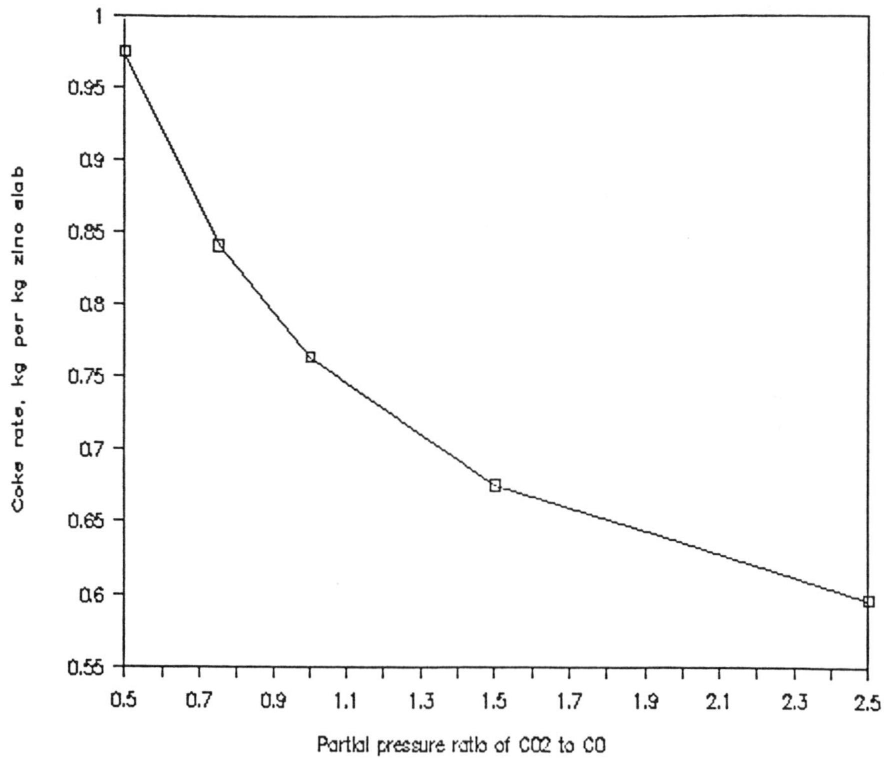

Figure 2.

Conclusions

A simple and convenient method has been described for
developing complex material and heat balances through the use of
the commercially available Outokumpu-HSC and the LOTUS 123
programs. The calculation of material and heat balances of an
envisaged zinc reduction process has shown that this method
greatly simplifies computational effort and programming; and
provides for the use of the considerable resources of the above
two programs in assessing the effect of various factors and
operating parameters on the performance of novel processes.

REFERENCES

1. HSC program, Outokumpu Oy, P.O.Box 60, SF-28101, Pori, Finland

2. LOTUS 123

3. Sources of thermodynamic data used by HSC

4. N.J. Themelis and G.M. Freeman, "Fluid Bed Behavior in Zinc Roasters", <u>J. of Metals</u>, Aug. 1984, pp. 52-57.

THE MEASUREMENT OF THE THERMAL CONDUCTIVITY AND SPECTRAL EMISSIVITY
OF POROUS SPHERES - COMPUTER SOFTWARE APPLICATIONS

K.S. Dominguez and J.R. Wynnyckyj
Department of Chemical Engineering
University of Waterloo
Waterloo, Ontario N2L 3G1

ABSTRACT

An experimental method to measure the thermal conductivity, and the total hemispherical and the spectral emissivities of porous spheres, using radiation pyrometry is described. Its uniqueness lies in the non-steady state measurement of a surface temperature. Techniques have been developed to interpret the experimental data such that process instrumentation time lags are eliminated from the raw results. Computer software has been developed to process the experimental data. This is a quick and easy method to measure these thermophysical properties without altering the integrity of the spheres.

KEYWORDS

Spectral and Hemispherical Emissivity; Thermal Conductivity; Pyrometry; Thermophysical Properties of porous spheres, Computer Software for determination of...

INTRODUCTION

In many chemical processes involving the reaction or drying of porous spheres, heat transfer is a dominant factor. Modelling such processes requires an accurate knowledge of their thermophysical properties. These properties may, moreover, vary during processing. The effective thermal conductivity of the solid is one such property. It must be known as a function of temperature and degree of reaction and/or drying for the very spheres undergoing processing.

In general the usefulness of correlations (Krupiczka, 1967) is limited. Even more limited are theoretically based models (Torquato, 1987) and, at this time anyway, one must resort to measurement.

Barua and Wynnyckyj (1986) have developed a method to measure the effective thermal conductivity k_e of spheres. Its uniqueness lies in the use of anon-contact spectral radiation pyrometer to measure the transient surface temperature. Inherent in the technique, however, is the need for an accurate value of the spectral emissivity, $\varepsilon_{\Delta\lambda}$, of the solid in the wavelength band $\Delta\lambda$ of the pyrometer. Hence, a second experiment was developed to measure this quantity (Barua, Wynnyckyj and Rankin, 1987).

In order to evaluate k_e, ε and $\varepsilon_{\Delta\lambda}$ from the experimental data, long and tedious calculations are necessary. Following a brief summary of the experimental methods, this paper presents the basis for the calculations, and describes the computer software developed to deal with the latter.

THE EXPERIMENTAL MEASUREMENTS

Spectral Emissivity

When a solid is in thermal equilibrium with its surroundings at a fixed temperature, it radiates as a black body. If this solid is suddenly isolated by a non-reflecting shield, it will radiate as into free space. The ratio of the latter radiant flux to the former flux is the emissivity. In the experimental method, a spectral pyrometer is used to measure the radiant flux emitted from a sphere before and during isolation. The signal from the pyrometer is monitored by a differential voltage chart recorder. Ideally, the recorder output from the spectral emissivity experiment would appear as shown in Fig. 1, where section A represents the black body flux of the sphere in thermal equilibrium with its surroundings. Section B would represent the flux upon isolation.

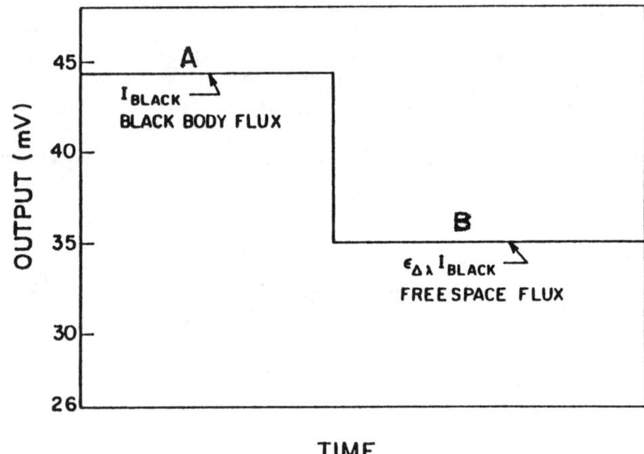

Figure 1: Ideal response of a pyrometer receiving radiation from a black body, following its isolation from the black body enclosure.

In practice, however, the pyrometer and recorder have finite response times and furthermore, the pellet begins to cool as soon as it is isolated from the furnace walls, ie., the black body enclosure. Figure 2 shows a typical experimental output. While section A still represents the black body flux of the sphere, it is obvious that section B has been affected by the process time lags and the cooling of the sphere upon isolation.

The units of this output are millivolts. They are converted to flux units (W/m^2) as part of the computer calculations. The equations constituting the mathematical operations necessary to evaluate the spectral emissivity are

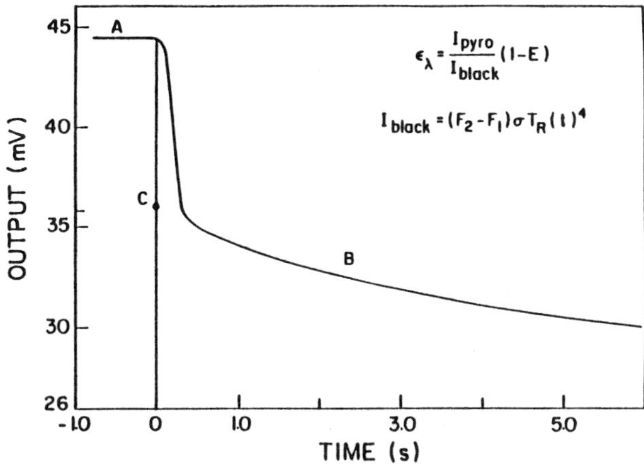

Figure 2: Pyrometer response in a typical spectral emissivity experiment.

summarized in Table 1. Equation (1), Table 1 relates the spectral emissivity of the pellet to the ratio of the two fluxes mentioned above. The value of E, a correction factor, is negligably small for the experimental conditions readily constructed and described by Barua, Wynnyckyj and Rankin (1987).

TABLE 1 Equations Required To Evaluate Spectral Emissivity

$$\epsilon_{\Delta\lambda} = \frac{I_{pyro}}{I_{Black}} (1-E) \tag{1}$$

where

$$I_{Black} = \{F_{\lambda 2} - F_{\lambda 1}\}\ \sigma\ T_R(t)^4 \tag{2}$$

$$E = (1 - \epsilon_{\Delta\lambda})(1 - \epsilon_{\Delta\lambda}^S)\ \left\{\frac{4R_p^2 L_T}{R_T(L_T^2 + 4R_T^2)}\right\} \tag{3}$$

In evaluating $\epsilon_{\Delta\lambda}$ the task is to determine I_{pyro} in equation (1), i.e. the ordinate of point C in Figure 2. To that end, section B of the curve must be extrapolated to zero time of the experiment. This is the instant that the shield has been moved over the pellet, where no cooling can be assumed to have occurred. To accomplish this, an improved method over that published by Barua, Wynnyckyj and Rankin (1987), has been reported recently by Dominguez and Wynnyckyj (1988). The response of the apparatus -- the pyrometer, recorder and shield-movement system -- to a step input is measured first. This is done by focussing the pyrometer on a pellet at a temperature of experimental interest, followed by rapidly blocking the pyrometer sight. From the flux (mV) versus time chart produced by this experiment a characteristic function G(s) is evaluated, based on the standard systems-response theory as discussed, for example, by Stephanopoulos (1984). It was found that a second-order overdamped transfer function served best.

To evaluate point C, an overall input function C(t) was written which fits the totality of the experimental curve in Figure 2; both in region B as well as in the transition region between A and B. This overall function is constructed such that it is a product of the transfer function G(s) and another function describing the true radiative flux received by the pyrometer, called the input function m(t). Since both C(t) and G(s) are obtainable from experiment m(t) can be evaluated. It is then m(t) which is extrapolated to zero time to obtain I_{pyro} of equation (1).

The equations pertaining to the extrapolation of curve B to point C (Figure 2) are summarized in Table 2. The transfer function G(s), equation (4), contains three constants L, T_1 and T_2. These are evaluated for the particular apparatus from the results of the response experiment as shown by Dominguez and Wynnyckyj (1988). The procedure described by Sundaresan et al. (1978) is used. The latter will not be detailed here.

TABLE 2 Equations Relating To Transfer Function Analysis

In Laplace Domain

$$G(s) = \frac{e^{-Ls}}{(T_1 s + 1)(T_2 s + 1)} \qquad (4)$$

In Real Time

$$C(t) = M + N(t-L) + P(t-L)^2 + Qe^{-(t-L)/T_1} + Re^{-(t-L)/T_2} \qquad (5)$$

$$m(t) = a + bt + ct^2 \qquad (6)$$

Auxiliary Equations

$$a = M + N(T_2 + T_1) + 2P(T_1 T_2) \qquad (7)$$

$$b = N + 2P(T_1 + T_2) \qquad (8)$$

$$c = P \qquad (9)$$

$$Q = (M + NT_2)\,T_1/(T_2 - T_1) \qquad (10)$$

$$R = (M + NT_1)\,T_2/(T_1 - T_2) \qquad (11)$$

Equation (6) is the imput function $m(t)$. It is simply a second order polynomial containing three constants a, b and c. The rigorous solution to the cooling problem involved in the emissivity experiment is given by the familiar infinite series solution, equation (15) in Table 3, to be discussed later. The present authors have shown (1988) that the second order polynomial represents this much more complex formallism quite adequately.

The overall input function $C(t)$, equation (5) in Table 2, is actually a product of $G(s)$ and $m(t)$. Since, however, $G(s)$ is in the Laplace domain and $m(t)$ is in real time the multiplication involved a prior conversion of $m(t)$ into the Laplace domain, and then a re-conversion of the product $C(s)$ back into real time.

The constants in equation (5) are not "adjustable", they are unique functions of L, T_1 and T_2 contained in $G(s)$, and of a, b, and c, contained in $m(t)$. The relationships between these constants are given as equations (7) to (11) in Table 2.

The value of the radiant flux from the pellet at point C, Figure 2, is given directly by equation (6) at $t=0$ as

$$m(0) = a \tag{12}$$

Thermal Conductivity

The experimental aspect of this measurement is particularly simple. A pellet is moved from a colder to a hotter portion of a tube furnace. The temperatures of both locations, T_I and T_W respectively are measured by thermocouples. The heat flux emitted by the pellet under the transient conditions is monitored using the same spectral radiation pyrometer as in the emissivity experiment just discussed. The rate of change is slow enough for the flux to be logged by an inexpensive analogue/digital converter, data logger unit. Its output may then be fed directly to a microcomputer for data storage, and if desired immediate processing.

In the earlier paper, Barua and Wynnyckyj (1986) showed that the transient surface temperature of the pellet under conditions of a long constant temperature zone of a tube furnace is given by:

$$T_R(t)^4 = \frac{I_{pyro} - (1 - \epsilon_{\Delta\lambda}) \, V^{pw} (F_2^W - F_1^W) \, \sigma \, T_W^4}{\sigma \, \epsilon_{\Delta\lambda} \, (F_2 - F_1)} \tag{13}$$

The spectral emissivity, $\epsilon_{\Delta\lambda}$, is available from the measurements discussed in the previous section.

The theory for evaluating the thermal conductivity is based on the analytical solution of Fourier's 2nd law. When a sphere is immersed into an environment at a temperature different from itself, the change in the <u>surface</u> temperature as a function of time is given by the familiar series solution, equation (15), Table 3 (Carlslaw and Jaeger, 1973). For values of $\tau < 0.12$ all terms in the series higher than the first are negligible and equation (15) reduces to equation (20) or, when logarithms are taken of both sides, to equation (21). Thus the logarithms of experimental values of $\theta(t)$ -- calculated from equations (13) and (18) -- plotted against time yield a straight line. The intercept θ_0 gives the value of ψ_1, equation (22). Using this value the effective thermal conductivity is calculated from the slope, via equation (23).

TABLE 3   Solution to the Unsteady State Heat Conduction Equation
For Sphere Surface

$$\theta(t) = 2 \sum_{n=1}^{\infty} \frac{\sin\psi_n - \psi_n\cos\psi_n}{\psi_n - \sin\psi_n\cos\psi_n} \; \frac{\sin\psi_n}{\psi_n} \quad \exp\left(-\psi_n^2 \tau\right) \tag{15}$$

$$\psi_n\cot\psi_n = 1 - 1/m \tag{16}$$

$$m = \frac{k_e}{UR_p} \tag{17}$$

$$\theta(t) = \frac{T_R(t) - T_w}{T_I - T_w} \tag{18}$$

$$\tau = \frac{k_e}{\rho C_p R_p^2} \, t \tag{19}$$

. . . .

for $\tau < 0.12$ the following apply

$$\theta(t) = \left(2 \, \frac{\sin\psi_1 - \psi_1\cos\psi_1}{\psi_1 - \sin\psi_1\cos\psi_1} \; \frac{\sin\psi_1}{\psi_1}\right) \exp\left(-\psi_1^2\tau\right) \tag{20}$$

$$\ln \theta(t) = \ln \theta_0 - \psi_1^2 \frac{k_e t}{\rho C_p R_p^2} \tag{21}$$

$$\theta_0 = 2 \, \frac{\sin\psi_1 - \psi_1\cos\psi_1}{\psi_1 - \sin\psi_1\cos\psi_1} \; \frac{\sin\psi_1}{\psi_1} = \text{intercept} \tag{22}$$

$$\text{slope} = -\psi_1^2 \, k_e/\rho C_p R_p^2 \tag{23}$$

Since the Biot number is uniquely related to ψ_1, equation (16), an experimental overall heat transfer coefficient U pertaining to the experimental conditions is also obtained from this measurement. From the value of U, the total emissivity of the pellet is, therefore, obtainable via the relationship:

$$U = \frac{mR_p}{k_e} = h_c + h_r = h_c + \varepsilon 4V^{pw} \sigma T_m^3 \tag{14}$$

Where $T_m = (T_I + T_w)/2$, provided that T_I and T_w do not differ greatly. Here ε is the total hemispherical emissivity.

COMPUTER SOFTWARE AND RESULTS

Figures 3 and 6 are flow sheets of the computer algorithms to calculate k_e and $\varepsilon_{\Delta\lambda}$ from the experimental data. The software is written in either BASIC (1984), or LOTUS 123 (1982).

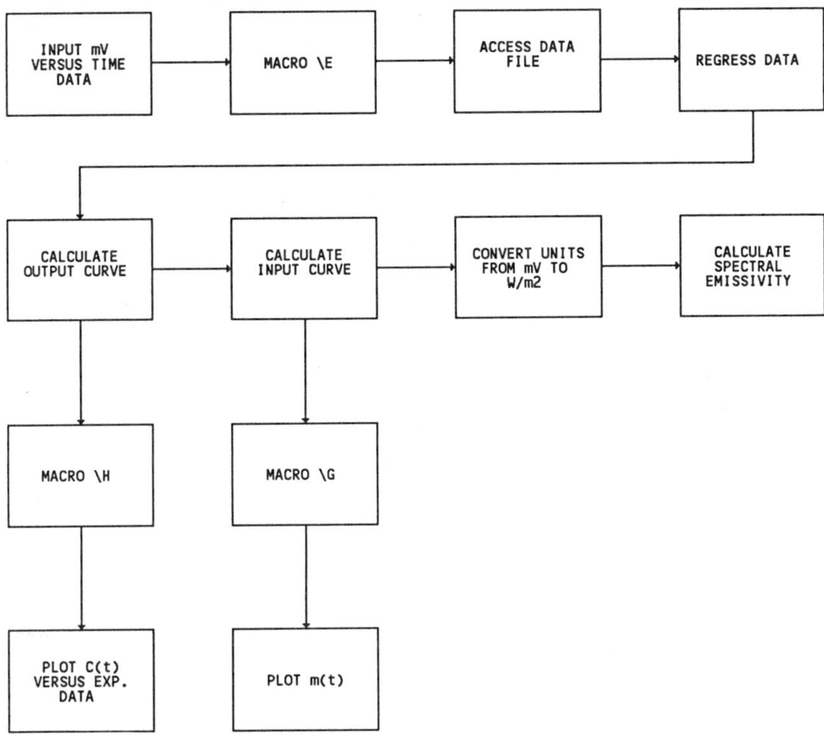

Figure 3: Computer algorithm for the evaluation of spectral emissivity
         from experimental data

Spectral Emissivity

The analogue chart record of flux versus time (such as Figure 2) is converted to a digital output and entered into a LOTUS 123 program called EMISS.WK1. (No attempt has been made to automate the data logging for the emissivity experiment, as was the case for the conductivity experiment. The rates of change are excessive for handling by reasonably priced A/D converters). Also entered into the program are the characteristic constants for the transfer function, L, T_1, and T_2 (equation 4), and the calibration data for the spectral pyrometer and for the thermocouple used to define the temperature of the black body cavity.

The program has been designed such that the user needs only to activate macro commands. By activating command macro/E, Figure 3, equation (5) is fitted to the flux vs. time data using the Lotus 123 data-regression software. This evaluates the constants M, N, and P making use of the constants L, T_1, and T_2 already in the file. The program then proceeds to evaluate the constants a, b, and c of the input function, equation (6). For this, use is made of the auxiliary equations in Table 2. At this point the user can call upon the program to display the results. These are either a column of smoothed out C(t) versus time data, or the program's plotting routine generates and prints flux (mV) versus time plots. Figure 4 is an example. (Dominiguez and Wynnyckyj, 1988). The solid curve was calculated by regression to the experimental points originally entered. The latter are shown as diamonds. The Macro/G command generates a plot of m(t) versus time, also in millivolts. This is shown in Figure 5.

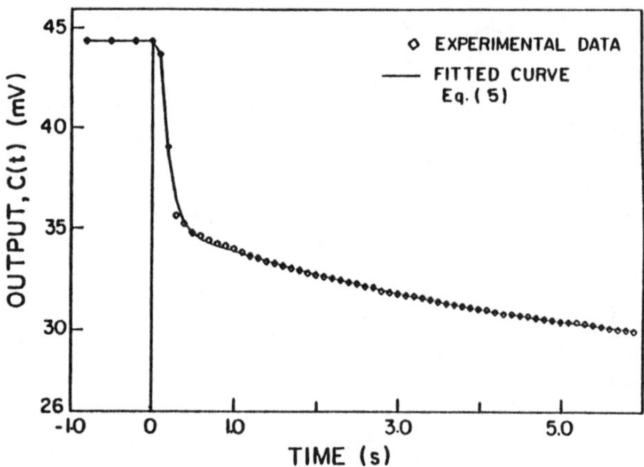

Figure 4: Emissivity experiment -- Typical fit between the regressed output curve, c(t), and experimental data.

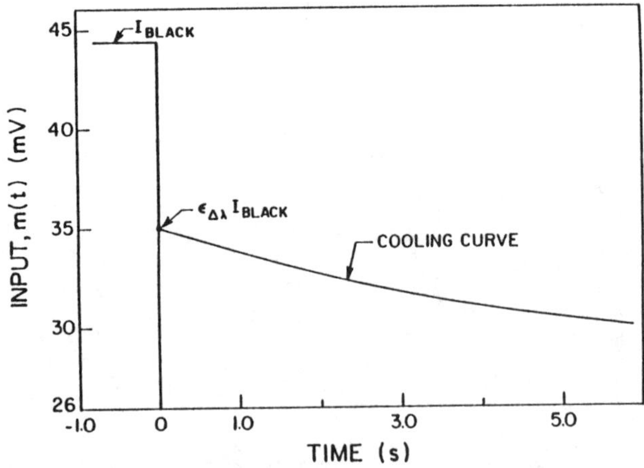

Figure 5: Computer printout of a typical calculated input curve (i.e., corrected pyrometer flux).

The pyrometer output in mV is non-linear, both versus flux (W/m^2) and versus temperature (K). To calculate $\varepsilon_{\Delta\lambda}$ from equation (1) the flux, both I_{pyro} and I_{black}, must be in units of W/m^2. The program first converts the thermocouple and the pyrometer outputs to temperature units (K) using the stored calibration curves and then, using equation (2), calculates flux in W/m^2. The values of the black body flux fractions F_1 and F_2 are required in this calculation. This data has been stored in the program in the form of the equations (25) to (27) as follows (Siegel and Howell, 1981):

$$F_{0-\lambda T} = \frac{15}{\pi^4} \sum_{n=1}^{\infty} \frac{e^{-nv}}{n^4} \{[(nv + 3)nv + 6]nv + 6\}; \; v > 2 \tag{25}$$

$$F_{0-\lambda T} = 1 - \frac{15}{\pi^4} v^3 \left(\frac{1}{3} - \frac{v}{8} + \frac{v^2}{60} - \frac{v^4}{5040} + \frac{v^6}{272160} - \frac{v^8}{13305600}\right); \; v < 2 \tag{26}$$

$$\text{where } v = 14,388/\lambda T \tag{27}$$

Finally the program calculates the spectral emissivity $\varepsilon_{\Delta\lambda}$ via equation (1) and displays the result.

Thermal Conductivity

The task of this program is to calculate the set of transient surface temperatures, $T_R(t)$, of the pellet from the pyrometer output using equation (13), convert it to $\theta(t)$ via equation (18), and to then regress $\theta(t)$ acording to equation (21). The values of k_e and ε are calculated using equations (23) and (14) respectively. Figure 6 gives the flow sheet of the calculations.

The first block in Figure 6 is data conditioning. The pellet radius, density and spectral emissivity, are entered. The raw experimental data from the logger is then entered into the program FINCON using software from the supplier (written in BASIC, with some modifications made to suit our application). This data is immediately stored. The collection of data is separate from its processing to allow for a shorter sampling time.

The raw temperature data is in units of mV. To convert to units of K, calibration equations are used. The initial pellet temperature, T_I equation (18), is an arithmetic average of the thermocouple readings logged prior to moving the pellet from the cooler region to the hotter region. The pyrometer is calibrated during the running of FINCON by correcting its mV output such that it predicts the pellet's initial and final temperatures as measured by independent thermocouples. The data logger used in the present experiment, Hewlett Packard Model HP 3497A, was operated at maximum speed. A data point was thus read and stored approximately every 2 seconds, except when unloading into the computer memory. The latter caused a delay of approximately 5 seconds. The program includes a command to display on the monitor average values of I_{pyro} (every 5 readings) during the progress of the experiment.

The second block in Figure 6 calculates the transient surface temperature T_R, the transient temperature-dependent heat capacity Cp, and the variables necessary to complete a linear regression of the data to equation (21). These calculations are contained within the WHILE loop. The calculations are limited to data collected after a specified initial time and until the pellet temperature has reached 95% of its final value. The initial time period is ignored due to its non-linearity, as discussed previously, while the final

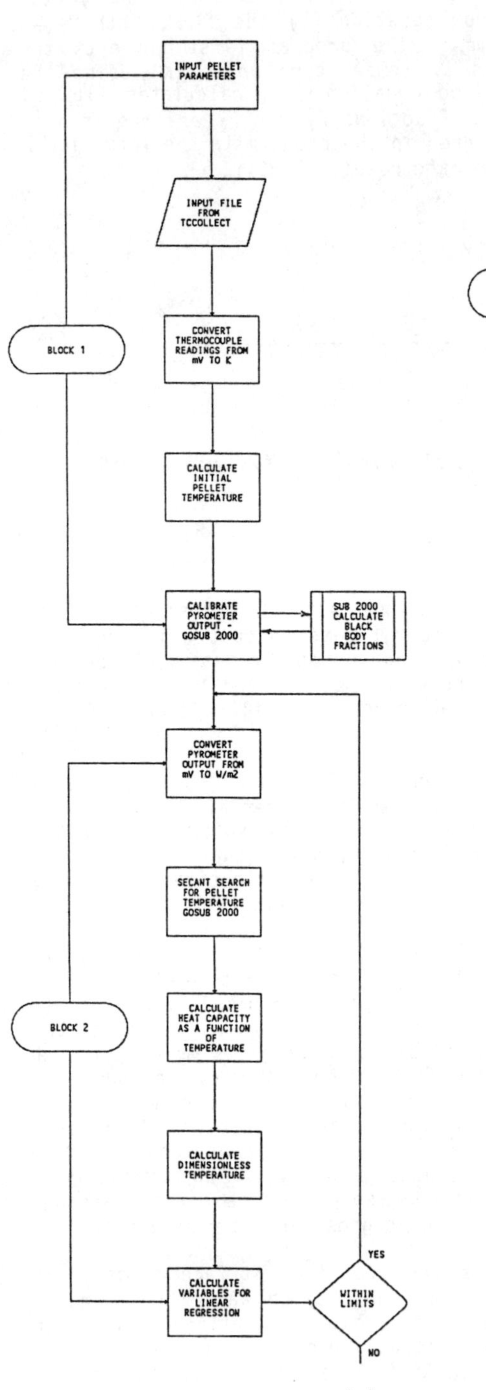

Figure 6: Computer Algorithm for the evaluation of the thermal conductivity from experimental data.

time period is ignored to minimize computation time. The data base for a typical conductivity experiment was some 200-300 digital points.

Within this loop, the calibrated pyrometer output is converted from flux units (mV) to temperature units (K) using a calibration equation supplied by the pyrometer manufacturer, and then from temperature units to I_{black} in flux units (W/m^2) using equation (2), Table 1. To complete this last conversion, the main program is left and a subroutine is entered to calculate the blackbody fractions according to equations (25) to (27) already referred to. Since the blackbody fractions, F_1 and F_2, are a function of the surface temperature an iterative technique is required.

Since equation (13) is not easily differentiated, the secant iterative technique (Horbeck, 1975) was chosen over the traditional Newton-Ralpson technique.   The secant method requires two initial guesses of T_R which bound the final or true value.   In FINCON, the initial guesses are the initial pellet temperature T_I and the wall temperature, T_W, measured during the same time interval as the pyrometer flux.   Equations (28) and (29) below relate two functions to the two temperature guesses, T_{R1} and T_{R2}.

$$FUNC1 = \frac{I - (1 - \epsilon_{\Delta\lambda}) \, \sigma \, VPW \, (F_{2W} - F_{1W}) \, T_W^4}{\sigma \, \epsilon_{\Delta\lambda} \, (F_{21} - F_{11})} - T_{R1}^4 \tag{28}$$

$$FUNC2 = \frac{I - (1 - \epsilon_{\Delta\lambda}) \, \sigma \, VPW \, (F_{2W} - F_{1W}) \, T_W^4}{\sigma \, \epsilon_{\Delta\lambda} \, (F_{22} - F_{12})} - T_{R2}^4 \tag{29}$$

As the values of the T_{R1} and T_{R2} approach the true surface temperature, T_R, equations (28) and (29) approach zero.

The two temperature guesses are updated as follows: T_{R2} takes on the value of T_{R1} and T_{R1} is now updated by the incremental value DELTA.

$$DELTA = \frac{-FUNC1}{(FUNC1 - FUNC2)} * DELTA \tag{30}$$

The above calculations are repeated until the value of DELTA is within user specified limits, usually 0.01 K.   This iterative method results in a converged value for the pellet temperature T_R within four to five iterations.

From T_R thus obtained the natural logarithms of the dimensionless temperatures are calculated.   To perform a least-squares cutting fit of the latter versus time, equations (31) and (32) below are used.

$$SLOPE = \frac{(\sum_1^n (time_i (ln \, \theta_i) \quad \frac{1}{n} \, [\sum_1^n time_i)(\sum_1^n ln \, \theta_i]}{\sum_1^n (time_i)^2 - \frac{1}{n} (\sum_1^n time_i)^2} \tag{31}$$

$$INTERCEPT = \frac{1}{n} (\sum_1^n ln \, \theta_i - (\sum_1^n time_i)(SLOPE) \tag{32}$$

The entire set of calculations is repeated for all the data collected within the time boundaries specified by the WHILE loop. Figure 7 is an example of the graph produced by regression of equation (21) to the logged $\theta(t)$ data. To produce this type of curve the following program operations are involved. Recall that the $\theta(t)$ data was filed in the FINCON program which is written in BASIC. This data is imported into the Lotus 123 program. Once in Lotus 123 a single command produces a graph such as Figure 7. The latter shows the usual Lurrie-Guerney pattern of ln $\theta(t)$; a linear portion at large times and a steep gradient towards $t = 0$ and $\theta = 1.0$. The data shown is for a partly-oxidized magnetite pellet (Dominguez and Wynnyckyj, 1988). The value of θ_0 is 0.80.

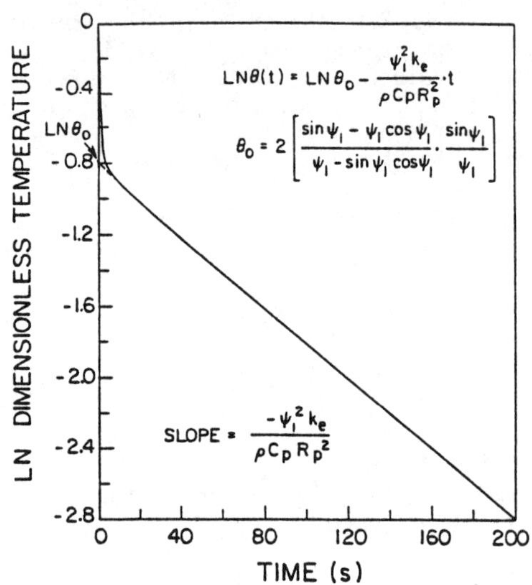

Figure 7: Computer printout of a plot of the dimensionless temperature versus time data from a typical conductivity experiment.

The third block in Figure 6 represents the calculation of the thermal conductivity and total emissivity. The value of ψ_1 is calculated from the intercept via equation (22) using the secant interative technique. The working equations for this calculation are:

$$FUN1 = \frac{2(\sin\psi_{11} - \psi_{11} \cos \psi_{11})}{(\psi_{11} - \sin \psi_{11} \cos \psi_{11})} * \frac{\sin \psi_{11}}{\psi_{11}} - \theta_0 \qquad (33)$$

$$FUN2 = \frac{2(\sin \psi_{12} - \psi_{12} \cos \psi_{12})}{(\psi_{12} - \sin \psi_{12} \cos \psi_{12})} * \frac{\sin \psi_{12}}{\psi_{12}} - \theta_0 \qquad (34)$$

$$DEL = \frac{-FUN1}{(FUN1 - FUN2)} * DEL \qquad (35)$$

The initial guesses of ψ_{11} and ψ_{12} are π-0.5 and 0.1 respectively. These were chosen to bound the true value of ψ_1. As described previously for the determination of the pellet temperature, two functions for the root of equation(22) are calculated using the initial guesses. From these functions an increment DEL, equation (35), is calculated. The guesses of ψ_{11} and ψ_{12} are updated as follows. ψ_{12} takes on the old value of ψ_{11} and ψ_{11} is updated by the incremental value, DEL. The entire process is repeated until DEL is less than 0.0001. Usually only five to six iterations are necessary to converge ψ_1 to within a 0.005% tolerance.

Finally the thermal conductivity is calculated using equation (23). Also the overall heat transfer coefficient U, and the total emissivity ε, are calculated using equation (14). All relevant data is filed.

DISCUSSION

It is possible to simulate the pyrometer output during the spectral emissivity experiment, i.e., the cooling curve shown in Fig. 4, by using the experimental data obtained from both the thermal conductivity and the spectral emissivity experiments (Barua, Wynnyckyj and Rankin, 1987). If the data from the two experiments are internally consistent, the simulated pyrometer output will match the cooling curve. If the two curves do not consistently agree it is possible to adjust the value of one of the parameters, k_e or $\varepsilon_{\Delta\lambda}$ until internal consistency is achieved. Work is in progress to complete a BASIC program, COOL, which will simulate the cooling curve. The data can then be entered into the LOTUS 123 program EMISS.WK1, and the curves can be compared.

The uniqueness and one of the attractive features of these experimental methods is that they can be applied to porous spheres without altering their integrity. Results are obtained on the very bodies which are to be subjected to other measurements or are to be used industrially, and for which accurate data may be required.

The methods themselves are experimentally very simple. An ordinary tube furnace, a commercial narrow-band infrared pyrometer and a standard laboratory chart recorder are all that is required. Added convenience accrues from the use of an A/D converter data-logger for the thermal conductivity portion of the experiments. This more sophisticated (and expensive) feature is not, however, essential. The methods require no special skills on the part of the experimenter. They are, moreover, very rapid. They are readily adapted to measurements in different gases and/or in vacuum. A wide range of temperatures can be covered, although to do the latter two or more pyrometers may be needed to insure accuracy.

It is evident from the body of this paper that tedious calculations are required to obtain the actual property values from the raw experimental results. To have to do these "by hand" might inhibit the methods finding a wider application. The principal task of the computer methods and software described in this paper, therefore, has been to mechanize and automate these calculations and to make the software user friendly. The calculations, indeed, can be performed by a person with only rudimentary computer skills. Presumably this makes the overall methods accessible for general use.

It is the intent of the authors to make this software available to the scientific community (at a nominal charge to cover distribution costs but otherwise royalty free) through the facilities of the Canadian Society for Chemical Engineering and of the Centre for Pyrometallurgy at the University of Missouri - Rolla.

CONCLUSIONS

(1) Recently proven, experimentally simple and rapid methods to measure thermophysical properties of porous spheres at high temperatures have been reviewed.

(2) Extensive calculations are required to evaluate the property values from the experimental results. Computer software is described which eliminates the need for special skills, and the considerable drudgery, in performing these calculations.

ACKNOWLEDGEMENTS

Financial support for this work has come from an Ontario Government Scholarship to K.S. Dominguez and from a National Sciences and Engineering Research Council operating grant. Mr. Bruce McVittie has contributed significantly to the initial steps of software development, as part of his 4th year design project in Chemical Engineering. Mr. Josef Slusarczyk is thanked for his help with apparatus construction.

REFERENCES

Arpaci, V., (1966). _Conduction Heat Transfer_, Addison-Wesley Publ. Co., Reading, Massachusetts.

Barua, S., and J.R. Wynnyckyj (1986). Measurement of high-temperature effective thermal conductivity of porous spheres. _Can. J.Chem. Eng._, 64, 695-701.

Barua, S., J.R. Wynnyckyj and W.J. Rankin (1987). Measurement of band emissivity of porous spheres at high temperatures. _Can. J.Chem. Eng._, 65, 328-334.

Basic (1984). International Business Machines Corporation. Boca Raton, Florida.

Carslaw, H.S. and J.C. Jaeger (1959). _Conduction of Heat in Solids_, 2nd Ed. Clarendon Press, Oxford.

Dominquez, K.S. and J.R. Wynnyckyj (1988). Spectral emissivity measurement, to appear in _Int.J.Heat Mass Transf._

Hornbeck, R.W. (1975). _Numerical Methods_, Prentice-Hall, N.J.

Krupiczka, R. (1967). Analysis of thermal conductivity in granular materials. _Int. Chem. Eng._, 7, 122-144.

Lotus 123 (1982). Lotus Development Corp.

Siegel, R. and J.R. Howell (1981). _Thermal Radiation Heat Transfer_. 2nd ed.Hemisphere Publishing, N.Y. p. 769.

Stephanopoulos, G. (1984). _Chemical Process Control_. Prentice-Hall N.J., pp. 186-211.

Torquato, S. (1987). Thermal Conductivity of disordered heterogenous media from the microstructure. _Reviews in Chem. Eng._, 4, 151-204.

Sundaresan, K.R., C. C. Prasad and P.R. Krishnaswami (1978).Evaluating parameters from process transients, _Ind.Eng.Chem_; Proc.Des.Dev. 17 (3), 237-241.

NOMENCLATURE

a,b,c constants in the imput function, eq. 6 (-)

C_p heat capacity (J/kg.K)

$C(t)$ output function (mV)

E correction factor for reflected energy from shield walls (-)

$F_2 - F_1$ fraction of the total energy emitted by a blackbody in the wavelength interval λ_1 to λ_2 (-)

$F_{1,2}^{w,w}$ radiant fractions emitted from the wall (-)

$G(s)$ transfer function; characteristic of the apparatus (-)

h_c convective heat transfer coefficient (W/m^{2}K)

h_r radiative heat transfer coefficient (W/m^{2}K)

I_{black} blackbody flux within wavelength range $\lambda_1 - \lambda_2$ (W/m^2)

I_{pyro} radiant energy flux received by pyrometer within wavelength range $\lambda_1 - \lambda_2$ (W/m^2)

k_e effective thermal conductivity (W/mK)

L_T length from centre to outside edge of shield tube (m)

L dead time of apparatus (s)

m inverse Biot number (-)

$m(t)$ input function (mV/s)

$M, N, P, Q, R,$ Auxiliary Constants, Table 2

R_T radius of shield (m)

R_p radius of pellet (m)

s Laplace parameter (s^{-1})

t time (s)

T_1, T_2 transfer function parameters (s)

$T_{I,s,R,w}$ temperature; initial, shield, pellet, wall, respectively (K)

U overall heat transfer coefficient (W/m^2.K)

V^{pw} view factor from pellet to wall (-)

<u>Greek Symbols</u>

ϵ total hemispherical emissivity of pellet (-)

$\epsilon_{\Delta\lambda}$ spectral emissivity of pellet in the range of $\Delta\lambda$

$\epsilon_{\Delta\lambda}^{s}$ spectral emissivity of shield (-)

$\theta(t)$ dimensionless temperature defined in eq. 18.

θ_0 intercept in Figure 7, defined by eq. 22 (-)

ρ density of pellet (kg/m^3)

ψ_n parameter in infinite series solution (-)

σ Stefan-Boltzmann constant ($5.6696 \times 10^{-8}\ W/m^2 K^4$)

τ dimensionless time or Fourier number (-)

<u>Auxiliary Parameters Used in Computer Programmes</u>

DELTA incremental value to adjust pellet temperature T_{R1} (K)

DEL incremental value to adjust ψ_1 (-)

$F_{o-\lambda T}$ fraction of total energy emitted by a black body in the wavelength range of $0-\lambda$

FUN1,2 name given in computer programme FINCON to equation 22, rearranged to find the roots of this equation (1 is present value, 2 is past value) (-)

FUNC1 2 name given in computer programme FINCON to equation 13, rearranged to find the roots of the equation (1 is present value, 2 is past value) (K^4)

n   counter

T_{R1} present estimate of pellet temperature (K)

T_{R2} past estimate of pellet temperature (K)

v   parameter used in fitting the fractional output of the blackbody curve (-)

Expert Systems and Artificial Intelligence

Co-Chairmen: W.T. THOMPSON
               Royal Military College
               Kingston, Ontario

               A. VAHED
               Inco Ltd.
               Sudbury, Ontario

REAL-TIME AND ARTIFICIAL INTELLIGENCE SYSTEMS
FOR CHEMICAL AND EXTRACTIVE METALLURGY

Stavros A. Argyropoulos   and   Osama T. Albaharna

Dept. of Metallurgy and Materials Science, University of Toronto
184 College street
Toronto, Ontario, Canada M5S 1A4

ABSTRACT

One of the theses of this paper is that the portion of software design and development dedicated to real-time activity of systems is very different from other programming activities. The first section of the paper presents theoretical and practical topics on real-time system design as well as two systems developed and used for chemical and extractive metallurgy. In the second section,an introduction to expert systems is presented. An example from extractive metallurgy application is used to illustrate some of the terms, ideas and techniques used in expert systems.

KEYWORDS

Real-time, Artificial Intelligence, Software, Expert Systems, Modification, Grain refining, Strontium, Aluminum, Silicon , Alloy.

REAL-TIME COMPUTING

Theoretical Topics

Whenever one hears the term *real* in *real time,* one suspects that this sort of computing contrasts with systems in which the time involved is in some sense less real (maybe even unreal, fictitious, phony, artificial, or the like). This, in turn, may imply to some that the programming done for real-time systems is also more genuine and the other kinds of programming are considerably less than respectable. A real-time system responds to external events as they occur. Since the precise time of these external events is an unknown factor, a real-time system is said to execute asynchronously. Unlike a batch system where one operation is completed before a new operation is started, a real-time system can delay the completion of one operation in order for another operation to be started, continued, or completed. This mechanism (where more than one operation is in progress at a given time) is called concurrent processing. Even though only one operation can be executed at a given time using a single central microprocessor unit, the concurrent processing mechanism of a real-time system gives the illusion of several operations executing simultaneously.

In every real-time microcomputer system, the microprocessor is in control of at least one device. Since the particular applications of systems may differ widely, it is difficult to generalize about the nature of this control. A computer controlled heating system for a refining furnace would be an observable case of control by the computer. The initial input to the system from the operator specifies that a given temperature is to be attained within the furnace, and the computer initially determines through repeated sampling of the actual temperature of the furnace that more heat is needed. When the desired temperature is attained, the microprocessor decides to discontinue adding new heat to the chamber until the moment that the heat drops once again below the specified threshold temperature.

It is important for the programmer of real-time control systems to feel something of the power that is inherent in the data structures and logic of the code "firmware" he or she will write for the microcomputer. The entire decision making process within the particular device controller will likely center around very few bits of information. For example, the processor determines from bits m through n of the temperature indicator register that the desired level has not yet been attained. Therefore the processor shall continue to add more heat to the furnace by leaving bit p set in high position. That single bit p may therefore ultimately result in a temperature of hundreds of degrees. The power that is left at the disposal of the programmer is made possible through the hardware, which essentially translates and amplifies firmware commands into the physical phenomena associated with the external device (i.e. heat, pressure, flow).

Although it is difficult to characterize all real-time programs with few brief generalizations, a suggested list of typical activities performed by a microprocessor in a real-time system, arranged by order of their priorities, is:

1) *Control of external devices.* The programmer must not simply allow for the microprocessor to monitor and control the external device. Rather, the entire design of the software must be focused on continual and timely service to the device.

2) *Perform low-priority (background) tasks.* In many applications, the speed of the real-time device is such that the microprocessor can satisfactorily service the device and still have most of its time available for other tasks.

3) *Perform null tasks.* It is characteristic of real-time event driven software that there is no *end* or *exit* point within the main routine. But the microprocessor must be kept runninig in order to service the devices properly. Therefore, quite often, the controller's software will simply execute a loop in the main routine, a loop that constitutes a *null task* to be performed whenever there is nothing else to do.

In the enormous variety of microcomputer designs of currently operating real-time systems there are two basic strategies for the microprocessor to communicate with external devices, polling and interrupts. The choice of one strategy over the other has important implications for the design of any system. A controlling system polls an external device when, and only as often as, the controlling system's original scheduling algorithm tells it to. However, a controlling system may be designed with sufficient *overhead capability* that it could respond immediately to any pertinent events or changes of state in the external device; we may say that such a system is *interrupt driven.*

System for Recovery Studies of Additives

The objective of an earlier study was to examine the effect of oxygen on the recovery and dissolution of ferroalloys in steelmaking. Conditions of steelmaking ladle metallurgy stations were simulated in an induction furnace (Sismanis and Argyropoulos, 1986). The steel bath was prepared by melting down 65 kg of ARMCO IRON. Figure 1 presents a schematic cross-section of the experimental apparatus. The solution speed characteristics of the ferroalloys were measured with a weight sensor. The particular application of this sensor to measure solution speed characteristics is given in detail in (Argyropoulos, 1983). The active oxygen and the temperature of the steel bath were measured with the Celox oxygen probe system made by ELECTRO-NITE Co. The analog outputs of the weight sensor and oxygen probe system were fed into the μMAC-5000 which is part of the intelligent measurement facility (INMEFA). A schematic layout of this facility is shown in Fig. 2. In this facility, the digitization of analog signals is done in the μMAC-5000 which is an intelligent satellite microperipheral (Analog, 1984). The μMAC-5000 has the Intel 8088 microprocessor, 80 k bytes of ROM and 64 k bytes of RAM. Analog to Digital conversion is performed on the μMAC-5000 via an integrating converter providing 14-bit resolution. The GIMIX system which is seen in Fig. 2 acts simply as a host computer.

A real-time software package called RECOVERY was written in the μMACBASIC for measuring the

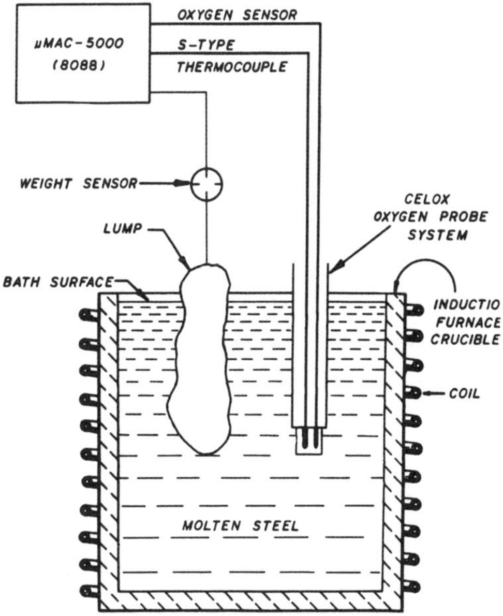

Fig. 1.   Schematic cross section of the induction furnace with ferroalloy
lump,weight sensor, Celox oxygen system,and the intelligent satellite
microperipheral µMAC-5000.

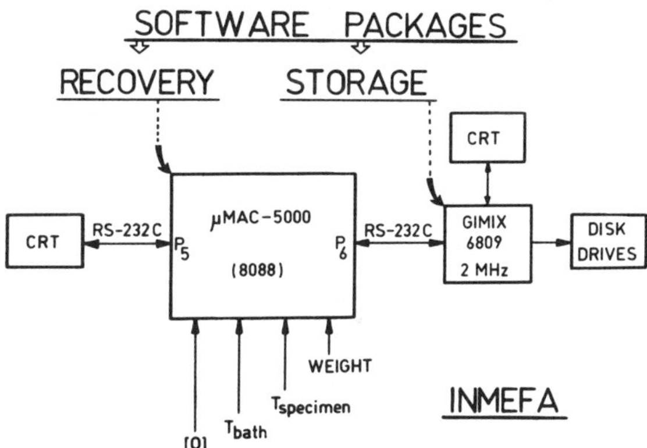

Fig. 2.   Schematic layout of a microprocessor based intelligent measurement
facility (INMEFA).

analog outputs every 250 ms. The program RECOVERY was runninig on the µMAC-5000 and the user had full control of its execution via a CRT which is connected with the serial port (P5) (Analog, 1984). A flowchart of the program RECOVERY is given in Fig. 3. Each time real-time measurements were made, the results were stored in the RAM of the µMAC-5000. At the end of each real-time measurement, the resultant data were sent to the host GIMIX system for storage on the disk. This operation was done through the serial port (P6) of µMAC-5000 which was connected to the GIMIX host. The program RECOVERY started running when the temperature of the steel bath first reached 1600^{0}C. This program initializes the timer on the µMAC-5000. The user can then select one of the following procedures which are also seen in Fig. 3.

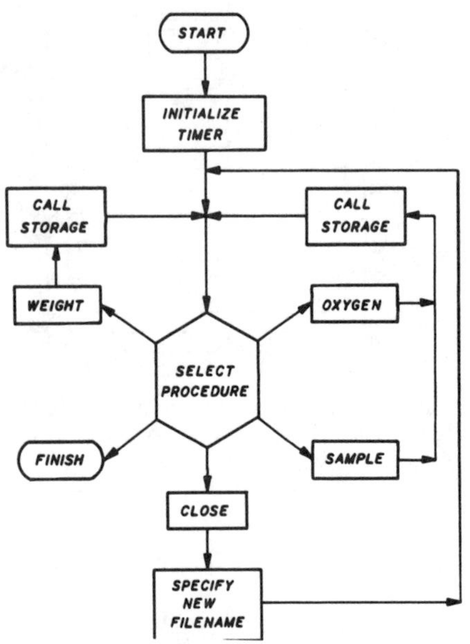

Fig. 3.   Flowchart of software package RECOVERY.

Procedure Oxygen: Measures active oxygen and temperature of the steel bath.
Procedure Weight: Measures solution speed of ferroalloy in liquid steel.
Procedure Sample: Records the time at which an aspirated sample is taken from the steel bath.
Procedure Close: Closes the file with the data on the GIMIX system.
Procedure Finish: Terminates the program RECOVERY.

Each time the procedure oxygen, or weight, or sample, is executed, the computer registers and stores the actual time from the beginning of the experiment.

Another software package called STORAGE was written in 6809 machine language. The program STORAGE was runninig on the GIMIX system. This software package enabled the GIMIX system to accept and store the data which were sent by the µMAC-5000. At the end of this procedure data are sent from the µMAC-5000 to the GIMIX followed by an end of transmission character (hex04). Upon receiving this character (hex04), the STORAGE software sends an XOFF character (hex13) to the µMAC-5000 and starts storing the data onto the disk. At the end of this operation, the software STORAGE sends an XON character (hex11) to the μMAC-5000 instructing it the μMAC-5000 to resume transmission. Another feature of the communication is that a file of data in the GIMIX system can be opened and closed from the μMAC-5000 using the RECOVERY and STORAGE software.

Figure 4 shows a typical experimental result obtained with the Celox oxygen probe system. Curve 1 presents the recorded temperature and curve 2 illustrates the measured active oxygen in the steel in (ppm). The INMEFA system started recording the analog signals from the oxygen cell and the S-type thermocouple a few seconds prior to immersion. Upon immersion, the recorded temperature increases very fast in segment AB. After that, it reaches equilibrium with the steel bath temperature in segment BC. As seen in curve 2, there is a steep rise in the recorded active oxygen in segment DE. After that, the measured active oxygen stabilizes as shown in segment EF. In this case, the represented active oxygen is given by that part of the curve to which arrow 2 points. Figure 5 presents a typical dissolution of standard ferroniobium into liquid steel. The results shown in Fig. 5 were obtained immediately after measurements, illustrated in Fig. 4. Slope of the segment DE is very small and this is an indication of a very low dissolution speed of standard ferroniobium in a steel bath with 1100 ppm active oxygen.

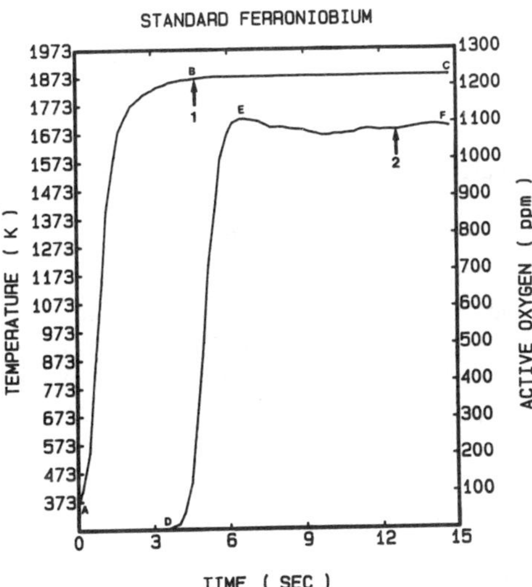

Fig. 4. Typical experimental results obtained by the Celox oxygen probe system, just prior to measuring the dissolution rate of standard ferroniobium. Curve 1 shows the measured temperature and curve 2 depicts the recorded active oxygen in liquid steel in ppm.

In the current configuration of this facility at the University of Toronto the GIMIX system has been replaced by the popular Macintosh Microcomputer. A new software package STORAGE has been written in the Macintosh Assembler language. This new software has the same features as the

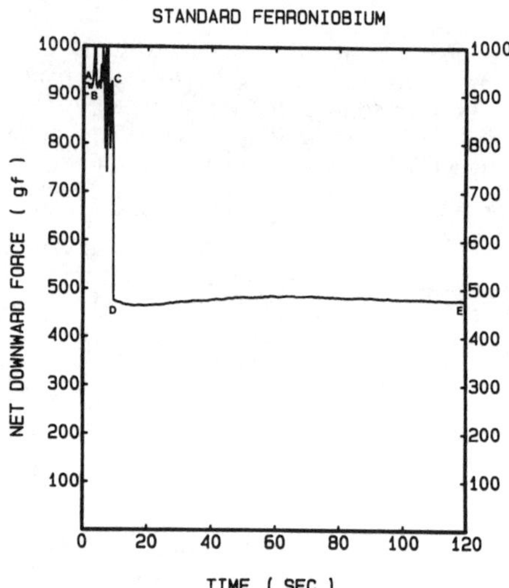

Fig. 5. Typical experimental results obtained from measuring dissolution of standard ferroniobium in liquid steel. The active oxygen of liquid steel for this case is shown in Fig. 4.

previous package developed for the GIMIX system. The interested reader can find more details in (Sismanis, Argyroppoulos, and Sommerville, 1988). The software package can also be adapted for personal computers. The software can be developed in any programming language provided that the compiler, or the interpreter, of this language can write to and read from the port of the personal computer to which the µMAC-5000 is connected.

System for Measuring Chemical Reaction

Microexothermic (Autoexothermic) ferroalloys is a new family of ferroalloys. They exhibit superior recovery and uniformity characteristics over conventional ferroalloys, for both cases where the conventional alloying or microalloying is implemented (Argyropoulos, 1984; Deeley and Argyropoulos, 1985; Argyropoulos and Deeley, 1984). In this family of ferroalloys the heat which is released from the intermetallic formation has beneficial effects on their assimilation into liquid steel. This is due to the fact that the heat release alters the assimilation process of ferroalloys from dissolution to melting (Argyropoulos, 1984). During the development of this family of ferroalloys an experimental procedure was devised to measure their "microexothermicity". This was done by heating samples from these additions in an air induction coil keeping the rate of temperature increase manually constant. The term "microexothermicity" was coined in an earlier work to indicate the type of exothermicity which will be released when the intermetallic formation in powder alloy compacts takes place (Argyropoulos, 1984).

The microprocessor based system presented schematically in Fig. 6 was used to replace the manual control with an automatic one. Figure 7 shows a schematic cross section of the composite sample

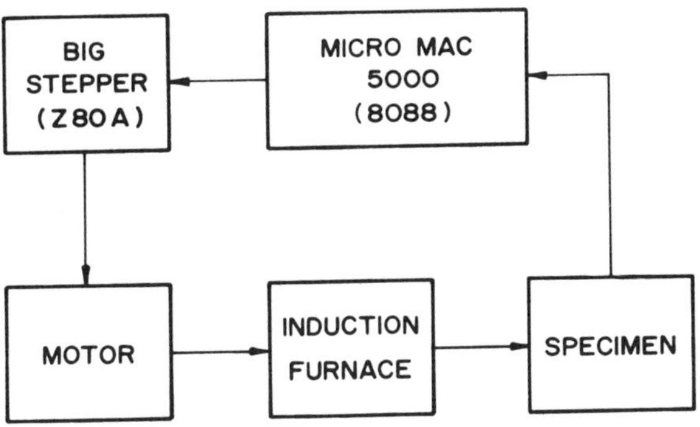

Fig. 6.   Schematic diagram of the microcomputer system used to control the induction furnace.

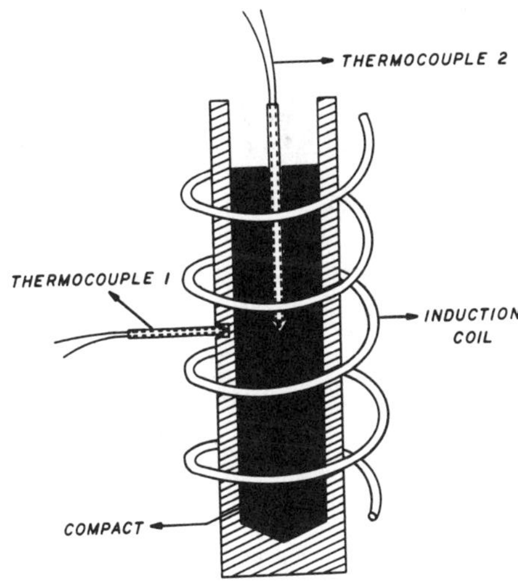

Fig. 7. Schematic cross section of the composite sample placed in an induction coil.

placed in the induction coil. The control thermocouple was located in the specimen's edge and in Fig. 7 is represented by thermocouple one. Thermocouple two is located in the center of the compact. As shown in Fig. 6 the μMAC-5000 is the host microcomputer and is connected with an intelligent motion controller, the BIG STEPPER, via a standard RS-232C interface. The BIG STEPPER is controlled via a series of ASCII command strings passed through the connecting RS-232C cable from the P6 port of the μMAC-5000 (Centre). Table I illustrates some of the commands which the BIG STEPPER understands. The stepping motor is a multiple pole DC permanent magnet motor with multiple windings. When a voltage pattern is applied to the windings in correct sequence, the motor can be made to rotate a specified number of steps in either direction or to stop in a given exact position. The unique feature of stepping motors is that they do not just rotate, but move instead in discrete repeatable steps (For this particular motor each step is 1.8 degree).

<u>TABLE I</u>

S,1,200,F,20	Step motor 1 forward 200 steps/sec for 20 steps.
S,1,200,R,10	Step motor 1 reverse 200 steps/sec for 10 steps.
A,1,600	Alter rate on motor 1 to 600 steps/sec on the fly.
F,1	Halt motor 1.

The μMAC-5000 measured the temperature of the specimen inside the coil. The objective of this controller was to apply a constant linear heating rate at the specimen's edge with simultaneous measurement of the temperature at the specimen's centre. When the temperature at the specimen's edge reached a given value, the controller kept this temperature of the specimen constant. The user of the controller specified the heating rate as well as the upper temperature limit. A proportional-integral-derivative (PID) algorithm was implemented in the software. The algorithm was designed to achieve true anticipation of heating changes resulting in tighter control and minimized setpoint overshoot. Figure 8 shows the block diagram of the induction furnace control. As shown this kind of feedback servo-control system performs three control actions. First, it measures the difference between the commanded and actual temperatures and uses a function of that difference to drive the error-reducing action of the furnace. Second, it introduces a "rate damping" to the system. Third, the addition of a quantity proportional to the integral of the controlled quantity reduces error when the system reaches equilibrium. The different values of K_1, K_2 and K_3 were determined with a series of tests which were carried out by heating a solid steel cylinder 3 cm in diameter. In these tests the control thermocouple was located 0.4 cm from the cylinder's edge. Different heating rates ranging from 200°C/min to 1000°C/min were used, with a constant upper temperature of 1000°C. Figure 9 shows a typical experimental result obtained from the controller. As shown, a perfect linear heating rate was imposed by the controller on the steel cylinder and the overshoot of the temperature when it reached the maximum quickly damped out.

The microprocessor based induction furnace controller was used to measure the "microexothermicity" of different types of ferrotungsten cylindrical compacts. Table II presents the chemical analysis of the materials used. Two types of compacts were made: first, the modified compacts, where the ferrotungsten powder was mixed with silicon powder; and second, the non-modified compacts where only ferrotungsten powder was used. The interested reader can find more details on the compact preparation in (Argyropoulos, 1984).

Figures 10 and 11 present typical experimental results for modified and non-modified ferrotungsten compacts, respectively. In both tests the microprocessor based controller imposed practically the same heating of rate 1200°C/min. In these figures, Curve 1 shows the temperature at the edge of the compact and Curve 2 shows the temperature in the center line of the compact. In both cases there is a delay in the temperature increase at the compact centerline (Curve 2) which is due to the fact that

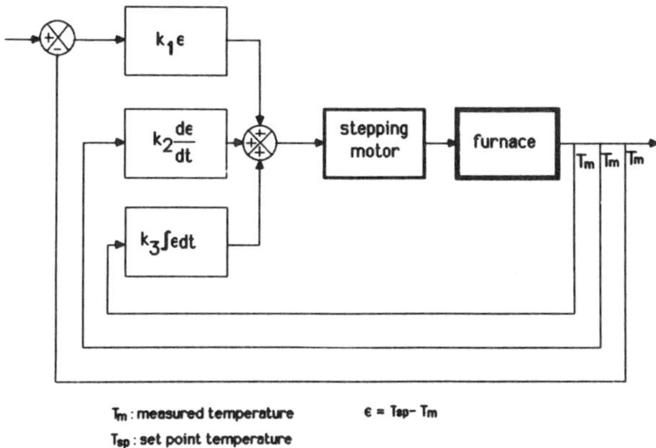

Fig. 8. Block diagram of the induction control system.

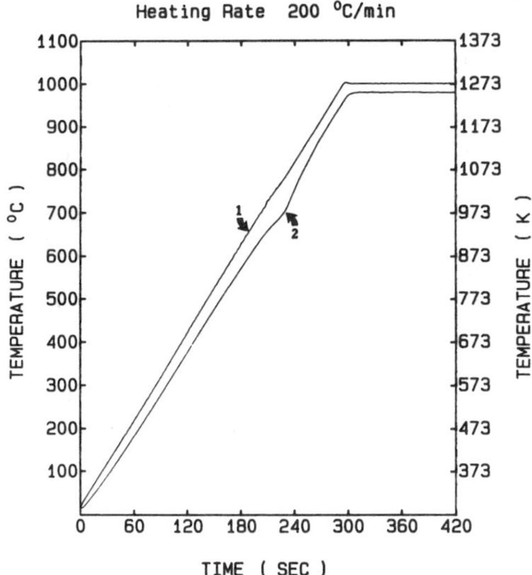

Fig. 9. Experimental results obtained by heating a solid steel cylinder inside the induction coil. Curve 1 shows the measured temperature at a location of 0.4 cm from the cylinder's egde. Curve 2 presents the temperature at cylinder's vertical axis.

<u>TABLE II</u>

Ferrotungsten powder (30 mesh x down)		
	W	79.40
	Si	0.55
	P	0.03
	C	0.01
	S	0.05
	Fe	Bal.
Silicon Powder (50 mesh x down)		
	Si	98.64
	Fe	0.55
	Ca	0.21
	Al	Bal.

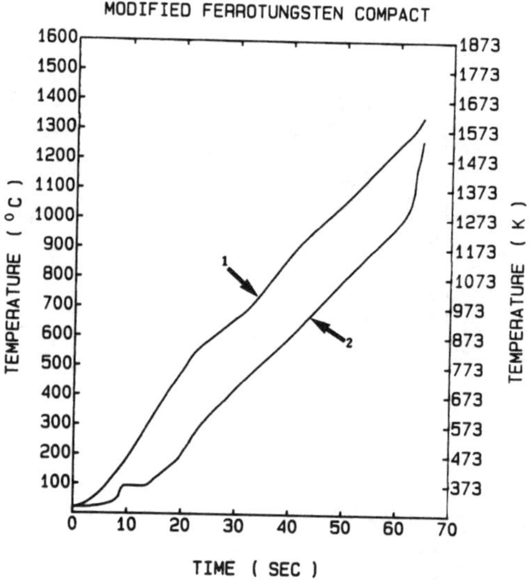

Fig. 10. Experimental results from the heating of modified ferrotungsten compact in air induction coil. Curve 1 is the controlled temperature at the edge of the compact. Curve 2 is the temperature at the centerline of the compact.

moisture exists in the compact and it takes time to evaporate. After this point both compacts follow similar temperature increases up to the 60th second. After this time, however, there is a dramatic increase in the slope of Curve 2 for the case of the modified compact. This abrupt change in the

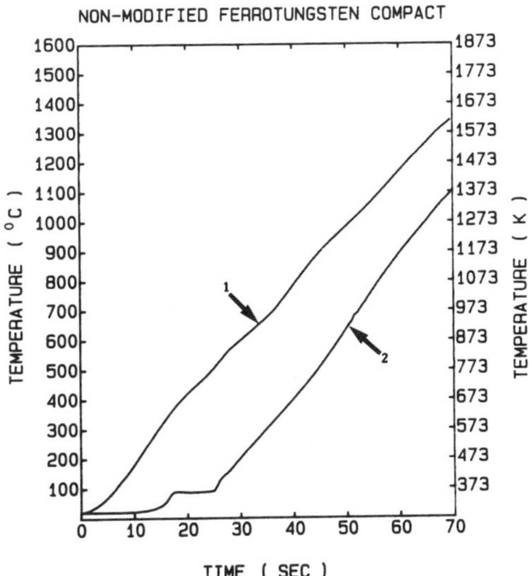

Fig. 11. Heating of a non-modified ferrotungsten compact in an induction coil with the microprocessor based controller. Curve 1 is the controlled temperature at the edge of the compact. Curve 2 is the temperature at the centerline of the compact.

temperature slope can only be explained on the basis of heat which is generated in the body of the compact. It is clear that the formation of different tungsten-silicon intermetallics is exothermic and consequently heat is generated from the interior of the compact. In the case of the non-modified ferrotungsten (Figure 11, Curve 2) where the compact was made only from ferrotungsten without any silicon, the above mentioned phenomenon is neither observed nor expected. Referring to Curve 2 in Fig. 10 there is an important feature to be focussed upon, the exothermic reaction is initiated when the centerline temperature is about 1000°C. After a few seconds, however, there is a corrosive attack on the thermocouple, and when the temperature rises to 1270°C the thermocouple is destroyed. The heat which is released from the intermetallic formation is enough to melt the tungsten-silicides. Consequently, this kind of addition can be dispersed into liquid steel following a melting pattern rather than a dissolution one. The whole mass transfer kinetics thus can be enhanced tremendously.

ARTIFICIAL INTELLIGENCE

The principal concern of Artificial Intelligence is the design of computer programs to undertake activities thought to require human intelligence. Expert Systems is a major research area within artificial intelligence. An expert system is basically a computer program that performs problem-solving tasks in a limited area of expertise and gives the same advice as a human expert.

In this section, an introduction to expert systems is presented. An example will be given to illustrate some of the terms, ideas, and techniques used in expert systems. This example is drawn from an existing data acquisition system developed to carry out thermal analysis (Closset and Gruzelski,

1983). It has been shown that the modification of heat treatable Al-Si-Mg hypoeutectic alloys significantly improves their mechanical properties (Charbonnier, Morice, and Portalier, 1979). Since the mechanical properties are strongly related to the form of the eutectic silicon, it is a matter of great interest to be able to determine the resultant microstructure prior to casting in order to ensure that the melt has been properly treated. Optical metallography is undoubtedly the best technique to evaluate the degree of modification. However, the major drawback of this method is the necessity to destroy a sample for the microstructural analysis. This is time-consuming and not always possible before casting. Obviously, a nondestructive method to control the eutectic silicon modification is desirable in order to give the foundry man a good tool to improve the casting quality.

The developed system is able to simultaneously perform simple thermal analysis, derived thermal analysis, and heat of solidification analysis. The results of this analysis can be quantitatively and qualitatively used to distinguish the microstructure (i.e. **acicular, lamellar, fibrous**) and the four main states of modification (**unmodified, undermodified, modified, and overmodified**).

<u>Expert Systems</u>

The general structure of an expert system is shown in Fig. 12. The two main ingredients of any expert system are the Knowledge Base and the Inference Engine. The knowledge Base is the information that an expert in the specialty addressed by the system, the domain, has or should have. It is a set of assertions of facts and rules. Facts about the world "Problem Domain" which include classifications and relationships between objects. Rules for manipulating those facts including information about when and how to apply the rules. The latter type of knowledge is referred to as procedural knowledge, while the former type is declarative knowledge. The boundary between the two is very flexible. Generally, the less knowledge we declare, the more procedural knowledge is required and vice versa (Waterman, 1986).

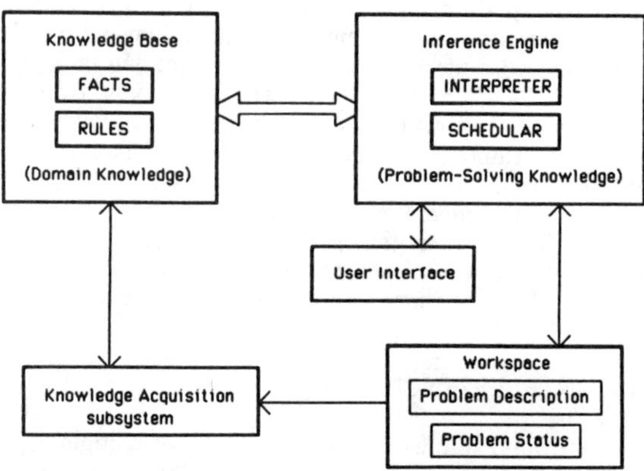

Fig. 12.   General structure of an expert system.

Despite advances in expert system technology, acquiring the knowledge needed to power an expert system and structuring that knowledge into a usable form is one of the primary bottlenecks in expert systems development (Kushner and colleagues, 1987). Among these difficulties is the availability of and quality of time of interaction with the experts, resolving conflicts between experts, and compiling the relevant expertise into problem-solving capabilities.

<u>Knowledge Representation</u>

There are many techniques to represent the experts' knowledge. An excellent summary of the most important knowledge representation techniques can be found in "The Handbook of Artificial Intelligence" by Barr and Feigenbaum (1981). The most popular of these are Production Rules, Frames, and Semantic Networks.

Production Rules are the most popular format for representing knowledge in a way that maintains its procedural character. They are simply statements of the form:

IF   condition   **THEN**   action.

The action part of the rule may be required to ask a question of the user, impliment some standard programming procedure, or even interact with some physical device to switch the system on and off, in addition to alteration of the knowledge base. A set of production rules define a set of allowed transformations which moves a problem from its initial statement to its solution. The current state of a solution is represented by the set of facts asserted in the knowledge base. When the current problem situation satisfies or matches the IF part of a rule, the action specified by the THEN part of the rule is performed.

Figure 13 gives a summary of some of the terms used in the thermal analysis system. Figure 14 shows a possible set of rules for a production rule representation of the problem of identifying the microstructure and degree of modification of Al-Si-Mg alloys. At each execution cycle, all rules whose conditions are satisfied are performed and the facts list modified. In cycle 1, only rule 4 fires. This rule adds the level of strontium to the facts list. In cycle 2, rule 1 fires and identifies the degree of modification. During cycle 3, rules 3, 6, and 7 are executable. Note that rule 7 sends the degree of modification to an I/O device. In cycle 4, rule 2 fires. This rule updates the value of $(T_e - T_c)$ in the facts list. Processing continues in the same manner until a final solution is reached. The inference engine is responsible for choosing which rule to perform next. The method could be a simple sequential search or a more involved heuristic method. More on the inference engine will be elaborated later.

Frames started with attempts to organize the properties of some objects or events. The strength of this system lies in the fact that those elements that are conventionally presented in the description of an object or event are grouped together and may thus be accessed and asserted as a unit. Frame-based knowledge representation uses a network of nodes (frames) connected by relations and organized into a hierarchy. Nodes low in the hierarchy automatically inherit properties of higher-level nodes.

Figure 15 gives an example of such a network of frames. Names in block letters represent other frames in the hierarchy. Thus frame SAMPLE-2 gives the objects associated with the second experimental sample of Al-Si-Mg. Some frames like Al-Si, and MODIFIER define general properties while others like SAMPLE-2-MODIF and ACICULAR&LAMELLAR are more specific and give precise details of certain objects.

Semantic networks are based on the notion of "associative memory". The basic functional unit of a semantic network is a structure consisting of two points of "nodes" linked by an "arc". Each node represents some concept and the arc represents a relation between pairs of concepts. Such pairs of related concepts may be thought of as representing a simple fact. Nodes are labelled with the name of the relevant concept. The arc itself is directed, thus presenting the "subject/object" relation between the concepts within the fact. Moreover, any node may be linked to any number of other nodes giving rise to a formation of a network of facts.

TERM	DEFENITION
%Sr	Level of Strontium in Al-Si-Mg melt
T_c	Eutectic Nucleation Temperature
T_e	Eutectic Growth Temperature
Q_t	Total Heat of Solidification
$\triangle T_c$	Variation in temperature for eutectic nucleation ($T_{ci} - T_{ci+1}$)
$\triangle T_e$	Variation in the eutectic growth temperature ($T_{ei} - T_{ei+1}$)
$\triangle \Theta$	Difference between the eutectic growth temperature and the eutectic nucleation temperature ($T_e - T_c$)

Fig. 13.   Terms and concepts referred in the text.

```
(1)  IF      %Sr between 0.0044 and 0.0079
     THEN    degree = modified.

(2)  IF      microstructure is fibrous
     THEN    △Θ between 3.2 and 3.4.

(3)  IF      degree = modified
     THEN    △Θ between 3.2 and 3.6.

(4)  IF      △Tc is 0.3
     THEN    % Sr is between 0.0044 and 0.0053.

(5)  IF      △Tc less than zero
     THEN    microstructure is fibrous and acicular.

(6)  IF      degree = modified and  Tc = Te
     THEN    microstructure is fibrous.

(7)  IF      degree is identified
     THEN    print degree.

(8)  IF      microstructure is identified
     THEN    print microstucture.
```

PROGRESS	FACTS
initial	$T_c = T_e = 0.3$
cycle 1 adds	%Sr between 0.0044 and 0.0053
cycle 2 adds	degree = modified
cycle 3 adds	△Θ between 3.2 and 3.6 microstructure is fibrous
cycle 4 puts	△Θ between 3.2 and 3.4

Fig. 14.   Example of a production rule representation scheme showing a possible set of rules with corresponding modifications to the knowledge facts.

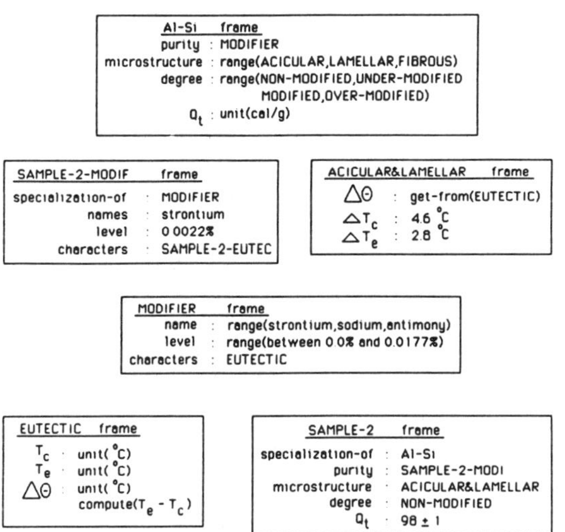

Fig. 15. Example of a frame representation scheme.

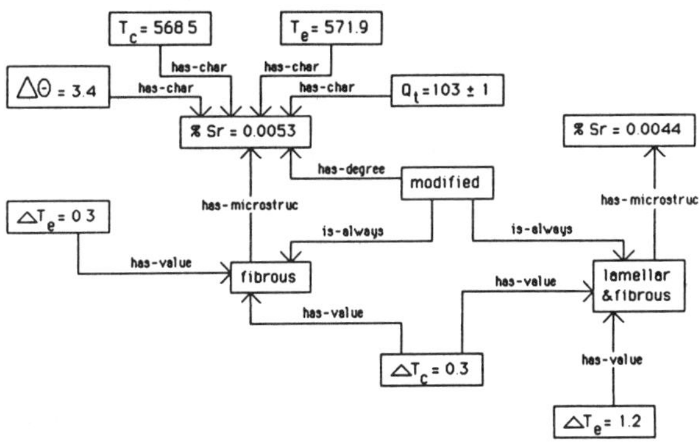

Fig. 16. Example of a semantic network showing part of the results obtained by simple thermal analysis.

Figure 16 is one example of a semantic network. Here, a central node is associated with the concept %Sr=0.0053. Other nodes give other concepts such as T_c and T_e. Arcs are directed into the %Sr=0.0053 node showing that this node has the characteristic of T_c=568.5 for example. Each node is labelled with the relationship it represents. It is clear how such a semantic network and the frame network in Fig. 15 can be used to represent an experts' knowledge. However, other problems do exist. In addition to the problem of how to construct such networks, that is the network building software, there is the more compelling and complicated problem of how to interpret and use the networks as well as the rules and facts in Fig. 14, to arrive at a correct solution.

Inference Engine

The Inference Engine is the controller in an Expert System. It is the software which uses information from the knowledge base together with information supplied by the user to arrive at possible solutions to the problem at hand. It contains an interpreter that decides how to apply the rules to infer new knowledge, and a scheduler that decides the order in which the rules should be applied.

The problem of control in an inference engine could be reduced to the problem of state-space search. In a conventional program, algorithmic methods are used to obtain a solution to the problem. These methods perform specific operations and tasks on each piece of information and data available. Thus, they perform exhaustive search of state-space for the best possible answer. While such methods are guaranteed to produce correct or optimal solutions to a problem, they are often inefficient and time consuming.

In an expert system's inference engine, often, the control structure is heuristic because the tasks these systems undertake are typically difficult and poorly understood. The heuristic methods would selectively examine some of the data and information based on certain conditions. Although this makes it easier to search for a solution, it nonetheless does not guarantee the best solution but produces an acceptable one most of the time. An inference engine could be domain-independent, that is, totally separate from and independent of the domain knowledge. In real-world problems however, this is not easy to achieve and may not be desirable in terms of efficiency and optimization. One possible control method is to perform the first executable rule at a given state. Alternatively, all rules executable at a given state could be performed at the same time before re-examining the rule sets. For such method, the question of ordering the control rules become of extreme importance. A thorough examination of problem-solving methods can be found in "Principles of Artificial Intelligence" by Nilsson (1980).

In addition to state-space search methods, it is also important to employ different reasoning techniques in the problem-solver. For example, in the thermal analysis problem discussed above, it is very useful to employ both qualitative and quantitative analysis techniques. Figures 17a, 17b, 17c, and 17d show the microstructure and graphical data for the simple thermal analysis obtained for four different strontium levels. The portion of the curves of interest in controlling the modification is the eutectic plateau region. It is observed that when the strontium level is increased from 0.0% to 0.0495%, the eutectic temperature decreases from 576.2^{o}C to 571.7^{o}C and then increases slightly to 572.5^{o}C. With proper data/knowledge representation, qualitative analysis comparing the successive curves is sufficient to distinguish the four main states of modification.

Figure 18 gives a plot of the total heat of solidification for a 0.0079% strontium level. Examining such plots for various levels, it is observed that the strontium modification results in the disappearence of the eutectic undercooling peak, S_1 and some sharpening of the ternary eutectic, Al-Mg$_2$Si-Si, peak. The liquidus undercooling peak, S_0, is not affected since no grain refining was used. Curves for the derived thermal analysis resemble those in Fig. 18. The study by Closset and Gruzleski (1983) shows that such curves are sensitive to grain refinement since the magnitude of the liquidus undercooling peak, S_0, is related to the ease of nucleation of primary grains. The peak S_0 disappears with grain refinement. Thus, given proper qualitative representation of the relative heights, differences between peaks, and inflection points, it is possible to differentiate the nonmodified structure from the other forms of structure and to control, nondestructively, the degree of modification.

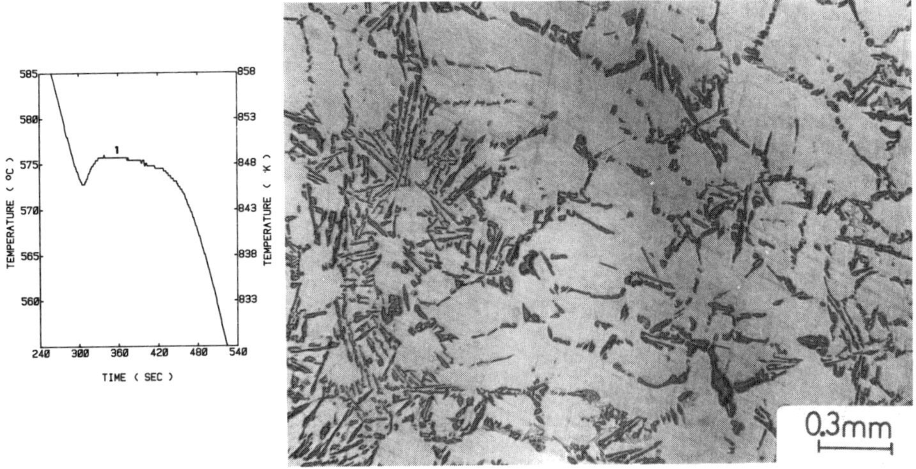

Fig. 17a.   Curve 1: Base alloy, no Sr , structure acicular.

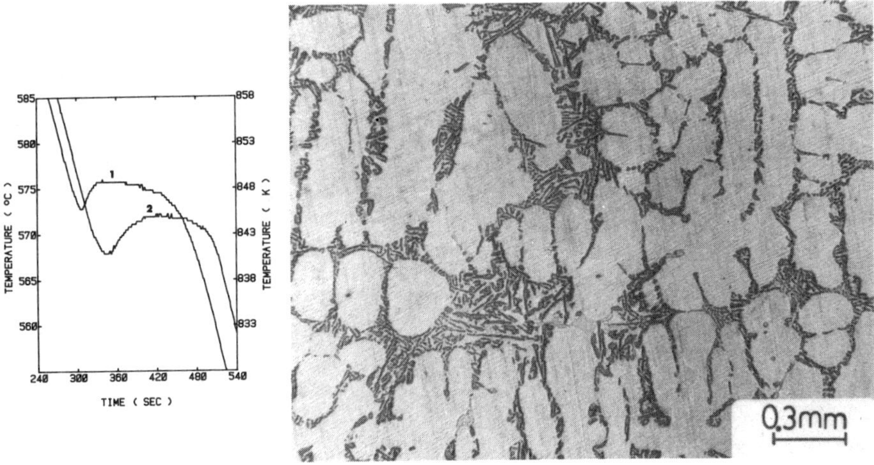

Fig. 17b.   Curve 2: 0.0038 wt.% Sr, structure lamellar.

The importance of combining both qualitative and quantitative analysis techniques can not be over emphasized. First of all, they solve simpler problems with drastically simpler techniques. In addition, even when qualitative analysis fails, it sets up specific plans which greatly simplifies the quantitative

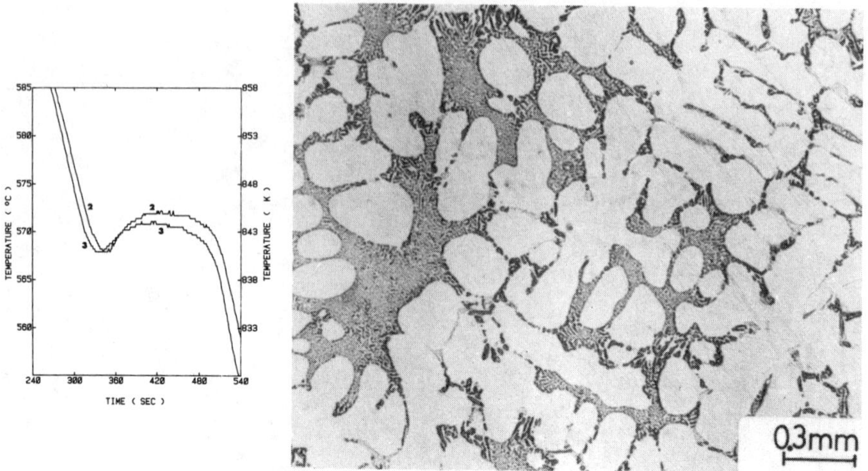

Fig. 17c.   Curve 3: 0.0089 wt.% Sr, structure fibrous.

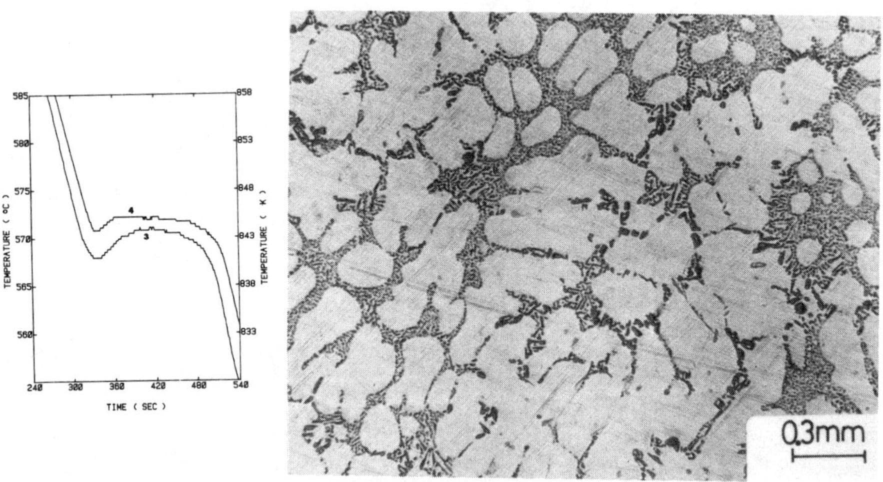

Fig. 17d.   Curve 4: 0.0495 wt.% Sr, structure fibrous and acicular.

analysis and state-space search for the problem. Furthermore, qualitative analysis can handle indeterminacies in problem specification (de Kleer, 1981).

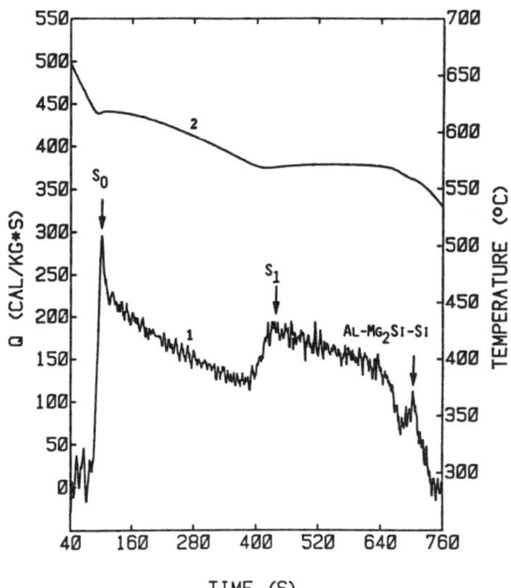

Fig. 18. Total heat of solidification for 0.0079% strontium level.

Real-Time Expert Systems

The complexity of real-time systems is increasing rapidly.The number of functions controlled, the rate at which the functions must be controlled, and the number of factors that have to be considered before a decision can be made are approaching a level with which direct human involvement and present software architectures can't cope (Laffey and colleagues, 1988). This increase in complexity has caused considerable interest in the use of expert systems for real-time applications. Such a system operating in a real-time environment is required to provide the necessary reasoning on rapidly changing data and dynamically changing requirements with limitations on time, hardware, and other system resources.

The large size and dynamic nature of the domain requires new approaches to the inference procedure, since exhaustive search procedures are not possible in real-time (Moore and colleagues, 1984). Proper application of problem-solving techniques can result in sophisticated strategies to handle intelligent interpretation of sensor data, diagnosis of problems, and to cope with process disturbances. Adaptation to complex applications such as process control, aerospace systems, and communications network monitoring and control is a prime research area.

Already, considerable work has been done to attempt and use expert systems in real-time applications. Laffey and colleagues (1984) give a comprehensive survey of real-time expert systems. Nonetheless, few applications have progressed beyond the prototype stage to be used everyday in a real-time domain. One of the main reasons for this situation is that expert system developers have often tried to apply traditional tools to applications for which they are not suited. Much research is needed to develop special expert system tools for real-time domains.

SUMMARY

Real-time implies meaningful dialogue between devices,with software as a necessary intermediary. The designers of such system should ideally be neorenaissance types who are as much as possible at home in both camps (i.e. Metallurgy and Computer Technology). This will enable system design, implementation , and integration to proceed more efficiently. In addition, inevitable future maintenance of the system can be performed with equal smoothness.

A simple example from the Extractive Metallurgy field was presented to emphasize and clarify the main aspects of expert systems. The presented material did not fully cover expert systems but concentrated on the main building blocks. Obviously, more environmental and application related issues such as user interface, expert systems tools, languages, and the hardware system require comprehensive study and pose a big challenge for expert system designers. Nevertheless, it is felt that the presented material is of sufficient detail so as to be a starting point for a more comprehensive study by users interested in specific expert systems applications.

REFERENCES

Analog Devices Inc. (1984). μMAC-5000, Operation, Concepts, Command Reference, Programming, *User's Manual.*, Norwood, Massachussets.

Argyropoulos S. A. (1983). Dissolution Characteristics of Ferroalloys in Liquid Steel. *Electric Furnace Proceedings, ISS-AIME*, <u>41</u>, 81-93.

Argyropoulos, S. A. (1984). The Effect of Microexothermicity and Macroexothermicity on the Dissolution of Ferroalloys in Liquid Steel. *Electric Furnace Proceedings, ISS-AIME*, <u>42</u>, 133-148.

Argyropoulos, S. A., and P. D. Deeley (1984). Exothermic Alloy for Addition of Alloying Ingredients to Steel. *United States Patent, Patent Number 4472196.*

Barr, A., and E. Feigenbaum (Eds.) (1981). *The Handbook of Artificial Intelligence,* vol. 1, William Kaufmann Inc.

Centre Computer Consultants. *Stepping Motor Cookbook.* P.O. Box 739, State College, PA 16804, U.S.A.

Charbonnier, J., J. Morice, and R. Portalier (1979). Thermal analysis of Aluminum alloys to determine their suitability for casting. *AFS International Cast Metals Journal,* 39-44.

Closset, B, J. E. Gruzleski , S. A. Argyropoulos, and H. Oger (1983). The quantitative control of modification in Al-Si foundary alloys using a thermal analysis technique. *Proceedings of AFS Transactions,* 351 -358 .

Deeley, P. D., and S. A. Argyropoulos (1985). Autoexothermic Ferroalloys and New Ladle Metallurgy Tool for Improved Continuous Casting Success. *Proceedings of Continuous Casting '85, The Institute of Metals, London,* 8.1 - 8.10 .

de Kleer, J. (1981). Qualitative and Quantitative resoning in classical mechanics. In P. Winston, and R. Brown (Eds.), *Artificial Intelligence: An MIT Perspective,* vol. 1. MIT Press. pp. 9 - 29.

Kushner, B. G., R. A. Geesey, P. A. Parrish, and S. G. Wax (1987). A knowledge acquisition tool for the intelligent processing of materials. In R. J. Harrison , and L. D. Roth (Eds.), *Artificial Intelligence Applications in Material Science.* The Metallurgical Society Inc. pp. 195-203 .

Laffey, T., P. Cox , J. Schmidt, S. Kao , and J. Read (1988). Real-Time Knowledge - Based Systems. *The AI Magazine,* 27 - 45.

Moore, R., L. Hawkinson, C. Knickerbocker, L. Churchman (1984). A Real-Time Expert System for Process Control. *Proc. IEEE First Conf. on Artificial Intelligence Applications*, 569-576.

Nilsson, N. (1980). *Principles of Artificial Intelligence*, Tioga, Palo Alto, California.

Sismanis, P. G., and S. A. Argyropoulos (1986). The Effect of Oxygen on the Recovery and Dissolution of Ferroalloys. *Fifth International Iron and Steel Congress, Steelmaking Proceedings*, 69, 315-326 .

Sismanis, P. G., S. A. Argyropoulos, and I. D.Sommerville (1988). Development of a Facility for Heat and Mass Transfer Studies. *Proceedings of W.O.Philbrook memorial symposium, ISS-AIME*.

Waterman, D. A. (1986). *A Guide to Expert Systems*, Addison-Wesley.

SIMSMART: SYNTHETIC INTELLIGENCE (SI): THE INTEGRATION OF AI AND SIMULATION

DON WAYE

ANDRE TERROUX

APPLIED HIGH TECHNOLOGY LIMITED

2000 MANSFIELD STREET, SUITE 506

MONTREAL, (QUEBEC), CANADA

H3A 2Z2

ABSTRACT

Process industries, whether batch, continuous or both; rely heavily upon:
1) The productive efficiency of its equipment;
2) The operating, supervisory and managerial skills of its staff.

Recent modernisation and automation of the industrial base aimed at increasing productivity and profitibility; have imposed much greater demands on all levels of human skills than those that were required to operate the older, simpler and clearly uneconomical installations. Satisfactory control of the day-to-day operation of this high-speed equipment is often beyond:
1) The reach of the humans reaction time (however skillful);
2) The capabilities of the process control systems (however fast).

It is the interaction of human skills and machine efficiencies that provide the basis for potential increases in profitability. The real-time optimisation of both the process (machines) and process control (humans) systems is required. This paper reviews the application of the integrated synthesis of AI (Artificial Intelligence) and Simulation technologies into a new technology we have labelled Synthetic Intelligence (SI). The paper will cover the concept and application of SIMSMART, the first proven Synthetic Intelligence (SI) machine; at the Clermont paper mill of Donohue Inc. The expert system includes complete simulation of not only the papermaking process, but also both process control systems; a Bailey Network 90 and an Allen-Bradley PLC-3. Examples will be given of the use of SI techniques for the optimisation of the process equipment (pump/impeller sizing, tank sizing, valve sizing, etc.) and the process control systems (display configuration, controller tuning parameters, control configuration,etc.)

INTRODUCTION

The secure operation of an industrial process depends on the human operator and the real-time process control system. Both of them must perform well together. The operator, in order to make decisions; depends upon the availability, the timeliness and the quality of the information presented to him by the process control system. If the process measurement and control system does not perform well, the operator, no matter how capable he might be, would have great difficulty in making the right decision. On the other hand, even if the control system does perform well, the operator may not be psychologically or technically prepared to make the right decision.

Most industrial processes use a combination of both batch/sequential and continuous manufacturing to convert the raw materials into products. The process is controlled by both Distributed Control Systems (DCS) for continuous (analog) control and Programmable Logic Controllers (PLC) for batch/sequential (digital) control. The expert system's realism is achieved, since both the control system emulators and the process simulator are exact duplicates of reality.

The first delivery of the SIMSMART expert system was for training the operators of the new Number 5 Valmet-Dominion paper machine at the Clermont, Quebec mill of Donohue Inc. The expert system included complete simulation of not only the papermaking process, but also both process control systems; a Bailey Network 90 (DCS) and an Allen-Bradley PLC-3 (PLC).

Based on SIMSMART's original design specifications, it was known that the expert system would have to incorporate some form of Artificial Intelligence (AI) as a means of deriving system models for various user-designed process scenarios. Also, it was assumed that the mathematical models representing the process equipment ICONS would be purely mathematical.

As a result of the experience gained at the Donohue installation, it was found that process equipment ICON models based solely on pure mathematical principles either do not meet the objectives of the SIMSMART expert system or are impossible to derive because the models required parameters which could not be quantified.

These findings led to a systems modelling approach that entailed synthesising the mathematics of the ICON to create «synthetic» models capable of reproducing the desired dynamic response. This level of computer automation could be said to possess a «Synthetic Intelligence» (SI), since the intelligence is based on data and rules derived from the knowledge of engineers and operators acquired over many years of experience.

SYNTHETIC INTELLIGENCE

Much has been published during the last decade on the different «levels» of the process control «models», such as regulatory, supervisory, etc. Is is important to note that each level of process control can be similarly represented by an equivalent process control simulation of that level. The process control simulation models are configured to include:

1.   Analog signals, normally connected to the DCS.
2.   Digital signals, normally connected to the PLC.

Process simulation models have traditionally consisted of sets of (non) linear (partial) differential equations. What had been suggested and has since been proven is that there exists a one-to-one correspondance between on the one hand, analog and digital control models, representing the process control model; and on the other hand, data and rule driven models, representing the process model. The key to SI is the integration of data-driven models (analog equivalent) with rule-driven models (digital equivalent). (see Figure 1).

Previous attempts at process simulation have been limited to either data or rule driven models, but not both at the same time. The detail of representation of the process model must be «sufficient» to accurately

represent the same «level» of knowledge required by the process control model. The engineering and operations knowledge considered «necessary» to build «sufficient» models is available only at the ICON level. The SIMSMART expert system's Synthetic Intelligence compiler automatically creates process models from the ICON model library and the process flowsheet network. (See Figure 2).

The concept of a model as a data base has not yet been exploited to its full potential. A model is defined as a comprehensive representation of captured intelligence. Synthetic Intelligence (SI) is a knowledge-based expert system

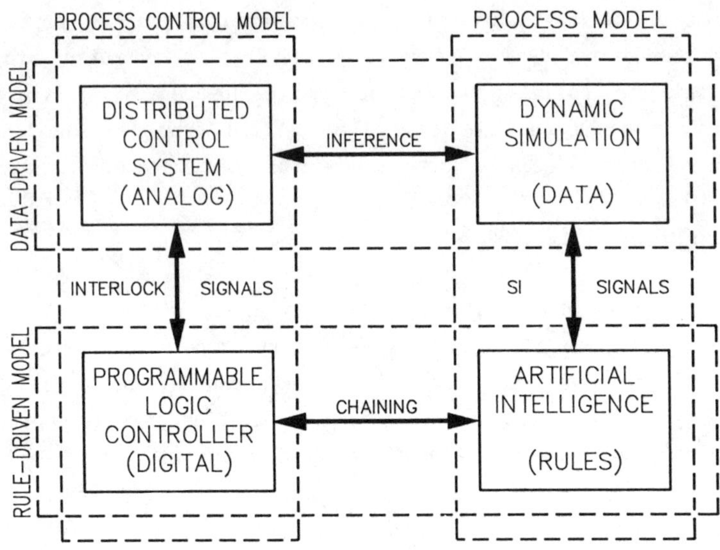

FIGURE 1 SIMSMART EXPERT SYSTEM
PROCESS / PROCESS CONTROL
MODELING ANALOGY

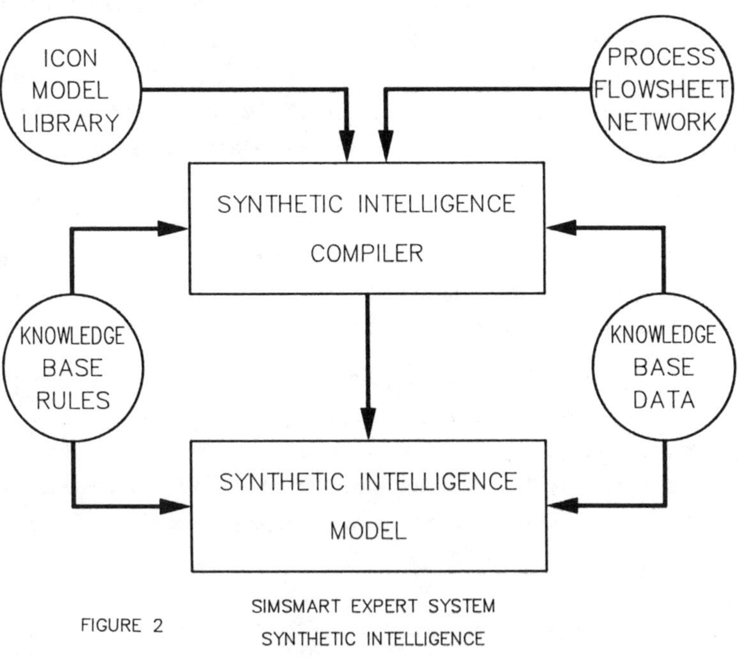

FIGURE 2 SIMSMART EXPERT SYSTEM
SYNTHETIC INTELLIGENCE
MODEL BUILDER

that is cognizant, or is capable of «thinking». Thinking is information processing. Information processing is computation, which is the manipulation of symbols. Symbols, because of their relationships and linkages, express something about the external world. An expert system exists when the symbols (ICONS) acquire meaning through the way they influence the behaviour of the system as a whole (Process Model). The connections (streams and signals) and the organisation (network) make the whole (process) into more than the sum of its parts (ICONS).

A knowledge-based expert system such as SIMSMART, assists the human user in solving problems in a restricted domain.

Knowledge-based expert systems are information-processing systems, using a combination of symbolic reasoning and data-processing. The central question in using knowledge-based expert systems to create the process models is whether the synthesis process (or part of it) can be adequately formalised and subsequently captured in rules, or whether the SI process shall always remain a human skill.

The intelligent aspects of a knowledge-based expert system must be able:

1.   *to FORMULATE HYPOTHESES*
2.   *to PRODUCE NEW DECISIONS*
3.   *to ATTEMPT TO ACHIEVE NEW GOALS*

Figure 3 shows the signals that link together the various technologies.

It is the integration or synthesis of these «stand-alone» technologies that provides the first step towards an autonomously «thinking» system.

It sould be noted that the process model, process control model and the «human» provide the required integration to create a symbolic reasoning process. The SI models will always be able to *«produce new decisions»* without the «human» expert; but the SI system cannot autonomously «think», since the human expert is required to *«formulate hypotheses»* and to *«attempt to achieve new goals»*. The first fusion of AI and Simulation was to create the prototype installation of SIMSMART at Donohue's Clermont mill. This delivery was limited to the creation of the knowledge base, composed of the process and process control models.

HISTORICAL BACKGROUND

Donohue's first major expansion occured in 1977 with the completion of the bleached kraft pulp mill in St-Félicien, Québec. The mill was «traditional» and used proven technology. With this in mind, the process control systems were pneumatic for continuous or analog control and electric relays for motor or digital control. The training of the operations personnel was also traditional, using standard classroom techniques such as: textbooks, exercises, exams, etc. At that time, simulation and its potential uses were virtually unknown.

The next, and possibly most significant expansion was the construction in 1981 of a major newsprint mill in Amos, Québec. The project employed state-of-the-art technology for the process and process control. [Ref.1, 2] The process control was accomplished using Distributed Control Systems (DCS) for analog control and Programmable Logic Controllers (PLC) for digital control. Apart from the use of videotapes, training of the operations personnel remained «traditional». However, the potential of simulation as a training tool was recognized. As a consequence and on an experimental basis, a «process and instrumentation» simulator (valves, pumps, tanks, transmitters, etc.) and a refiner simulator (control console) were purchased as operator-training tools. The experience gained by Donohue personnel justified the use of simulation in a training environment.

In 1985, Donohue extended their pulp production facility at the Clermont newsprint mill. For the first time, a software based simulator was used for operator training.
Despite some difficulties, Donohue remained committed to the on-going use of this technology.

TRAINING SIMULATOR OBJECTIVES

Two years later, Donohue, added a high-speed Valmet-Dominion newsprint machine at the Clermont mill. Both the process equipment and the process control were considered to be «state-of-the-art». In fact, the DCS or analog control was provided by a Bailey Network 90 and the PLC or digital control was through two Allen-Bradley PLC-3's. Donohue opted to purchase the prototype (first installation) of SIMSMART from Applied High Technology for operator training. [Ref. 3] The SIMSMART expert system was to simulate both the process and process control systems; and was to provide as much as possible, an exact duplicate of reality . For the process simulation, this meant dynamic simulation (transient response) of all the process equipment from the 3 high-density storage towers to the reel-up on the paper machine. The simulator was to be modular, based on the equipment tag-numbers and in addition to the standard mass and energy balances, was also to be capable of full hydraulic modelling (force balance). The additional requirement is mandatory to provide the required degree of reality for the process control systems. Most measurement and control devices (instrumentation) function on the basis of pressure/flow characteristics (force balance). To adequately simulate the process and the process control, therefore required mass, energy and force balance dynamics. [Ref. 4]

IMPLEMENTATION OF SIMSMART

Donohue and their consulting engineer provided the required engineering detail for the process and process equipment. This process engineering information included tank drawings, piping isometrics, pump calculation sheets, control valve characteristics, etc. All of this information was required to construct the simulator portion of SIMSMART. [Ref.5]

To match the level of detail (granularity) to that of the process models, the process control models were designed to be exact duplicates of the Bailey and Allen-Bradley control systems. The process control engineering contained in the control simulators included control configuration editors, display configuration editors, graphic editors, input/output cards, wiring and terminal blocks, etc. Again, as before, the consulting engineers and the control system vendors provided the required control engineering detail to configure an exact duplicate of the real process control systems. [Ref. 6]

INSTALLATION OF SIMSMART

The process engineering and process control engineering information were loaded into the simulator, exactly as received from either Donohue, the consultant or the control system vendors. Most of this information was installed in SIMSMART by July, in anticipation of training the operators during the months of August and September. Most of the process simulator had been debugged and was operational by the end of July. At that time the process control simulators, were «connected» to the process simulator so that training could begin.

The first indication of impending malfunctions (bugs) occured when the main TMP feedstock supply pump was started. It shouldn't have started without the agitator running. But it did! Then it was discovered that the main TMP feedline to the mixing chest could supply about 20,000 litres/minute, but the flow-measuring instrument had only been sized for 12,000 litres/minute. The demand signal for the TMP fibres was a flow-ratio signal from the level of the mixed stock chest. The logic (ladder diagram) programming of the PLC «locked-out» (CO-tracking) the controller in the DCS if the level in the mixed stock chest was normal. Therefore, with the agitator off and the TMP pump running, and a «normal» level in the mixed-stock chest, the operator could not operate the TMP flow controller. After reprogramming the ladder diagram to fix the mistake in the ladder logic, the TMP flow controller opened the control valve and saturated the flow-measurement signal (100% output was 12,000 Lpm) at about 60% of the available flow (maximum flow was 20,000 Lpm).

And so on, and so on...

The next two months were spent «debugging» the control systems, the field instrumentation and the process engineering so that the process and process control operated in a way consistent with both Donohue and the consulting engineers design specifications. Daily reports were sent to the construction personnel such that they could correct, for example:

1.   Errors in the DCS and the PLC «wiring» between the measurement/control devices and the input/output cards.
2.   Errors in the hardware addresses on the I/O cards and their respective software addresses in the control configuration, etc.
3.   Errors in the sizing of field measurement and control devices.
4.   Errors in the range and calibration of the field instrumentation.
5.   Errors in the various display and graphic configurations, etc.

RESULTS

The construction and mill personnel folowed up on the various memos and corrected the errors in both the process engineering and process control configurations prior to start-up. Thus, when it came time to actually start the paper machine, everything worked; and the 30-day budgeted start-up period was reduced to less than one week, resulting in considerable savings to Donohue.

Clermont's business plan calls for the production of saleable grade newsprint while maintaining optimum quality and production. All available resources must be concentrated and directed toward achieving the primary goal. Any «forced» expenditures, such as correcting and debugging the programming of the process control systems must be viewed as a negative influence towards the primary goal. Since the process and process control systems had been «debugged» prior to start-up; the operations personnel were able to spend more time solving the papermaking problems, and much less time debugging the process and process control systems than has been experienced on similar projects. Consequently, Donohue achieved «nominal» production (the goal) in less than one week, which is somewhat of a record for a grass-roots installation such as this.

PROTOTYPE & PRODUCT EVOLUTION

Donohue-Clermont was the prototype installation for SIMSMART. The goal was to research and develop a dynamic process and process control simulator for the pulp and paper industry, that included sufficient engineering detail to be able to accurately reproduce reality. In our research, we discovered that it is the integration of Artificial Intelligence (AI) techniques with traditional dynamic simulation that provided both the reality and orders-of-magnitude computational speed increases that have made SIMSMART look so real to the engineering and operations personnel. In our quest for reality, we not only duplicated the process engineering, but also the process control engineering. SIMSMART is so real that we were able to duplicate errors in engineering, which due to project time contraints, forced us to debug all the control systems and some of the process. There is a very big difference between how the process/process control was designed (supposed) to work and how it was actually programmed (configured) to work.

Therefore, our experience from this fast-track project, leads us to believe that in the SIMSMART «product», a «generic» control system should be provided, to overcome the potential lateness and possible error-prone delivery of the detailed process control engineering. In this way, the controls can be configured to control the process as intended by the client; and training can proceed without interference by errors and omissions. Once the operator training has been completed, the real control system engineering can be loaded into SIMSMART, and the operators could compare the intended operation of the process, on which they had been previously trained, versus the programmed operation of the process, which would be used at start-up. Corrections could then be made to the various control and display configurations to achieve the intended operating procedures, and this information could be loaded directly into the real control system just prior to start-up.

The consequences of this product procedure would ensure a successful operator training, no matter how late or how correct the programming of the control systems is. In addition, the provision of «generic» controls (based on ICONS) would allow both the client and the consulting engineer to better understand the dynamic «war» between the process and process controls before committing the design to detailed engineering. In this way, more efficient training and start-up could be expected.

ON-LINE SIMULATION AND OPTIMISATION

SIMSMART is now evolving from a prototype to a product, based on the experience of the first installation at Donohue-Clermont. The future value of SIMSMART will only become apparent once the simulator's ability to accurately reproduce reality is used for on-line simulation and then optimization. Unlike other simulators, SIMSMART should not be «scrapped» after the training, since all knowledge acquired during the engineering and training periods is stored in SIMSMART.

The objective of all knowledge-based expert systems is to provide a simple «WHAT IF» package to the process operations personnel (operators, supervisors and managers). The accuracy (success rate) of the analysis is closely tied to the «detail of representation» of the process and process control models. Therefore, it is most important that the higher level models be correct before attempting a «WHAT IF» analysis.

The SIMSMART expert system uses on-line simulation to collect, analyse and classify data (See Figure 4). The good data is used to develop the «missing» higher level models required for the «WHAT IF» analysis. The first step in this procedure is to attach the SIMSMART expert system to the real-time information networks of the various process control systems installed in the process/plant. Next, the collection of measured data and simulated data must be synchronized, so that the sampled data from the process control systems is lined up in time with the simulated data values. The data sets, once synchronised in time, can be analysed by an acceptable mathematical method, such as Gaussian elimination. The analysed data can then be classified as either «good» data or «bad» data. Good data are those process variables and control signals whose values fall within acceptable limits, as determined by the chosen mathematical method, and as set by the operations personnel (during training, at start-up, etc.)

The «bad» data should be examined by the maintenance personnel to identify why the data is bad. Bad data is attributable to several possible sources:

1.   Instrument calibration (zero and span).
2.   Instrument location (physical attachment to the process).
3.   Signal filtering (lead/lag at the transmitter and/or at the input card).

It should be noted that installations in the Scandinavian pulp and paper industry have shown that on average, 37% of all measurement signals were classified as «bad» for the aforementioned reasons. [Ref. 7] The bad data should never be used for closed-loop control, either analog or digital; and the operator should be warned to put the affected loop(s) or circuit(s) in manual, and not to rely on the bad data to effect manual control.

It is the responsibility of the mechanical, instrument and electrical technicians to repair or replace the bad data sources as quickly as possible. It should be noted that some «bad» data points will remain «bad», no matter how much maintenance is carried out; due to:

1.   The technology (gamma ray, infrared, microwave, etc.) used to «measure» the process variable is inappropriate.
2.   The process variable cannot be measured accurately, under any operating conditions.

The «good» data should be used to develop better process and process control models, based on statistical analysis of the data. The improved process models can then be substituted back into the SIMSMART ICON library to improve the «detail of representation» of the simulated model.

It should be noted that in addition to the process model refinement above, it is also possible to create control models of higher control levels, such as those that would be required for «target optimization». Targets (such as Kappa number in the pulp and paper industry and matte grade in the base metals industry) are usually implemented by changing various controller setpoints, coordinated in time by a high level algorithm (the so-called Statistical Process Control or SPC model). Most of the «quality» target models are based on «unmeasureable» process variables, due to:

1.   Measurement device too expensive (capital investment and maintenance costs).

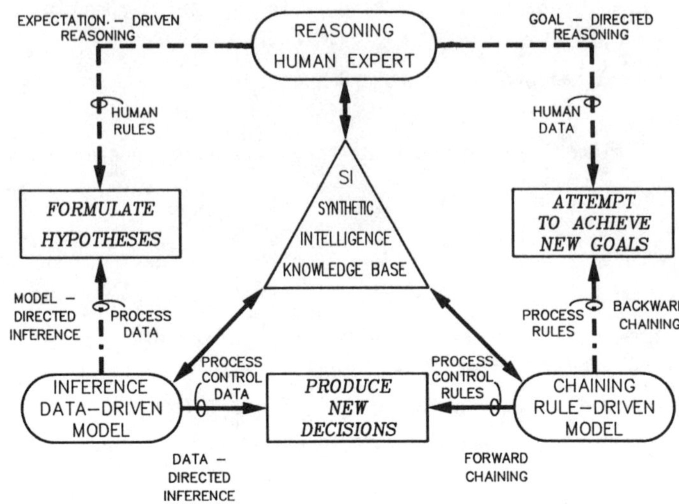

FIGURE 3   SIMSMART EXPERT SYSTEM
SYNTHETIC INTELLIGENCE
INTEGRATION

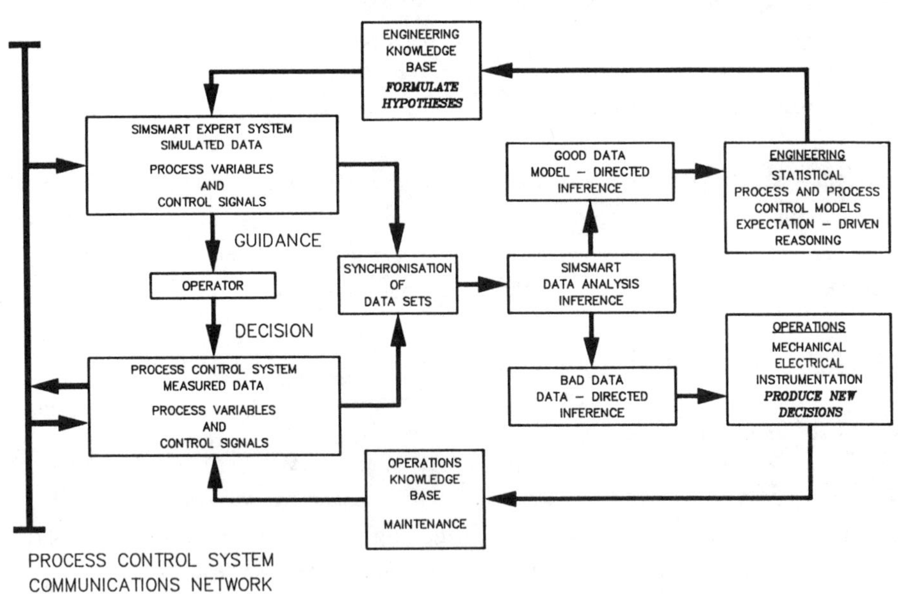

PROCESS CONTROL SYSTEM
COMMUNICATIONS NETWORK

SIMSMART EXPERT SYSTEM
FIGURE 4   REAL – TIME
DATA ANALYSIS

2.   Unreliable repeatibility of the measurement (calibration, drift, etc.)

3.   No available measurement device (lack of fundamental research, exhorbitant development costs, etc.)

However, the possibility of predicting the values of unmeasureable process variables through on-line simulation should have some very positive benefits. The large number of correct, and yet identical solutions to the dynamic mass, energy and force balances clearly demonstrates that the simulated problem is over-specified.

It should be noted that even with 37% or so of the measurement devices producing eroneous data, the operators were still able to run the plant. This means that data points (measureable and unmeasureable) can be accurately predicted by on-line simulation. While some of theseredundant measurements can be used to verify the values at other points in the process, it is also true to say that some of the measurement devices can be eliminated altogether, resulting in reduced capital expenditures and decreased maintenance costs. Then, once the operations personnel have accepted the validity of the simulated data, some of the «unmeasureable» process variables could be «controlled» basedon the simulated measurements, even if there are NO real-time measurements.

Eventually, as the ICON models are refined and the higher level control models are created and then improved, the amount of bad data will be reduced to a minimum. This is an on-going exercise that will require time and close cooperation between the operators, the engineers and the technicians. It should also be noted that these upper level models are process specific, and can be developed only for a specific installation, and are NOT transportable to another installation.

«WHAT IF» ANALYSIS FOR OPERATOR GUIDANCE

A «WHAT IF» analysis requires a faster-than-real-time simulation. The ability to accurately predict reality requires complete, detailed and accepted process and process control models. Therefore the engineering and operations knowledge base developed during the real-time data analysis form the backbone of the faster-than-real-time data analysis package. (See Figure 5).

The first step to perform a «WHAT IF» is to define the following:

1. Start-time and initial values of all process variables and control signals.
2. Duration and clock rate.
3. Intermediate changes to selected values of process variables and control signals (ramp, on, off, setpoint, etc.).

The next step is to run SIMSMART in faster-than-real-time to generate the simulated data sets. It should be noted that if SIMSMART is run 60 times faster-than-real-time; then 1 hour of future data can be generated in 1 minute of real-time. Each simulated data set contains the values of all process variables and control signals for each simulated clock «tick». A «tick» of the simulation clock is set by the master clock control program which also regulates the execution and loop times of both the DCS and the PLC. [Ref.6]

The next event in the procedure is to search through all the simulated data sets (using «forward chaining» techniques) for any «unacceptable» values of either the process variables or the control signals. The «unacceptable» values were previously entered into the engineering knowledgebase during the training of the operations personnel. The operator is warned of all impending problem(s), based on the time of occurance and duration of excursion of the unacceptable values. These events, once detected by the «forward-chaining» data analysis package, initiate a «backward chaining» data analysis package. The operator is warned of the possible cause(s) of impending problem(s) and the real-time time(s) of the initial onset of the cause(s) of the problem(s).

The objective of the «WHAT IF» analysis is to forewarn the operator before he implements any control changes, of exactly what would happen, when it would happen, and possibly why it would happen on a 24 hours-a-day, 7 days-a-week basis. This means that the knowledge of both the best engineers and the best operators (those who know) would be available to the operators (those who do not know), no matter what degree of technical or psychological skills the operator might have. It should be noted that once the operations personnel gain confidence in the simulated consequences (i.e. future data), the «quality» targets could be directly «controlled» by either manual or automatic implementation of the time-dependent setpoints as prescribed by the optimized plan (forecast of the consequences). [Ref. *8*]

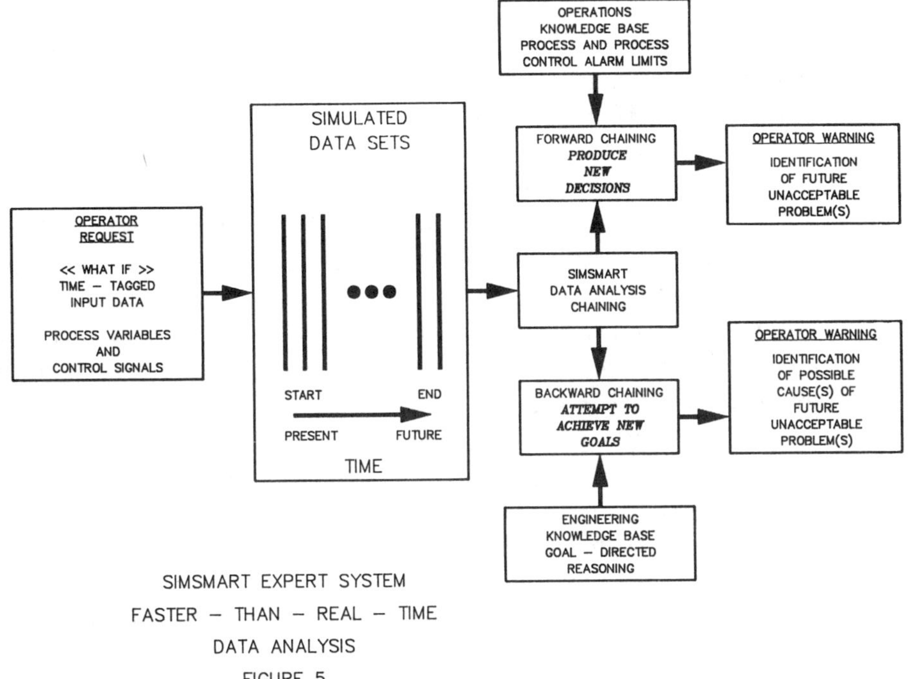

SIMSMART EXPERT SYSTEM
FASTER — THAN — REAL — TIME
DATA ANALYSIS
FIGURE 5

CONCLUSIONS

Synthetic Intelligence (SI); the integration of AI and Simulation, has been
shown to provide an almost exact duplicate of reality, with computational
speeds of over 1000 ICONS (representing abour 40,000 process variables) in
less than one second of real-time on a standard DEC VAX/VMS computer. The
success obtained with a great variety of process simulations to date would
tend to indicate that the SIMSMART SI approach is indeed cost-effective for
dynamic process and process control simulations of large complex networks in
real-time and faster-than-real-time.

REFERENCES

1.   WAYE, D.: «An approach to Distributed Computer Control in an Integrated
      Pulp Mill» - TAPPI Journal, Vol. 64, No.6, June 1981.
2.   WAYE, D., O'Reilly R., Caron S. ,Barrussaud J., Laviolette, J.G.,
      Massicotte G. and Bujold, J.:   «Mill-Wide Hierarchical Distributed

Control at Donohue-Normick Inc.» - <u>Pulp and Paper Canada</u>, Vol.86, No.6, June 1985.

3.   WAYE, Don; Dynamic Simulator Aids Operator Training at Donohue's Clermont Mill; <u>Pulp and Paper</u>, March 1987.

4.   WAYE, D.:   «SIMSMART: Dynamic Simulation for Engineering Design, Operator Training and Automated Control of Industrial Processes» - <u>TAPPI Journal</u>, Vol.10, No.8, August 1987.

5.   TERROUX, A., and Waye, D.:   «Modelling Made Easy: The Synthetic Intelligence (SI) Approach» - <u>1988 Eastern Simulation Conference</u>, Orlando, Florida.

6.   TERROUX, A. and Waye, D.:   «The Integration of Multiple Paradigms: The Synthetic Intelligence (SI) Approach» - <u>1988 Eastern Simulation Conference</u>, Orlando, Florida.

7.   WAYE, D., Editor: «How to Implement a Production Management System through Integrated Simulation and Mill-Wide Control», <u>1983 TAPPI Process Simulation Seminar</u>, March 1983, Atlanta, Georgia.

8.   WAYE, D. and Terroux, A.:   «SIMSMART for Process and Process Control Optimization: An Application Using the Synthetic Intelligence (SI) Approach» - <u>1988 Eastern Simulation Conference</u>, Orlando, Florida.

COMPUTERIZED PROCESS PLANT SCHEDULING

W. M. Sztrimbely*, P. J. Weymouth** and S. N. Smith**

*Applied High Technology Systems Ltd., Toronto, Ontario
**Advanced Dynamic Systems Inc., Toronto, Ontario

ABSTRACT

Industries have been scheduling their manufacturing processes for many years. The
common approach being followed is that of having an expert in plant operations,
usually a senior operator, draw up a schedule of tasks for each day making use of
a pen, paper and his intuition. The job of the scheduler is to plan when each
required task should be done, what equipment will be used and approximately how
long the task should take. Optimizing the schedule in terms of minimizing costs
and maximizing throughput, in even the simplest of plants, is not an easy task.

The SIMSMART Scheduling Advisor is an advanced dynamic scheduling tool designed to
relieve the human scheduler of the mundane tasks associated with developing the
"first cut" schedule, so that the human may spend his time and effort refining and
optimizing the proposed schedule. Over time, and through extensive usage, the
"first cut" schedule produced by The SIMSMART Scheduling Advisor will improve, with
the inclusion of more sophisticated rules, as the system "learns" more about the
operation and the associated interactions.

KEYWORDS

Computerized Scheduling; Advisory Intelligence; Expert Systems Methodologies; Batch
and Continuous Processing.

INTRODUCTION

The complexity of process operations and equipment interactions in the daily
operation of a steel producing Meltshop or a Nickel Smelter Converter Aisle is
overwhelming, even in the simplest of facilities. A computerized scheduling system
incorporating Advisory Intelligence capabilities based on proven Expert Systems
Methodologies, extensive data processing, extremely user friendly interactive menu
driven interfaces and colour graphic displays, known as The SIMSMART Scheduling
Advisor, has been developed for the scheduling of both Batch and continuous processing operations.

The scheduling system is designed to reside on the shop floor, in a PC-AT computer
with automatic data transfer between the scheduling system and the Process Control
Environment. The decision making modules make use of forward and backward reasoning, conditional reasoning algorithms and sophisticated search routines in order to

optimize the schedule of operations.

The level of operational detail is so refined that it takes into consideration the impact on production of having an average or better than average operator controlling various pieces of equipment. The effect of equipment maintenance, both scheduled and unscheduled, shift change and resource availability is taken into account when producing the schedule.

The following sections of this paper will describe the steps for customizing and implementing The SIMSMART Scheduling Advisor into an existing or greenfield processing plant environment.

PROJECT STRUCTURE

The projects are conducted in phases, in order that progress may be monitored and the scope of work reviewed after the completion of each phase. The two key factors that ensure the success of the scheduling system are: i) the commitment of the Plant Management to the project, in both their actual support and in allowing absolute freedom of access by the Knowledge Engineer to the operating personnel, and ii) the involvement of the operating personnel throughout each phase of the project.

The major phases in completing a scheduling project will now be presented.

Phase I: Definition of Scope of Work

This is the most crucial phase since the foundation of the entire project is developed through a series of both formal and informal meetings and discussions between the Knowledge Engineer and all levels of the plant management and operating personnel.

The objectives of this phase are:
- Selection of process equipment and plant activities to be modelled.
- Definition of the project scope, anticipated results and scheduling output presentation format.
- Collection of process sequence information for model generation.
- Generation of the "Process Logic Diagrams" (PLD's) which reflect material flow, operating practice and all decision making in the daily operation of the plant.
- Specification of process data requirements for the scheduling system's data base.

Phase II: Computer Software Development & Calibration

- Development of computer programs for specific modules required to model the plant operation.
- Data collection and manipulation to generate program data base.
- Review and evaluation of individual modules.
- Testing of assembled scheduling system.

Phase III: Installation of Scheduling System on Client Hardware

- Installation of The SIMSMART Scheduling Advisor onto client hardware (PC-AT).
- Development of interfaces to Process Control Environment for automatic data retrieval.

<u>Phase IV: Up-Dating of Advisory Intelligence Modules</u>

- Ongoing introduction of new operating rules into Advisory Intelligence Modules
  that evolve through changes in the standard operating practice as a result of
  equipment changes, metallurgical changes, etc.

PROGRAM ARCHITECTURE

The scheduling system consists of three distinct elements: i) Interactive User
Friendly Menu Driven Front End, ii) Advisory Intelligence Modules, and iii) Graphic
Colour Display and Summary Reports.

<u>Interactive User Friendly Menu Driven Front End</u>

In order to facilitate the initialization of the scheduling system to the current
status of the plant (prior to linking the scheduling system into the Process Control
Environment), a series of menus are presented to the user. These menus appear in
succession, automatically, after the previous menu's information requirements have
been satisfied.

In an attempt to speed up the initialization process, default conditions are built-
in which are invoked if the user chooses not to answer a question directly.

<u>Advisory Intelligence Modules</u>

The PLD's developed during the first phase of the scheduling project are used to
present the operating logistics and decision making process followed in the daily
operation of the plant. These PLD's are the scheduling model, which is translated
into the rule-based Advisory Intelligence Modules that are activated by the
scheduling system, in order to develop and optimize a schedule.

The modularity of the Advisory Intelligence component of The SIMSMART Scheduling
Advisor enables a trained computer programmer to easily enter new operating rules or
to modify existing rules. This feature will allow the system to "learn" as well as
enhance the level of sophistication in the decision making process.

<u>Graphic Colour Display and Summary Reports</u>

The production schedule produced by The SIMSMART Scheduling Advisor indicates all
major material flow, processing activities, maintenance activities and equipment
interactions.

The terminal display of the schedule is in colour as this allows for easy cross-
referencing of material flow from source to sink (e.g. furnace matte to a converter
or the tracking of specific grades or heats of steel). The system provides the user
with the option of viewing the schedule on the screen in one hour, three hour, eight
hour, twelve hour or twenty-four hour display formats.

The user can scroll forward or backward through the schedule and question the system
about different activities displayed on the screen.

The schedule may be modified using the edit functions provided by the system and
after approval, the schedule may be displayed throughout the plant via the Process
Control Environment. Hard copies of the schedule may be printed on standard

printers as the print function is adaptable to any commercially available printer.

Summary reports are generated by the system and are customized to the format that exists within the plant (i.e. shift summary, daily summary, etc.). Information such as equipment utilization, delays, production rates, etc., are available and can be printed upon request.

CONCLUSION

The SIMSMART Scheduling Advisor has been developed for use by non-computer trained operating personnel. Because of the techniques used in designing and developing the system, any person familiar with the plant operation can generate schedules after less than one hour of training.

Acceptance of the results by operating personnel is more likely if they have been involved throughout all phases of the development and implementation. An operator who has had input into the system design is more likely to use and rely on the schedule because he will understand how the schedule was generated.

ACKNOWLEDGEMENTS

The authors would like to acknowledge financial support by the National Research Council of Canada, under the Industrial Research Assistance Program, for the development of numerous libraries utilized in The SIMSMART Scheduling Advisor.

Alloy Corrosion Data Bases Combined with Thermochemical Analyses

R. C. John*, W. T. Thompson** and I. Karakaya***

*Shell Development Company, Houston, Texas, USA
**Royal Military College of Canada, Kingston, Ontario, Canada
***Middle East Technical University, Ankara, Turkey

ABSTRACT

Experimental data and data analyses methods for high temperature alloy corrosion by gases have been incorporated into a computer-based information system. The system considers corrosion mechanisms that form mixtures of sulfides and oxides (sulfidation and mixed sulfidation/oxidation). Applications of the system can include equipment design, alloy selection, component failure analysis, and equipment corrosion studies. Corrosion rate data are correlated with temperature, alloy composition, exposure time, and gas composition. Data from a variety of sources are examined, archived, statistically analyzed, and then built into equations that have forms based upon theoretical expressions that are applicable to corrosion of simple alloys by gases. Interpolations and extrapolations of the conditions associated with the databases and the predictions by the system are obviously indicated. There is a large number of corrosion data observations for commercially available alloys in the growing databases. These alloys are a significant collection of those commonly incorporated into equipment used in high temperature processes.

KEYWORDS

Alloy corrosion data bases; sulfidation; sulfidation/oxidation; corrosion rate prediction/correlation.

INTRODUCTION

High temperature corrosion in processes can limit the service lifetime of equipment. Corrosion specialists often need access to current methods of data retrieval and analysis in order to evaluate these lifetimes. The complex interactions between corrosive environment composition, alloy composition, temperature, and exposure time are among the factors that influence corrosion rates, and must be considered in the identification of suitable materials of construction. The corrosion engineer uses an enormous amount of data that have been generated for a diversity of process conditions (environment), corrosion mechanisms and alloys. These data are scattered over documents, such as, technical journals, private reports, symposia volumes, books, monogeraphs, and private databases and in various formats, and they certainly represent varying degrees of thoroughness in the technical analyses of the data.

Use of the extensive corrosion literature is limited by its accessibility, the thoroughness of the studies, and the diversity of the methods of reporting. The value of the data depends upon the expertise of the corrosion engineer plus his/her ability to acquire, to interpret, and to exploit the data. Recent years have shown a high interest in a class of computer-based systems arising from studies in artificial intelligence that are termed "expert systems". Expert systems are often considered to include some portion of the methodology of a human expert into an electronic information system. The reasons for the creation of an expert system include: (1) increasing the accessibility of the expert's expertise to the user community, (2) improving the exploitation of the expertise by making its application more reproducible and verifiable, and (3) archiving the expertise before access to the expert is lost. The technology must be well enough established that the information can be adequately organized into and then exploited via the electronic information system.

Building an expert system is challenging because portions of the expertise may evade incorporation into the computer system, due to the intuitive nature of the decision making process of the human. Also, the technology may not be sufficiently mature. An expert system alternative is a system that supplements, rather than replaces the human expert. This approach can ease the burden on the expert by building the tedious portion of the technical work, such as data archival, data retrieval, data analyses, and data correlation into the system. This "advice system" can allow the expert to devote more time to the creative aspects of the job.

The engineering discipline that has been built into this advice system is the correlation and prediction of alloy corrosion rates by high temperature gases that contain mixtures of H_2, H_2O, H_2S, S_2, CO, CO_2, COS, O_2, CH_4, and inert gases. The alloy corrosion mechanisms (sulfidation and sulfidation/oxidation) that are possible in these gas mixtures are discussed in the next section of this article. For substantial variations of gas compositions, exposure times, and temperatures, many commercial alloys based upon combinations of Fe-Cr-Ni-Co-W-Mo-Si form sulfides or mixtures of sulfides plus oxides. Long-term (tens of thousands of hours) corrosion rates are of key interest in most commercial applications of materials and equipment. These rates may be limited by diffusion through the growing corrosion product layers. The nature and growth of these layers are determined by the exposure conditions and the characteristics of the corroding alloys.

Understanding of these corrosion phenomena is important in equipment design and operation of both commercial and developmental processes. This paper describes how this understanding is built into a computer-based information for alloy gas phase corrosion [ASSET – Alloy Selection System for Elevated Temperatives (John and coworkers, 1988)]. Processes that have corrosive environments within the framework of ASSET are: coal gasification, oil gasification, refinery processes with H_2/H_2S gases, fluid bed coal combustion, and other processes that contain reducing, sulfur-containing gases. These processes use large amounts of metal equipment which corrode in high temperature reducing, sulfidizing gases. This area of corrosion science has been studied for several decades and has produced voluminous amounts of information. Unfortunately, the utilization of the total body of information does no approach a fraction of its potential due to the diversity of the types of corrosion studies, the variation in the rigor with which test conditions were defined and the dispersion of the data among various reporting means.

This study has created a central computer system that is accessible to a group of expert users and has eliminated many of the shortcomings that have been mentioned. The following capabilities: data archival, data analysis, and data exploitation have been built into ASSET. The system uses statistical analysis methods and experimentally determined data. System capabilities include: (1) thermochemical analyses of the corrosive environments, (2) determinations of corrosion product stabilities (incorporating commonly used dependences of corrosion upon time, temperature, and gas composition), and (3) statistical correlations of corrosion with exposure conditions, as guided by recognized empirical trends. Corrosion data that can be used by the system represent metal penetrations resulting from experiments conducted under properly controlled conditions, for adequately long exposure times, with well described prepared alloy specimens, and employing adequate replication.

ASSET uses include: interpretations of corrosion-related failure analyses, analyses of the corrosive environment, selections of materials for equipment design, test program designs to minimize the amount of testing required, designs of corrosion-resistant alloys, archival of corrosion data, data sharing, comparison of interactions between changing exposure conditions and the corrosion-limited lifetimes of equipment, and comparison of data from different studies. The essential characteristics of the corrosion mechanisms that have been addressed and the capabilities of the ASSET are described below.

CORROSION MECHANISMS IN ASSET

The system consideres two corrosion processes which are well known to occur in reducing, sulfidizing gases: sulfidation and sulfidation/oxidation. The alloy, the corrosion product, and the corrosive gaseous environment, for both mechanisms, are illustrated in Figure 1. Corrosion by sulfidation is the formation of sulfide corrosion products. Gases that can provide the sulfur are: H_2S, COS, and S_2. Sulfidation/oxidation is the growth of corrosion products that contain both sulfides and oxides and is generally associated with alloys that contain more than 15%w Cr. Gaseous environments that promote these corrosion mechanisms contain mixtures of species such as: H_2, H_2O, H_2S, CO, CO_2, COS, CH_4, S_2, O_2 and inert gases. Corrosion in ASSET is expressed as metal penetration, which is the sum of uniform metal loss plus internal corrosion (*e.g.*, grain boundary corrosion or internal precipitates).

Corrosion by Sulfidation/Oxidation

Sulfidation/oxidation is the formation of corrosion products that contain mixtures of sulfide and oxide compounds. Alloys found to exhibiting this corrosion mechanism contain Cr concentrations greater than 15%w. Mathematical description of the relations between the environmental variables, the alloy characteristics, and the rate of corrosion that would be applicable to a variety of alloys over wide variations of gas composition, temperature and time has remained elusive in spite of considerable effort. Corrosion rates are more rapid as the P_{S_2} and temperature increase, and as the P_{O_2} decreases has been known from studies using complex gases. Only recently was quantitative correlation of these variables with corrosion rates accomplished (John, 1985, 1986), and this correlation method has been incorporated into ASSET.

Traditional interpretations of the influences of the gas composition and temperature upon alloy corrosion have included thermochemical phase stability diagrams (TPSD's), as illustrated in Figure 2, and interpretations of the phases that develop on the corroding alloys. Use of the TPSD's is limited by simplifying assumptions, such as attainment of equilibrium and pure immiscible condensed phases. These assumptions permit their routine calculation. Equilibrium between the gas and the corrosion products is not guaranteed and phases do often mix. The widespread use of the TPSD's to diagnose corrosion problems has clarified the importance of complex gas composition interactions on the corrosion mechanisms on many alloys and has repeatedly encouraged emphasis upon the strong relations between thermochemistry and corrosion behavior. The diagrams can be used to compare the thermochemical characteristics of the environments, but they offer no guidance on how corrosion rates can be quantitatively influenced by variables such as temperatures, times, and gas compositions.

The quantitative relations between gas composition, corrosion product stabilities, alloy composition and sulfidation/oxidation corrosion rates of chromium-containing alloys have been explained previously (John, 1985, 1986). For purposes of this paper, it is sufficient to say that the primary characteristics of the mixed gas in the case of sulfidation/oxidation are the partial pressures of O_2 and S_2, plus the temperature. These characteristics determine the stabilities of the chromium oxides and sulfides. These are the primary corrosion products for alloys containing significant concentrations of chromium. The relation between the corrosion products and the gas composition are illustrated below. The relative stabilities of Cr_2O_3 and CrS as corrosion products in S_2–O_2 gases are defined by the reaction:

$$2CrS + 3/2O_2 \leftrightarrows Cr_2O_3 + S_2 \tag{1}$$

The equilibrium constant for the above reaction is:

$$K = \frac{a_{Cr_2O_3} P_{S_2}}{(a_{CrS})^2 (P_{O_2})^{3/2}} \tag{2}$$

Equation (2) illustrates how the relative corrosion product stabilities can be directly calculated from the gas composition, independent of any knowledge about the alloy composition or other characteristics. Since the relative amounts of oxides and sulfides in a corrosion product can be used to suggest the relative corrosion rate, this knowledge has been used to formulate the equation that is used to correlate the gas composition and the corrosion rates:

$$K_p = (constant)\left(\frac{P_{S_2}}{P_{O_2}^{3/2}}\right) exp\,(Q/RT) \tag{3}$$

where K_p is the parabolic rate constant for the corrosion reaction. This equation is consistent with the observed trends of higher P_{S_2}'s, higher temperatures and lower P_{O_2}'s, all tending to increase the corrosion rates, as measured by the value of K_p. The conversion from equation 2 to equation 3 is not direct. However, since alloy corrosion parabolic rates depend upon characteristics of the CrS and Cr_2O_3 phases in the corrosion products and the relative growth rates of CrS and Cr_2O_3 vary by orders of magnitude (Danielewski and Natesan, 1977), it is reasonable that alloy corrosion should correlate with the ratio of formation tendencies (activities) of the principal corrosion product phases.

Equation 5 may seem consistent with the work by Yurek *et al.* (Yurek and co-workers, 1985, 1986) in that they find the ratio of P_{H_2O}/P_{H_2S} to determine corrosion product stabilities and growth rates in $H_2/H_2O/H_2S$ gases. The ratio of P_{H_2O}/P_{H_2S} is similar to the term $\left(\dfrac{P_{S_2}}{P_{O_2}^{3/2}}\right)$, but the background information is different.

Conditions in Yurek's tests that last several hours cause gas phase diffusion to be the rate limiting step, as shown by linear dependences of corrosion rate upon P_{H_2S} in the gas, gas velocity, and exposure time. However, the data (John, 1985, 1986) supporting correlation of corrosion with

$$\frac{a_{Cr_2O_3}}{(a_{CrS})^2} \quad or \quad \left(\frac{P_{S_2}}{P_{O_2}^{3/2}}\right)$$

clearly show that corrosion kinetics are independent of gas flow rate, independent of concentrations of gas species (except for P_{O_2} and P_{S_2}), and a dependent for linearly penetration with (time)$^{1/2}$ for times up to at least thousands of hours and temperatures of 300-1000°C. Diffusion of undetermined species in the corrosion product layer is the rate limiting step for commercially significant conditions of long exposure times.

Previous studies by have shown that the influences of temperature and gas composition can be quantitatively correlated with the stabilities of the dominant corrosion products of CrS and Cr_2O_3. An example as shown in Figure 3 (John, 1985, 1986) for three gas compositions shown in Table 1. Stabilities of corrosion products that are found in the surface corrosion product layer are often used as indicators of the likelihood of various corrosion mechanisms and relative corrosion rates. Sulfide corrosion products are associated with faster corrosion than oxide corrosion products. Correlation of the exposure conditions with corrosion is shown in Figure 4 for the commercially available Alloy 800 (21% Cr, 33% Ni, and balance Fe) for variations in gas compositions and temperatures.

Data correlation by this approach have been found to be applicable for a variety of alloy types, such as: Fe, Ni-Cr, Fe-Cr, Fe-Cr-Ni, Co-Ni-Cr, Ni-Cr-Fe, and Co-Cr. The method has been previously explained (John, 1985, 1986). The method is briefly described here. A parameter that measures the relative stabilities of the dominant corrosion products of chromium (CrS and Cr_2O_3) is calculated on the basis of the gas composition and the temperature. This parameter is the ratio of the thermodynamic activities of CrS and Cr_2O_3, $(a_{CrS})^2/(a_{Cr_2O_3})$, which includes the influences of temperature and gas composition. Correlations of alloy corrosion rates with this parameter expressed in terms of P_{O_2} and P_{S_2}, as shown in equation 3 can also be exploited as a function of the alloy composition.

The demonstrated limits of successful correlation of sulfidation/oxidation corrosion with the exposure conditions for alloys that contain at least 15%w Cr are:

- $-12 < \log(a_{CrS})^2/a_{Cr_2O_3} < +2$
- $300°C < \text{temperature} < 1000°C$
- $50\ \text{hours} < \text{exposure time} < 3000\ \text{hours}$

The combination of the first two limits can be re-expressed in terms of P_{O_2} and P_{S_2}:

- $-30 < \log P_{O_2}(\text{atm}) < -21$
- $-12 < \log P_{S_2}(\text{atm}) < -7$

for a temperature range of 300-1000°C.

The actual limits for use of the above parameters with corrosion rate correlations with environment are not known and may be expanded upon future consideration of more extensive data collections.

Corrosion by Sulfidation

Corrosion of metals by sulfidation produces in a sulfide layer separating the metal from the sulfur-bearing gas. The extent and rate of corrosion depend upon several factors including: alloy, time, temperature, and the concentrations of the corrosive gas species present. A generalized kinetic relation between sulfidation and the important variables in the environment is:

$$K_1 = A(P_{H_2S})^{(B + C/T)}(P_{H_2})^D exp(Q/RT) \qquad (4)$$

where K_1 = *(metal consumption by corrosion)/(time)^N* and the values of A, B, C, D, N and Q are empirical constants, while T is the absolute temperature, and R is the gas constant. Linear regression analysis fits the data to equation 1, freely determining values of A, B, C, D and Q. Repeated analyses with sulfidation data for various steels have shown that values of C are small, relative to Q and that the signs of B and D are opposite (one is negative and the other is positive). Therefore, fitting of real data to equation (1) produces an equation which is in fact similar to equation (2).

Sulfidation of iron and low alloy steels has been repeatedly demonstrated to involve kinetics to be diffusion limited by transport in the growing sulfide scale (Mrowec, 1980; Danielewski, 1977). When the sulfidation rate is due to the diffusion limited transport of species through the growing sulfide layer, equation (1) can be reduced to:

$$K_1 = constant(P_{S_2})^n exp(Q/RT) \qquad (5)$$

where K_1 = *a parabolic rate constant (metal consumption by corrosion)/(time)$^{0.5}$*, and n is a constant.

CAPABILITIES OF ASSET

The information system that facilitates predicting of corrosion rates is called the "Alloy Selection System for Elevated Temperatures" or ASSET. Thermochemical characteristics of the corrosive gas phase are combined with empirical correlations of corrosion rates with exposure conditions. The system contains user-friendly modules which perform these roles:

1.   Store experimental data;

2.   Statistically analyze the data to produce "best-fit" equations that can predict corrosion;

3.   Scan the data base;

4.   Calculate corrosion and corrosion rates for alloys in the data base at selected exposure conditions and rank the alloys in the order of increasing corrosion;

5.   Predict corrosion for alloy compositions that are not yet experimentally examined.

The basic features will be explained. Figure 5 is an example of alloys for which data are archived currently. Data in the data base can be displayed by entering the alloy name. Following the display of data in Figure 6 on this alloy, the user has the option to display the individual experimental points and the exposure conditions which gave rise to the regression equation shown in Figure 6. Each data point is referenced to the original bibliographic source, as indicated in the column titled "Ref". The uncertainty factors (the average of the absolute deviations between the correlated {predicted} and the experimental corrosion rates, expressed as a fraction of the best fit values) are also shown. The quality of the fit for the forms of equations 3 and 5 is perhaps best appreciated by examining Figures 7 to 11.

Predicting corrosion rates requires the use of an accessible procedure to compute the gas phase equilibria. For example, with a gas mixture with a nominal composition of 80% H_2, 15% H_2O, and 5% H_2S at 800°C, it is first necessary to know the partial pressures of H_2S, H_2, O_2, and S_2, before one can apply equations (1) and (5). One of the many programs available that can provide this information is the EQUILIB program of F*A*C*T (Thompson, Bale, Pelton, 1985). F*A*C*T provides the equilibrium gas composition from the input composition, temperature and pressure, as shown in Figure 12. The reactants and the numbers of moles are shown in the first line of Figure 12. The program then determines the total number of moles of gas and the mole fractions of each product species for a specified temperature and pressure. The calculated partial pressures of H_2S, H_2, O_2, and S_2 are automatically supplied as the first line of numerical input of ASST, as shown near the top of Figure 12. It may be inappropriate to consider that the gas is at equilibrium. Alternately, one may supply other values of P_{O_2}, P_{S_2}, P_{H_2}, and P_{H_2S}. Following the entry of exposure time in Figure 13, the result is the listing shown near the bottom of the figure. The corrosion rates of all the alloys in the data base are individually calculated and then ranked in order of increasing corrosion. The uncertainty factors and the temperature ranges that are represented by the data bases for each alloy are indicated. Asterisks are used in the extreme right-hand columns to indicate when the selected conditions of gas composition and/or temperature represent extrapolations relative to the alloy data bases. AF is the activity function defined as:

$$AF = log \frac{(a_{CrS})^2}{(a_{Cr_2O_3})} \tag{6}$$

and is shown at the bottom of the figure. It is simply related through equation (2) to the partial pressures of S_2 and O_2. For some alloys there are insufficient data to permit meaningful statistical analyses of the data and sample designations of the names of such alloys are listed at the bottom of the output.

The real gas composition and the equilibrium gas should be compared and the influences upon corrosion were discussed previously (John, 1987). If the gas is not at equilibrium, then the actual values of P_{O_2} and P_{S_2} (or AF) must be used in calculating corrosion rates. However, non-equilibrium gases may not have unique values of AF. The interchangeability of equilibrium and non-equilibrium gases in the corrosion rate correlations based upon equation 5 was previously demonstrated (John, 1986).

Demonstration of the validity of ASSET is shown by the ability to predict corrosion rates that are close to the original measured values. Examples of the comparisons between original and predicted values for the corrosion mechanisms of sulfidation and sulfidation/ oxidation are shown in Figures 7 to 12. Close study of the figures shows that over several orders of magnitude of variation in the corrosion rates, the actual and the correlated (predicted) values are indeed close. The width of the dispersion of points around the line illustrtaes the uncertainty factor that is reported with the corrosion rate predictions. This is true for simultaneous use of a large number of data from several independent public data sources (Sharp, Haycock, 1959; Haycock, 1959; Hosier, Harris, 1980; Lai, 1985; Lloyd, Cooke, 1980; Bakker, Perkins, Liene, 1985; Perkins, Morse, Coons, 1982; Stoklosa, Stringer, 1977; Lai, 1984) and a large amount of private data, for temperature variations of several hundred °C, test times spanning several thousand hours, and different combinations of corrosive gas species and their concentrations. The diverse alloy types (*e.g.*, carbon steel, Fe-Ni-Cr {Alloy 800}, Ni-Cr-Fe {Alloy 600}, Ni-Cr-Mo-Cb {Alloy 625}, Fe-Cr-Al, Fe-CR {AISI 446}, and Co-Cr-Ni-Mo) are equally well described by ASSET.

SUMMARY

A computer-based information system has been created to assist corrosion engineers with questions on high temperature alloy corrosion by gases. Some aspects that are similar to those incorporated in expert systems hae been included. But, rather than attempting to replace the corrosion engineer, the system aims to supplement, and is therefore termed an "advice system". The abilities to archive, analyze, and exploit (via interpolation/extrapolation) measured corrosion data have been based upon real data combined with theoretical understanding.

The system was developed to reduce the need to measure corrosion rates, to make better use of existing data, to supplement the availability of the corrosion expert, to share a central collection of data, and to provide advice based upon consistent data analysis methods.

The present uses of the advice system include equipment design, materials selection, equipment corrosion lifetime prediction (including considerations of changing exposure conditions), alloy development, experimental selection of the test matrix for a corrosion research program, and failure analyses.

REFERENCES

Bakker, W. T., R. A. Perkins, and J. V. Lierre, (1985). *Corrosion in syngas coolers of coal gasification, combined cycle power plants*, **Mater. Perform.**, January, 9.

Danielewski, M., and K. Natesan, (1977). **Oxid. Met.**, 12, 227.

Danielewski, M., S. Mrowec, and A. Stoklosa, (1982). **Oxid. Met.**, 17, 77.

Haycock, E. W., (1959). *High temperature sulfiding of iron alloys in hydrogen sulfide-hydrogen mixtures*, **Journ. Electrochem. Soc.**, 106, 764.

Hosier, J. C., and J. A. Harris, (1980). *Sulfidation resistance of high temperature alloys tested in reducing and oxidizing atmospheres*, **Indust. Heat.**, June, 14.

John, R. C., (1985). *Alloy corrosion in reducing plus sulfidizing gases at 600-950°C*, in **High Temperature Corrosion in Energy Systems**, AIME, Warrendale, PA, 161.

John, R.C., (1986). **Simultaneous sulfidation and oxidation of alloy 800 in mixed gases**, paper 358 of the May 4-9, 1986 meeting of the Electrochemical Society.

John, R. C., (1987). *Rapporteur's review on symposium on materials for coal gasification*, American Society for Metals meeting, Oct. 11-15, 1987.

John, R. C., W. T. Thompson and I. Karakaya, (1988). *ASSET Advice System on High Temperature Alloy Corrosion*, National Association for Corrosion Engineers – Corrosion 88, paper 136.

Lai, G. Y., (1984). *Materials behavior in high temperature, sulfidizing environments*, paper 73 NACE Corrosion/84.

Lai, G. Y., (1985). *Sulfidation resistance of various high-temperature alloys in low oxygen potential atmospheres*, in **High Temperature Corrosion in Energy Systems**, AIME, Warrendale, PA, 227.

Lloyd, D. M., and M. J. Cooke, (1980). *Corrosion of alloys at high temperature in coal gasification atmospheres*, **Behaviour of High Temperature Alloys in Aggressive Environments**, Metals Society, London, 799.

Mrowec, S., (1980). **Werk. Korros., 31**, 371.

Perkins, R. A., G. Morse, and W. C. Coons, (1982). *Materials for syngas coolers*, report AP-2518, EPRI Research Project 1654-5, Palo Alto, August.

Sharp, W. H., and E. W. Haycock, (1959). *Sulfide scaling under hydrorefining conditions*, paper from 24th Midyear Meeting of API, May 27.

Stoklosa, A., and J. Stringer, (1977). *Studies of the kinetics of Ni sulfidation in H_2-H_2S mixtures in the temperature range 450-600°C*, **Oxid. Met., 11**, 263.

Thompson, W. T., C. W. Bale, and A. D. Pelton, (1985). *Facility for the analysis of chemical thermodynamics (FACT)*, McGill University Montreal, Royal Military College of Canada in Kingston, Ecole Polytechnique, Montreal.

Yurek, G. J., M. H. LaBranche, and Y. K. Kim, (1985). *Oxidation of Cr and an Fe-25 Cr alloy in H_2-H_2O-H_2S gas mixtures at 900°C*, Proc. Symp. High Temperature Corrosion in Energy Systems, TMS-AIME, Warrendale, PA, 295.

Yurek, G. J., and K. Przybylski, (1986). **Journ. Mat. Sci. Eng.**, June.

Table 1. Example Gas Compositions Used to Calculate the Variation of $a^2_{CrS}/a_{Cr_2O_3}$ as a Function of Gas Composition and Temperature

Species	Gas Compositions*		
	Gas I	Gas II	Gas III
H_2O	0.37	6.14	6.18
CO_2	–	1.95	1.96
CO	0.04	13.74	13.84
H_2S	1.53	0.78	0.11
CH_4	0.88	6.56	6.61
P_{O_2} (atm)	2.1×10^{-26}	1.1×10^{-23}	1.1×10^{-23}
P_{S_2} (atm)	7.2×10^{-9}	3.5×10^{-9}	7.2×10^{-11}

* Equilibrium gases at 1 atm pressure and a temperature of 700°C, with concentrations in volume %, and H_2 is the balance species.

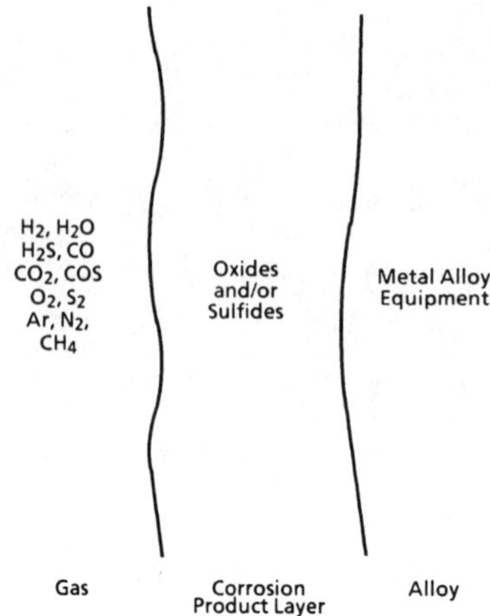

Fig. 1. Arrangement of alloy, corrosion product layer, and of corrosive environment that is considered by ASSET.

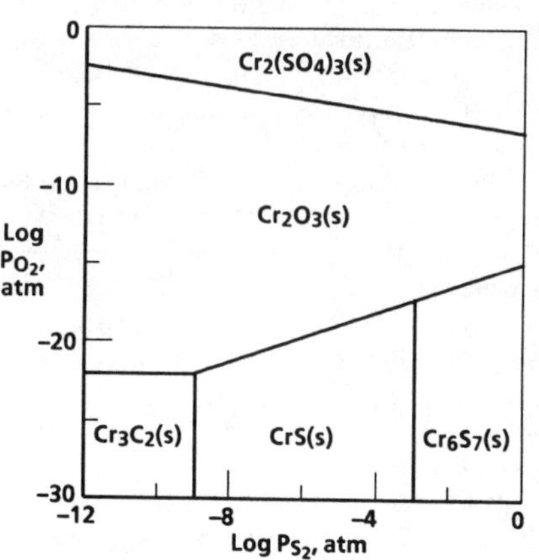

Fig. 2. Calculated thermochemical phase stability diagram for the Cr–S–O–C system at 900°C and a carbon activity of 1.1.

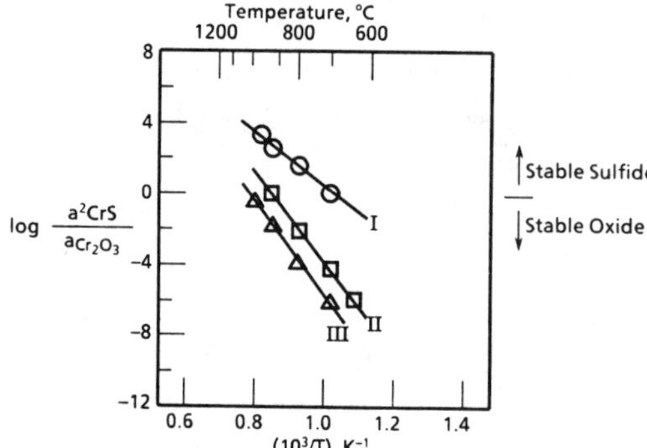

Fig. 3. Variation of CrS/Cr₂O₃ corrosion product stability parameter dependence upon a function of (temperature)⁻¹ for gases I, II, and III.

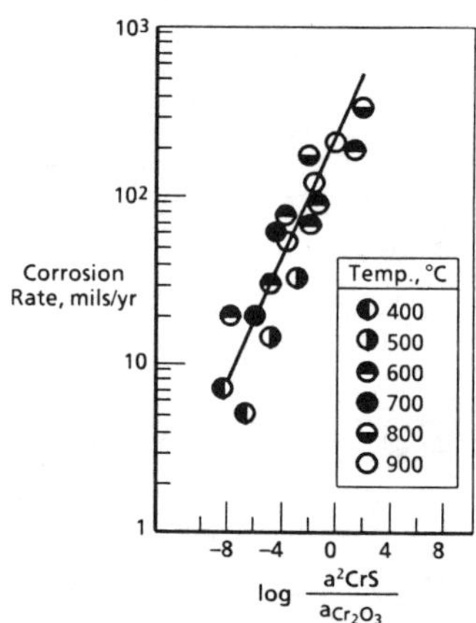

Fig. 4. Corrosion rate of Alloy 800 dependence upon gas compositions and temperatures after 720 hours of exposure.

Enter Alloy Name or Press "Return" to List the Alloys

–	–
–	HK40
1Cr–1/2Mo	Alloy 617
9Cr–1Mo	AISI 446
AISI 310	AISI 304
AISI 410	AISI 316
Alloy 800	Alloy 625
Alloy 825	Carbon Steel
Alloy 600	Alloy 25
Pure Cu	Pure Ni
Alloy 188	Alloy 601
Alloy 556	Alloy 671
–	AISI 317
	–

Fig. 5. Example listing of alloys with corrosion data in the ASSET system.

Alloy: AISI 310

Composition (%w):

Fe	Cr	Ni	Co	Mo	Al	Si	Ti	W	Mn
55.00	25.00	20.00	0.0	0.0	0.0	0.0	0.0	0.0	0.0

Data Fit   : (Sulfidation/Oxidation Mechanism)

$$\text{Log}(KP) = 0.478E+01 + (0.1158E+00)[\log(P(S2)) - 1.5\log(P(O2))] + (-0.8562E+04/T(K))$$

Where $K = \text{Penetration}/t ** 0.5$

Standard Error of Estimate for Log (K) = 0.2700E + 00
Average Uncertainty Factor for Penetration (mils) = 0.48

The Equation is Applicable Within the Ranges of
    773.000   to   1255.00 for Temperature (K)
     –4.00   to   1.00 for Log $(a^2_{CrS}/a_{Cr_2O_3})$

Experimental Data for Sulfidation/Oxidation Mechanism

Temp., °K	Pressure, atm		Penetration, mils	Time, hr	Grain Size	Ref. No.
	O₂	S₂				
1073.	0.16E–23	0.58E–07	24.15	720.	0	1
1073.	0.16E–23	0.58E–07	20.47	336.	0	1
1073.	0.16E–23	0.58E–07	22.84	336.	0	1
1073.	0.16E–23	0.58E–07	37.95	720.	0	1
773.	0.20E–32	0.98E–10	0.63	500.	0	2
773.	0.20E–32	0.98E–10	0.51	500.	0	2
1000.	0.10E–20	0.10E–06	1.08	96.	0	7
–	–	–	–	–	–	–
–	–	–	–	–	–	–
–	–	–	–	–	–	–

Fig. 6. Sample of corrosion data archived in the ASSET system. The example shown is AISI 310.

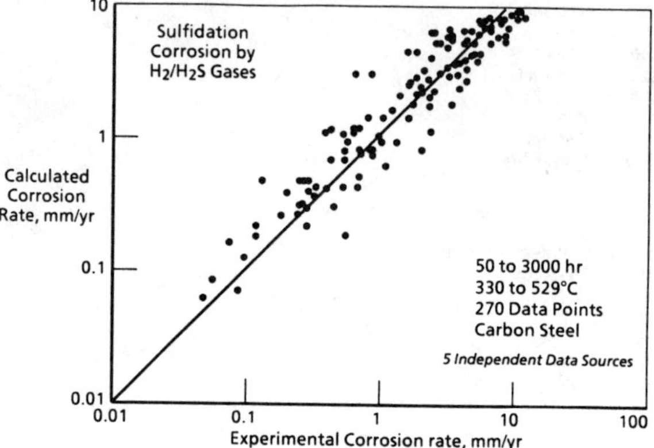

Fig. 7. Comparison of calculated and experimental
corrosion rates, by the ASSET system.

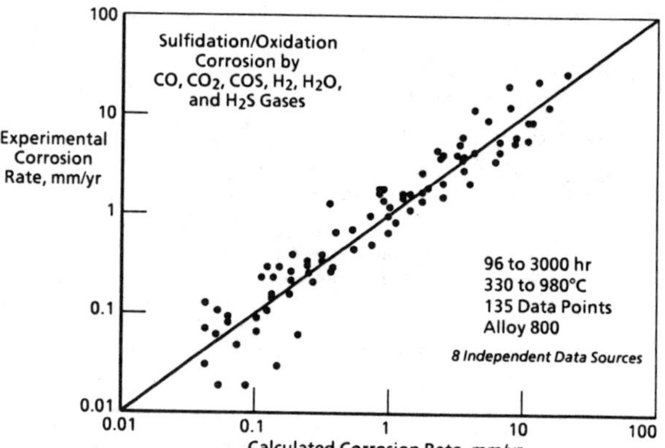

Fig. 8. Comparison of experimental and calculated
corrosion rates, by the ASSET system.

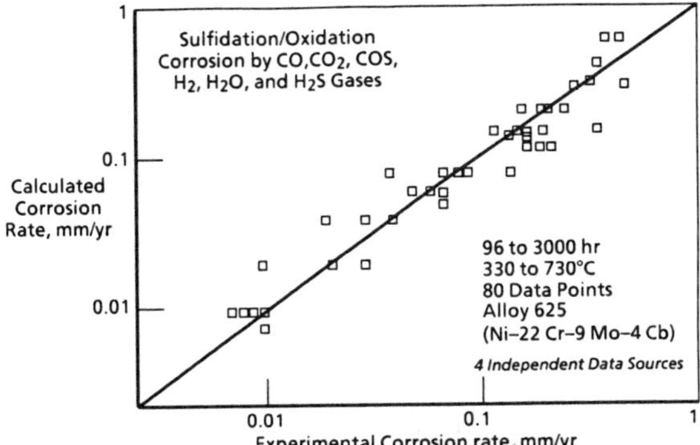

Fig. 9. Comparison of calculated and experimental corrosion rates, by the ASSET system.

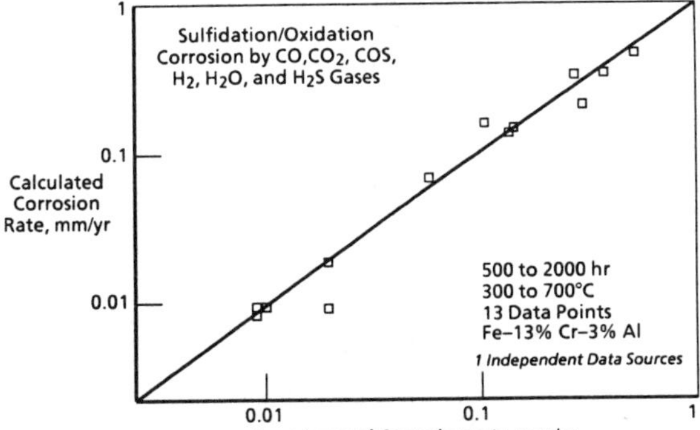

Fig. 10. Comparison of calculated and experimental corrosion rates, by the ASSET system.

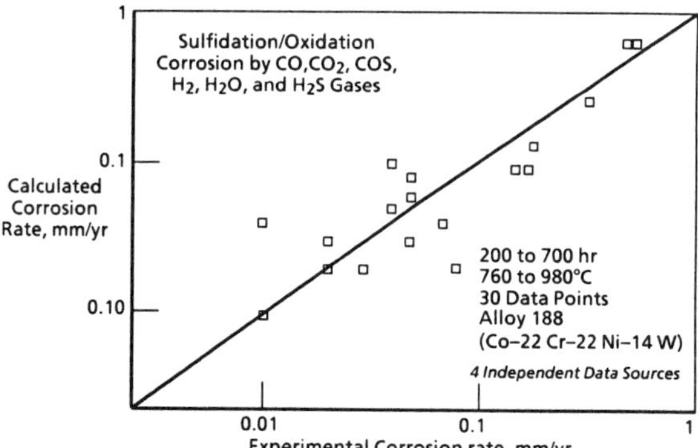

Fig. 11. Comparison of calculated and experimental corrosion rates, by the ASSET system.

$$< 0.500E + 02 > H_2S \; + \; < 0.800E + 03 > H_2 \; + \; < 0.150E + 03 > H_2O$$

1000.0 mol	(	0.80000	H_2
	+	0.15000	H_2O
	+	0.49993E–01	H_2S
	+	0.21720E–05	H_2S_2
	+	0.89579E–06	S_2
	+	0.70161E–06	SH
	+	0.12442E–07	H
	+	0.77246E–09	SO_2
	+	0.13072E–09	SO
	+	0.63866E–10	S_3
	+	0.28987E–10	S
	+	0.83099E–11	S_2O
	+	0.76984E–11	OH
	+	0.60284E–15	S_4
	+	0.27906E–17	S_5
	+	0.14149E–18	O
	+	0.75292E–19	SO_3
	+	0.61373E–19	H_2O_2
	+	0.13238E–19	O_2
	+	0.13544E–20	S_6
	+	0.11552E–22	H_2SO_4
	+	0.43474E–23	HO_2
	+	0.19472E–24	S_7

Fig. 12. Example of gas composition calculated using F*A*C*T, based upon the reactants shown at the top.
The temperature is 1073°K and the pressure is 1 atm. The mole fractions precede each species. These are 1000 moles of gas. Partial pressures are given by: (mole fractions) (total pressure). Exponential format for molar fractions is used. (eg. 0.500E–01 is 0.05).

Exposure Conditions
P (O$_2$), P (S$_2$), P (H$_2$S), P (H$_2$), P (SO$_3$), A (C) and Temp. (K)
  .132E–19 .896E–06 .500E–01 .800   .772E–09 .753E–19 .0   1073.
Enter the Time (hr) (1 Year = 8760 hr)
?
5000

Corrosion Mechanisms:
(1) Sulfidation/Oxidation
(2) Sulfidation

Enter Model Number 1, 2
?

Alloy	Data Points	Penetration, mils	Annual Rate, mils/yr	Uncertainty Factor	AE = –3.08 Extrapolated Variable(s)*	
AISI 446	12	3.6	6.3	0.61	–	
AISI 304	3	4.1	7.2	–1.00	AF	T
Alloy 188	30	5.9	10.3	0.80	–	
AISI 317	9	9.0	15.7	0.14	–	
Alloy 310	14	25.3	44.3	0.48	–	
Alloy 718	38	26.3	46.1	0.30	–	T
Alloy 825	86	27.6	48.3	0.33	–	T
Alloy 800	123	27.7	48.5	0.48	–	
9Cr–1Mo	23	30.0	52.6	0.30	AF	T
Alloy 617	5	36.2	63.3	0.18	–	
Alloy 625	80	38.7	67.8	0.26	–	T
Fe–13Cr–3A1	13	76.8	134.	0.17	–	T
Alloy 600	23	89.9	157.55	0.33	–	

No Rate Equation for the Following Alloys
    Copper
    AISI 316
    HK 30
    AISI 314
    Pure Ni

Where AF = Log (a$^2_{CrS}$/a$_{Cr_2O_3}$)
The Calcuation is Not Statistically Based When the Uncertainty Factor = –1.00

Fig. 13. Example of ASSET predicted corrosion and corrosion rates, using the gas composition displayed in Figure 12.

COMPUTER TRAINING SIMULATION AT ALCAN LTEE

G. Mak[*], D. Poulin[*], and P. Harrington[**].

[*]SACDA Inc. London, Ontario.
[**]Alcan Ltee. Jonquiere, Quebec.

ABSTRACT

Alcan Ltee. is undertaking a significant modification of its facilities at
Jonquiere Quebec. This project includes the addition of a Fisher Provox
distributed control system to provide monitoring and control of several new
decanters and a mud washing system. SACDA Inc. has been contracted to provide an
off-line real time dynamic training simulator which will accurately simulate the
plant facilities. The simulation interfaces with a Fisher Provue console that the
operators will use in the plant. For operator training purposes, upset conditions
can be initiated by the instructor or they can be programmed to take effect at a
given time in the training exercise. As well, the simulation can be set to run
faster than real time to facilitate training.

KEYWORDS

Dynamic simulation, training, decanters, high fidelity, software

INTRODUCTION

Alcan Ltee. is undertaking a significant modernization of its facilities at
Jonquiere, Quebec. This project will enable Alcan to enhance the operation of
its ore extraction facilities, and improve its efficiency and profitability. A
new Fisher Provox distributed control system is being installed to provide
monitoring and control of several new decanters, a mud washing system, and a
flocculant preparation area.

Because of the extent of the plant modernization, training of the operators is
an important aspect that can not be overlooked. For this purpose SACDA has been
contracted to provide an off-line real time dynamic training simulator which
interfaces with a Fisher Provue console identical to those used in the plant.

The computer simulation provided by SACDA models the process, its controls and
motor logic so that high fidelity responses for startup, shutdown, bypass, and
emergency exercises are provided. Thus intensive no-risk training on both the
DCS and the plant are provided.

For operator training purposes, an instructor can initiate up to fifteen
preprogrammed upsets from a menu of fifty, or he can cause an instantaneous

upset by stopping a motor, plugging a valve, or modifying the contents of a tank. If desired, the instructor can also set the simulation to run faster than real time.

PROCESS DESCRIPTION

The process included in the first training simulation provided to Alcan encompasses three areas of the plant. The first area involves the preparation of the various flocculants used in the decanters and mud washing units. This is a batch process where dry powdered flocculant is mixed with a precise amount of water at the right temperature. It is then followed by agitation for a predetermined time. During this time, the operator can monitor the status of the blowers, motors, pumps, agitators, flows, and the viscosity of the prepared flocculant in the mix tank. When the cycle is complete, the contents of the mix tank are automatically pumped to the distribution tank, if sufficient space is available. The operator may at any given time suspend the sequence of any one of the flocculant preparation systems while the others proceed normally.

The second area of the simulation includes the decanters where the bauxite ore in suspension is concentrated from 2 - 3 percent solids to 40 percent solids. In this step of the process the rate of flocculant addition must be carefully controlled to provide the maximum separation possible. If the process is upset because of improper flocculant addition, adding the wrong type of flocculant, temperature variations, or excessive flow, the solid / liquid interface will cease to be clearly defined. If the upset condition is not corrected in time, it is possible that the decanter underflow solids content will decrease, and later the overflow clarity will be reduced. All this results in an expensive decrease of efficiency.

The third area is the mud washing area. When the thickened bauxite mud leaves the decanters, it proceeds to the mud washing area. Here it is washed in a series of counterflow decanters. The mud is fed into the first decanter while clean water is added to the last of the decanters. As the underflow of the first decanter is fed to the second, it is mixed with the overflow of the third decanter, and so the process is repeated for each step. It can thus be seen that an upset in any one of the washing decanters can quickly affect all of the others.

TRAINING SIMULATOR

The successful startup of any plant is a function of extensive checking of the process, controls, and instrumentation combined with comprehensive training of the operators. Manuals and lectures, slides, films, video disks and tapes are good, but with the use of increasingly complex systems, just not sufficient. Hands-on experience has become a vital ingredient in ensuring adequate training in a safety-conscience and profit minded industry.

With SACDA's dynamic training simulator called "TRAINER", operators have the opportunity to learn how to operate the DCS and the process, at absolutely no risk to the personnel or the plant equipment. It also provides operators with the possibility of encountering emergency situations which would rarely occur in the plant. The trained response to such an emergency can frequently prevent a minor problem from becoming a major disaster.

SACDA's dynamic training simulator consists of a number of computer routines or modules, each one representing a specific unit operation. In order to simulate a given process, the engineer combines these modules to model the desired process.

TRAINER SOFTWARE SCHEMATIC
DATA FILE
DATA FILE INTERPRETER
DISK
INTERACTIVE EXECUTIVE PROGRAM
PROCESS DATABASE
PROCESS SIMULATION PROGRAM
DCS INTERFACE PROGRAM
INSTRUCTOR CONSOLE
DCS OPERATOR CONSOLE

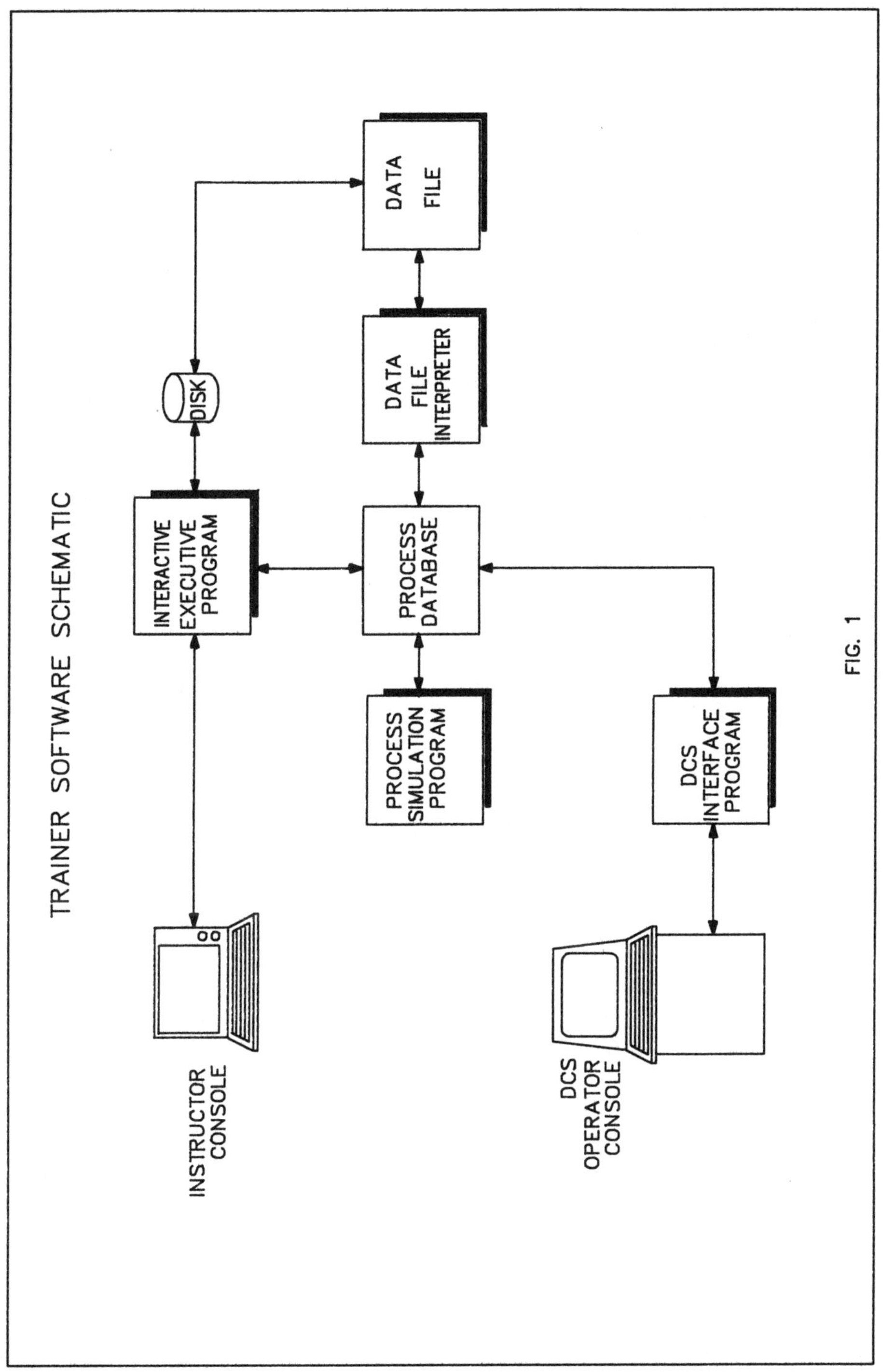

FIG. 1

Some of the modules available include pumps, control valves, tanks, solenoid valves, controllers, indicators, and logic modules. Using these modules, the engineer can simulate the process, its controls and all associated logic, such as might be contained in a PLC.

The model is prepared by gathering all relevant material from piping and instrumentation drawings, process equipment specification data, manufacturer's operating manuals, control configurations, SAMA drawings and logic drawings or ladder logic. This data is then used to determine which of the modules are to be used for the simulation modules. The desired modules are then combined in a data file which is read by the TRAINER executive program, and interpreted to provide the dynamic updates to the control console. This configuration of the TRAINER data file is currently being augmented with an automatic CAD input system.

TRAINER SOFTWARE

The TRAINER uses the same DCS console, the same graphics, and the same keyboard as used in the plant's control room. The interface which enables the communication between the operator console and the minicomputer which contains the simulation software consists of a highway device and a highway cable. The TRAINER software which resides on the minicomputer (typically a MicroVax II) consists of an executive program, a model interpretation program, a process simulation program, and a distributed control system interface program. (See Fig. 1)

The executive program coordinates the operation of the TRAINER system, and through menu structured commands, enables the instructor to control the course of the training session. The model interpretation program processes the configuration file representing the process and initializes the TRAINER database. The process simulation program performs the dynamic simulation of the unit operations, regulatory control, and motor logic. This program updates the TRAINER system database with the current state of the simulated process as the calculations proceed. The distributed control system interface program handles the communications between the TRAINER system database and the DCS console. (See Fig. 2)

During a training exercise, process upsets can be initiated in one of two ways. First, the operator trainee can make an error. This may cause an upset to occur if the error is severe enough. Secondly, the instructor can initiate an upset from the instructor terminal. The instructor can select up to fifteen malfunctions to occur within a given training exercise. This choice is made from a menu consisting of a maximum of fifty malfunctions. The malfunctions can be either digital, such as the tripping of a pump, or analog such as the plugging of a control valve over a period of time. In any case the instructor sets the time for the malfunction to occur and the interval over which it is to take place. A summary of some of the various commands that the instructor has at his disposal includes:

    CONDUCT - conduct a training session
    REVIEW - examine historical data from a previously conducted training session
    ENABLE (DISABLE) - enable or disable a previously defined malfunction
    BACKTRACK - backtrack in time to a previous process condition
    SNAPSHOT- create a snapshot of the existing process condition
    PAGE, FORWARD - display the next monitoring display page, or the previous page
    (PAGE, BACKWARD)
    FRZ/UNFRZ - freeze the simulation at the current time /unfreeze the simulation

FISHER PROVOX
DCS PROCESS CONTROL
TRAINER SIMULATOR
LCON
BCON
or PROVUE
OPERATOR
CONTROL
STATION
OPERATOR
CONTROL
STATION
LCON
BCON
or PROVUE
DCU
DATA
HIGHWAY
UOC
COMPUTER
INTERFACE
UNIT
MOD
DHI
CARD
DATA
HIGHWAY
PCIU
MICROVAX II
COMPUTER
IFC
LTD
INSTRUCTOR
CONSOLE
LTD
FISHER PROVOX COMMUNICATION NETWORK
AND TRAINER HARDWARE CONFIGURATION

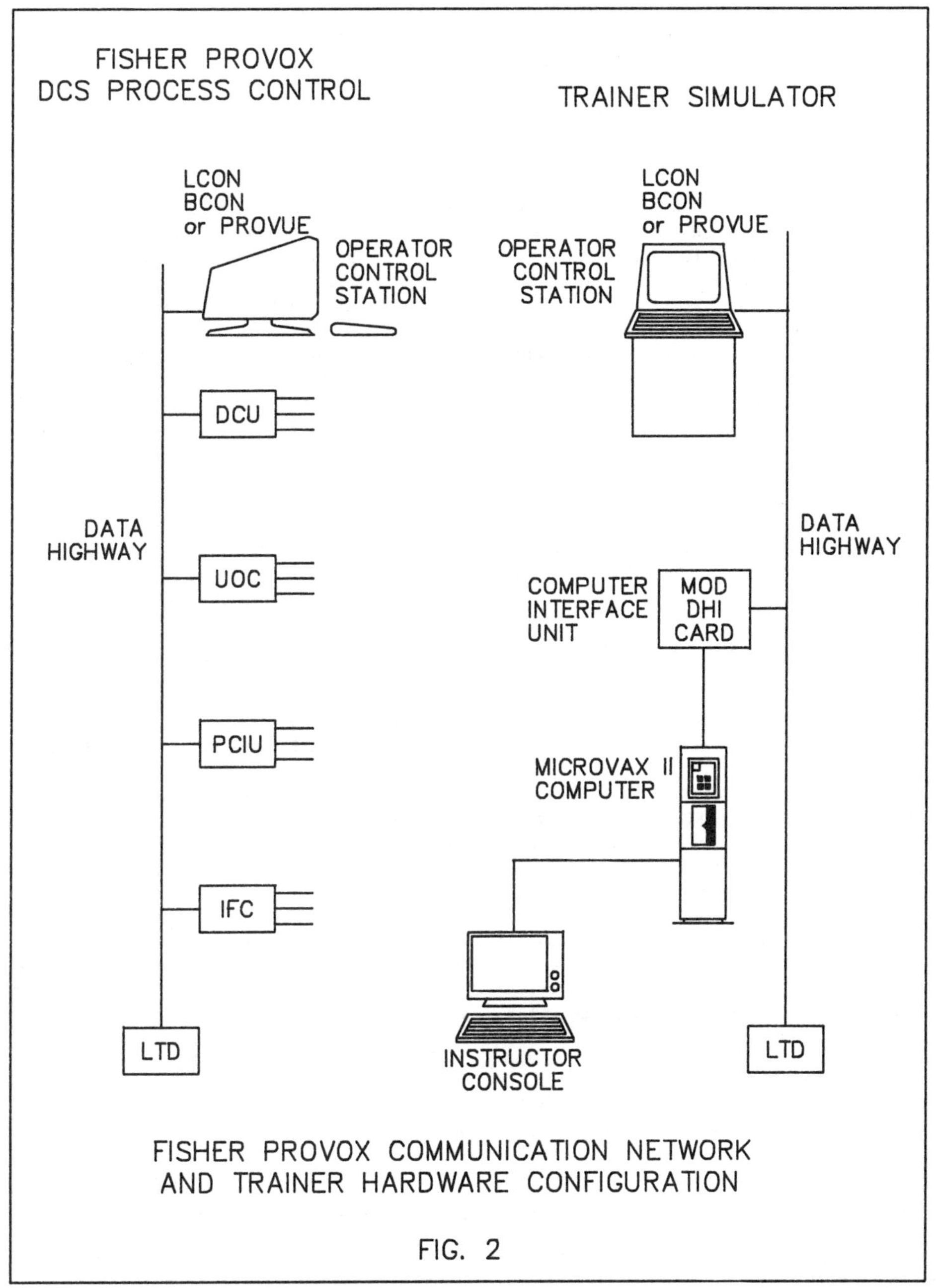

FIG. 2

```
MONITOR - monitor the trainee's performance
PREVIOUS - display the previous menu of commands
HELP     - display a help menu for the current command menu
RESET TIME - reset the simulation time
GAIN     - examine or change the specified control loop gain
RESET RATE - examine or change the control loop reset rate
DERIVATIVE - examine or change the control loop derivative constant
```

The control of bauxite decanters requires that the operator properly prepares the flocculants, adds them in the correct proportions, and carefully maintains the interface level in the decanter. If the flocculant pump were to fail, or the temperature of the feed were to rise too high, the distinct level interface would cease to be well defined. In this case the only indication that the operator would see is an apparent increase in the interface level. However, if no corrective action is taken, the underflow solids concentration would soon decrease as the interface spread continues. Eventually, the overflow clarity will decrease as solids make their way to the top of the decanter. Once this state has been reached, it can take a long time for the process to return to the normal steady state.

TRAINING PROGRAM

Training of operators is considered to be very important at Alcan. Because the process is so slow, upsets can take up to eight hours to occur, efficient training is a difficult task. The objective of the training requires that operators learn to recognize that an upset is in progress, and that the proper response is taken within the minimum time. The training program undertaken consists of four phases.

The first phase consists of the theoretical aspects of the plant. The operators are taught the process principles, unit operations, and the process flowsheets.

In the second phase they are taken into the plant where the equipment is identified and related to the unit operations learned in phase one. Here the operators gain an understanding of the actual physical layout of the plant and an appreciation for the various pieces of process equipment such as valves, pumps, flowmeters, etc.

In the third phase of training the operators are introduced to the control system console. Using a bench board simulation, they learn what constitutes a control loop eg. flow element, transmitter, controller, transducer, and control valve. Next, using the console they learn to operate the keyboard to call up loops, and graphic displays, and eventually to control the bench board simulation consisting of a pump, a tank, control valves, and a few controllers.

It is in the fourth and final phase of training where the operators are instructed on the actual plant operation using the full plant simulation built by SACDA. Here they are taught how the various plant units interact with each other, how the process is controlled, and most importantly how to recognise and rectify process upsets. During normal plant operation it may take up to eight hours for an upset to occur, and another twelve hours to restore the process to the desired steady state after correcting the cause of the upset. However, using the faster than real time feature of the SACDA simulator, an operator may experience several process upsets during his training session which may last from two to three hours.

After training was commenced, it was discovered that initially the trainees had some difficulties in determining the exact cause of an upset condition. After observing that an upset was indeed taking place, several incorrect causes were pointed out. For example, a flowrate which is lower than expected could be caused by a plugged valve, instrument calibration, plugged lines, or a closed hand valve. After several sessions with the training simulator, the operators learned to correctly identify the source of the upset condition, and to take appropriate remedial action.

CONCLUSIONS

The training of operators is an important task that requires considerable effort. Using a high fidelity plant simulator such as the one provided by SACDA enables operators to learn on the same console as the one they will use in the plant and allows them to learn at their own speed with absolutely no risk to the plant equipment or personnel.

REFERENCES

Bergeron, L., and A. Trudeau (1987). Formation des operateurs de la machine no. 4 a l'aide d'un logiciel de simulation dynamique "TRAINER' a la Consolidated-Bathurst, Inc. Division Port-Alfred. ACPPP, 24-35

Jackson, A., and C. F. Shewchuk (1985). Computer simulation as applied to process design, operator training and production management. Proceedings of 1985 Pulp and Paper Technology Conference, J. TAPPI/TS-CPPA, Tokyo, Japan

Kendrick, A. (1984). Simulation program saves steam at Petrosar. Canadian Chemical Processing, 34-40

Shewchuk, C. F. and E. W. Leaver (1986). New operator training system called ideal for many uses. PIMA Magazine, 40-43

Shewchuk, C. F. (1987). New modular implementation expands the role of dynamic simulation. Presentation at SPCI'87, Stockholm, Sweden